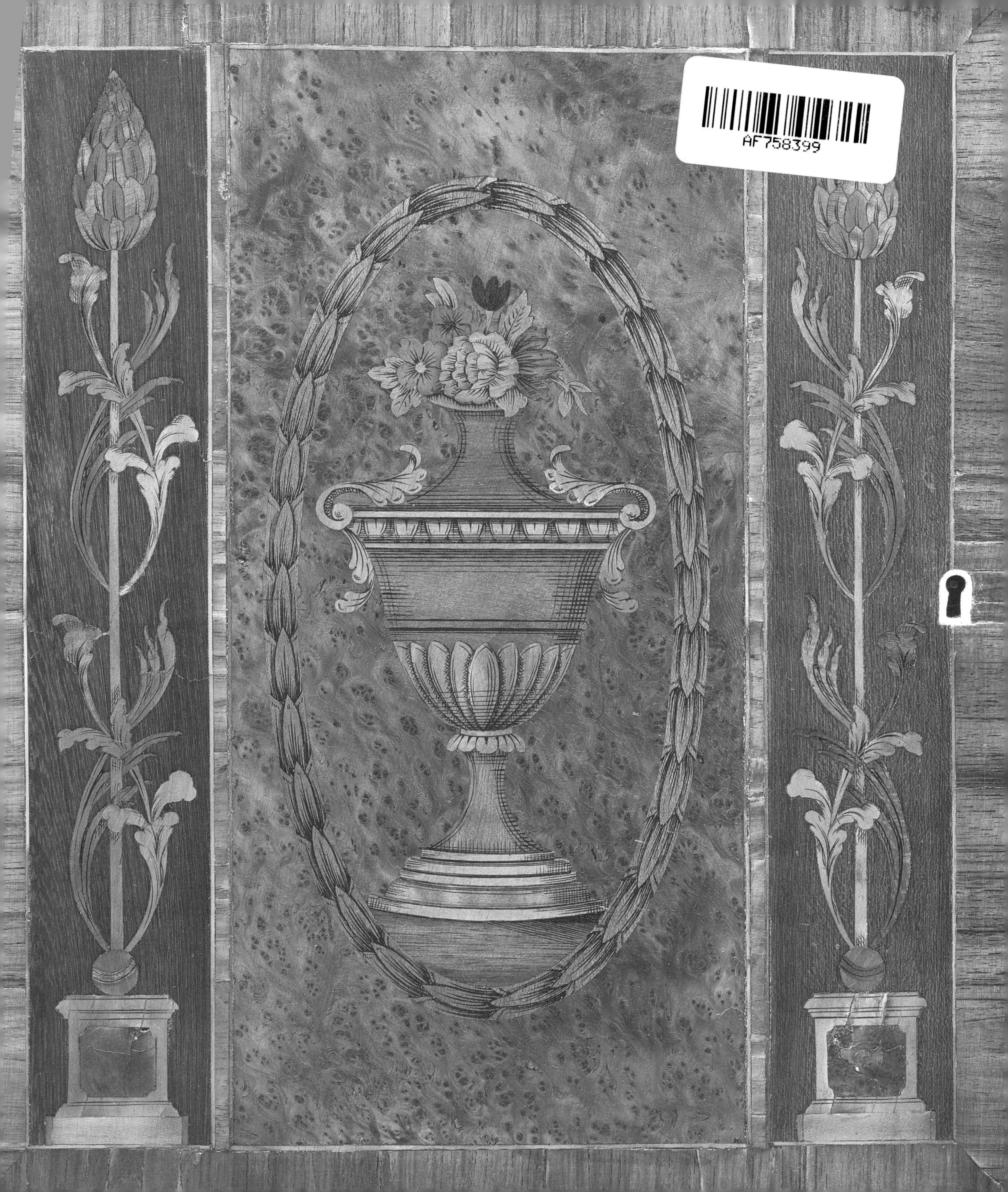

VASEMANIA

"An epidemical madness reigns for Vases, which must be gratified."
Josiah Wedgwood to Thomas Bentley, January 1769

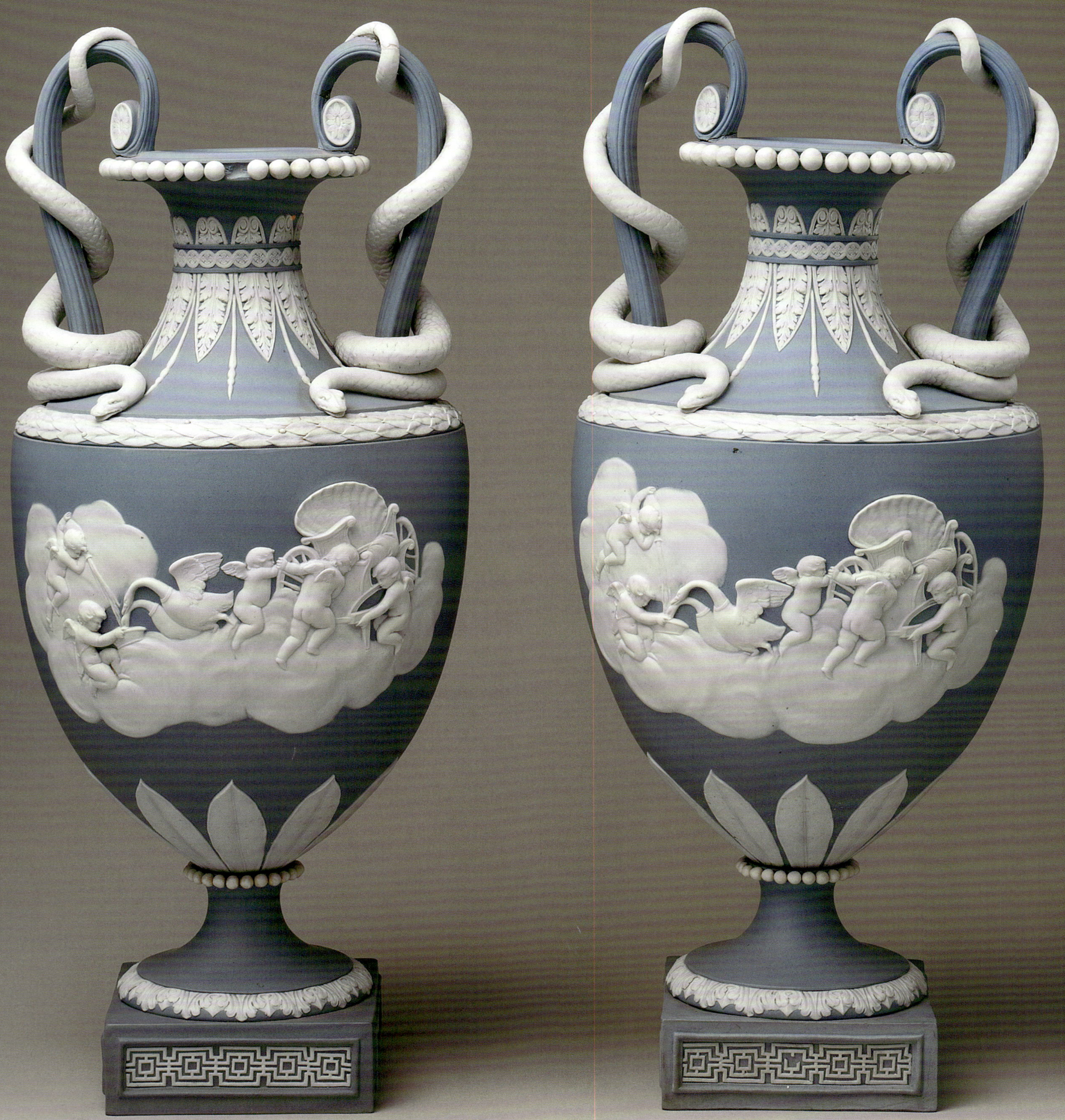

VASEMANIA

NEOCLASSICAL FORM AND ORNAMENT IN EUROPE

SELECTIONS FROM THE METROPOLITAN MUSEUM OF ART

Heather Jane McCormick
Hans Ottomeyer

With Contributions by
Master's and Doctoral Students
of the Bard Graduate Center

Stefanie Walker, Editor

Published for The Bard Graduate Center for Studies in the Decorative Arts, Design, and Culture, New York and The Metropolitan Museum of Art, New York, by Yale University Press, New Haven and London.

This catalogue is published in conjunction with the exhibition
V A S E M A N I A — *Neoclassical Form and Ornament in Europe: Selections from The Metropolitan Museum of Art*,
held at The Bard Graduate Center for Studies in the Decorative Arts, Design and Culture from July 22 through October 17, 2004.

Project director, editor, and project coordinator at the Bard Graduate Center: *Stefanie Walker*
Exhibition curator and project coordinator at The Metropolitan Museum of Art: *William Rieder*
Catalogue production: *Sally Salvesen*, London
Editor: *Susan F. Rossen*
German-language translator: *Terry D. Prince*

Director of Exhibitions and Executive Editor, Exhibition Catalogues, Bard Graduate Center: *Nina Stritzler-Levine*

Catalogue and jacket design: Cummings & Good, Chester, Connecticut
Printed and bound by Conti Tipocolor, Italy

Library of Congress control number: 2004107910

Front cover: *The Greek Bride*, Ennemond-Alexandre Petitot, 1771, Harris Brisbane Dick Fund, 1940; cat. no. 92.

Back cover: *Spring*, one of four overdoor panels depicting the Four Seasons, possibly made under the direction of François-Joseph Bélanger, about 1777–81, Gift of J. Pierpont Morgan, 1906; cat. no. 74.

Endpapers: Detail of *Lady's Writing Desk with Marquetry Panels*, about 1780–90, From the Marion E. and Leonard A. Cohn Collection, Bequest of Marion E. Cohn, 1966; cat. no. 72.

Frontispiece: *Pair of Vases with Serpent Handles*, Josiah Wedgwood, about 1787, Gift of Frank K. Sturgis, 1932; cat. no. 40.

VASEMANIA

Neoclassical Form and Ornament in Europe:
Selections from The Metropolitan Museum of Art

has been made possible with the generous support of
Mrs. Charles Wrightsman and the Samuel H. Kress Foundation

* * *

This book is dedicated to the memory of
Clare Le Corbeiller (1931–2003)

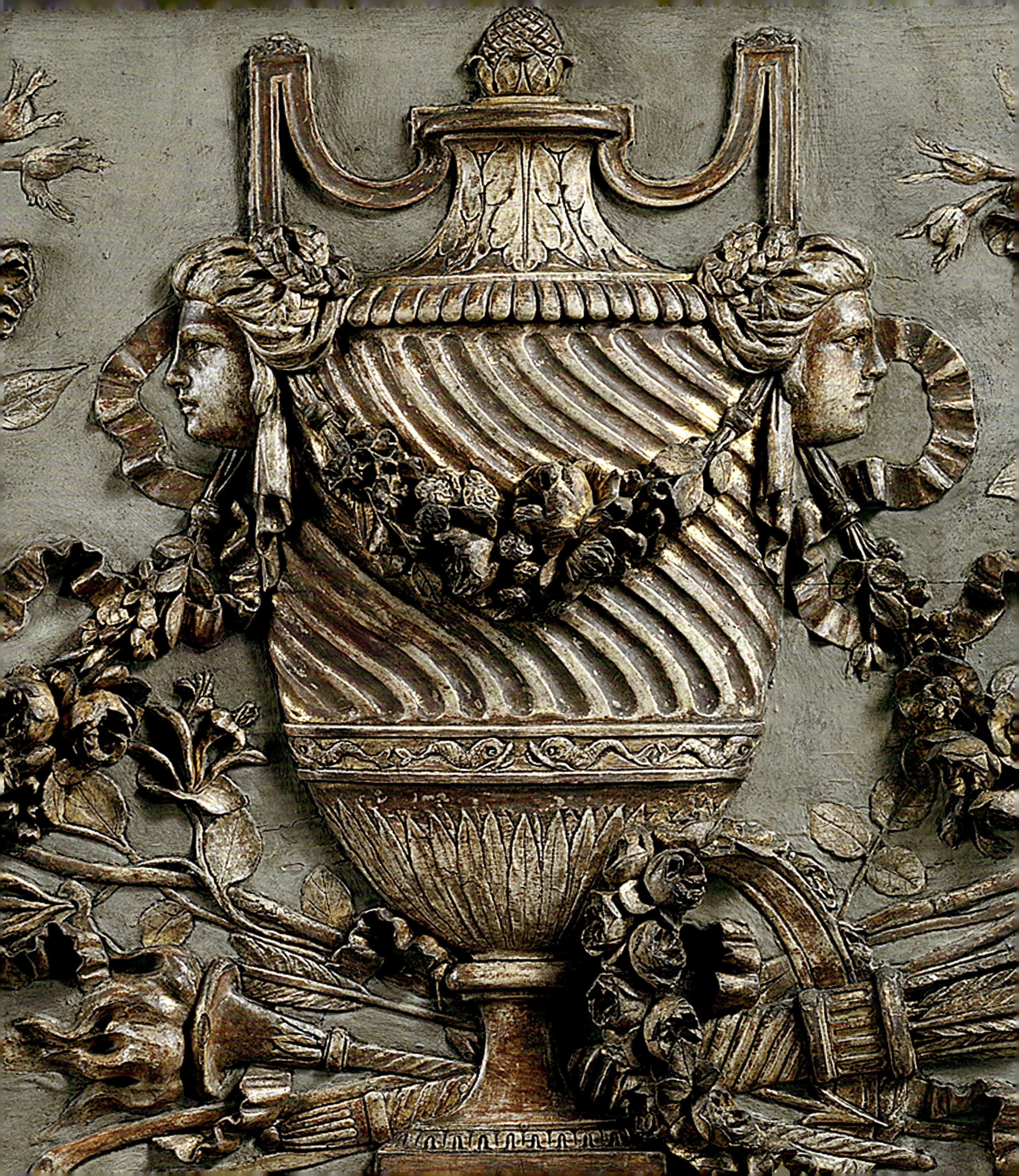

Contents

Detail of cat. no. 74

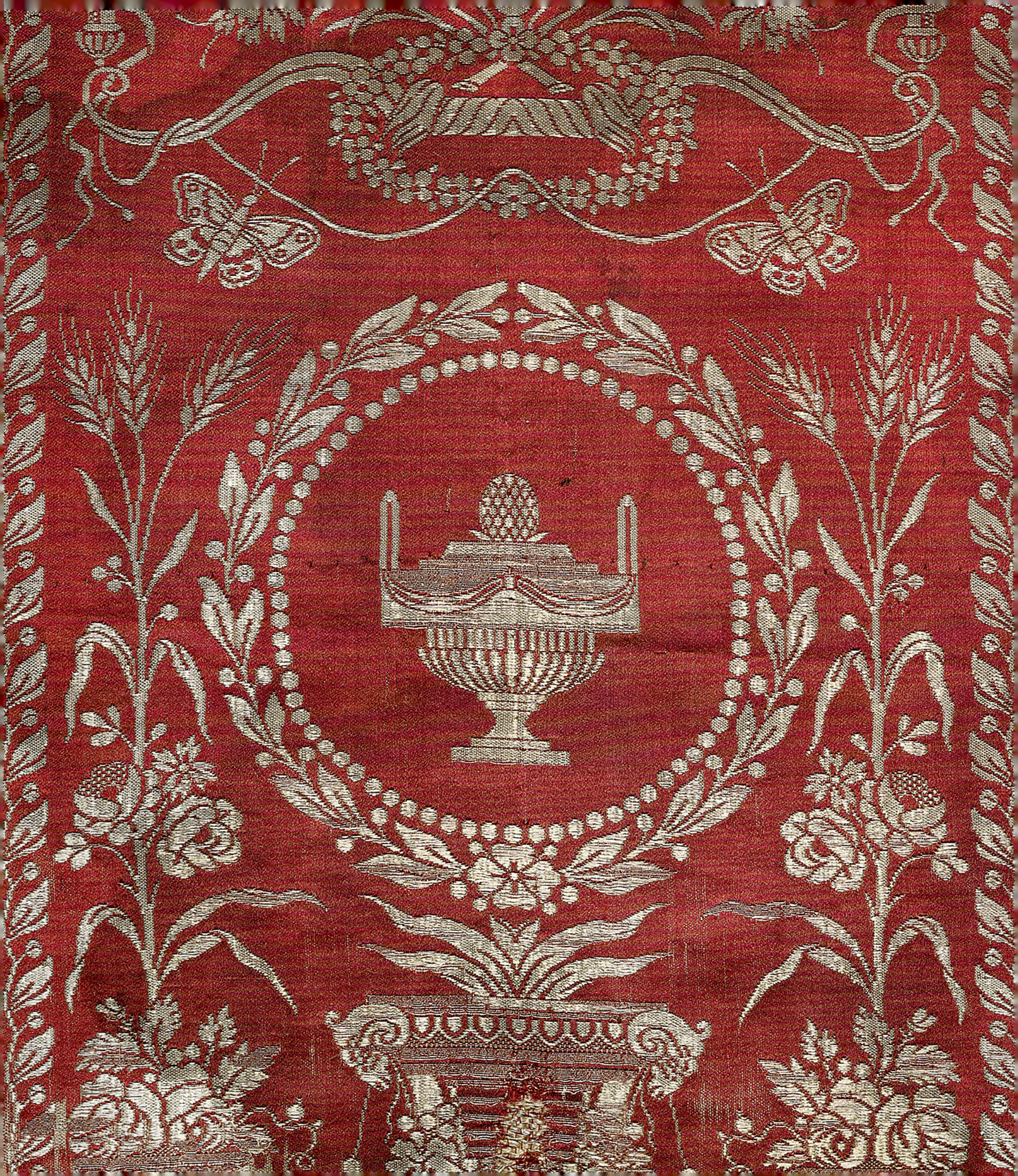

Foreword

Vasemania—Neoclassical Form and Ornament in Europe: Selections from The Metropolitan Museum of Art examines a magnificent array of ceramics, furniture, metalwork, textiles, woodwork, and works on paper that illuminates the pervasive use of the vase motif in European decorative arts from approximately 1750 to 1810. Ornament has been an important issue in design discourse, but it has been less common in the English-speaking world to consider the function and meaning of a single motif as a means of decoration. The remarkable diversity of vase forms revealed in this exhibition and book demonstrates the importance of ornament as an area of scholarly inquiry in the decorative arts.

Vasemania is the last of three exhibitions, which began in the fall of 2003, celebrating the ten-year anniversary of the Bard Graduate Center for Studies in the Decorative Arts, Design and Culture. The project inaugurates a unique educational initiative with The Metropolitan Museum of Art. Every two years, graduate students at the Bard Graduate Center will participate in curating an exhibition composed of selections from the Metropolitan that are not readily accessible to the public. Working under the direction of a faculty member at the Bard Graduate Center and a curator at the Metropolitan Museum, students will research objects in relationship to a particular subject or theme that becomes the focus of the exhibition. The goal of this collaboration is to train future curators of the decorative arts and to convey the essential role of art museums as repositories of objects and knowledge in the study of decorative arts, design, and culture.

The academic program and graduate students at the Bard Graduate Center are at the heart of its mission to enhance scholarship in and foster public awareness of the decorative arts and related fields. I am deeply grateful to Philippe de Montebello, director of the Metropolitan Museum, for providing the opportunity to expand our mission by making the museum a laboratory where our graduate students can study some of the world's greatest works of art. I am especially proud that this undertaking has received generous support from Mrs. Charles Wrightsman, one of the foremost champions of the decorative arts. I am also grateful to The Samuel H. Kress Foundation for their generous contribution to this unique educational endeavor.

A collaboration of this scope and ambition required the contribution of many members of the curatorial and administrative staffs at The Metropolitan Museum of Art. Under the leadership of Ian Wardropper, Iris and B. Gerald Cantor curator-in-charge, the Department of European Sculpture and Decorative Arts provided this project with many resources, both in staff and collections. William Rieder, curator and administrator in the department and a member of the Bard Graduate Center's Executive Planning Committee, proposed the idea of an exhibition about the vase motif in neoclassicism for a Bard–Metropolitan Museum collaboration. Mr. Rieder's extensive knowledge of the decorative arts is evident in the initial checklist he compiled, drawing upon the holdings of various departments, in addition to his own. He also attended to administrative aspects, taught in the Bard graduate program, and participated in reviewing the students' texts. Also in the department, Thomas Campbell, curator; Wolfram Koeppe, Jessie McNab, and Jeffrey Munger, associate curators; and Melinda Watt, research assistant, all devoted much time and energy to the project. I would like to thank Roger Haapala, associate for administration, and Stephanie Post, administrative assistant. Eric Peluso and William Kopp facilitated the study of objects in the museum's storerooms.

In the Department of Drawings and Prints, I am grateful to George R. Goldner, Drue Heinz Chairman, for his support. Special thanks go to Perrin V. Stein, associate curator, who played a major role in the exhibition. Elizabeth E. Barker, assistant curator, and Constance McPhee, study room supervisor, were also helpful. In the Department of Greek and Roman Art, Carlos A. Picón, curator-in-charge, supported loans to the exhibition and curator Joan R. Mertens reviewed entries. I am grateful as well to Kenneth Soehner, Arthur K. Watson Chief Librarian of the Thomas J. Watson Library. Herbert M. Moskowitz, chief registrar, and Marceline McKee, manager of special projects, coordinated loans from the various departments. Doralynn Pines, associate director for administration at the Metropolitan, shared her expertise and has been very helpful to the Bard Graduate Center. The exhibition is enhanced by one object from the private collection of Paul Micio, who also taught a class on silver. *Vasemania* introduces an ambitious digital photography initiative at the Metropolitan Museum, which Barbara Bridgers, general manager for imaging and photography, coordinated. Joseph Coscia, Jr., associate chief photographer in charge of collections photography, made the excellent digital captures.

Many individuals at the Bard Graduate Center contributed to the success of *Vasemania*. Foremost among them is Stefanie Walker, special exhibitions curator, who coordinated the project

Detail of cat. no. 77

with Mr. Rieder, served as general editor of the catalogue, defined the project's interpretative themes, and supervised the students.

In 2002 the Center offered a graduate course, "Neoclassicism in Europe," which served as a foundation for the exhibition. Participating, in addition to Stefanie Walker and myself, were faculty members Peter Miller, Andrew Morrall, and Elisabeth Simpson, along with Paul Micio and John Ward (a graduate of the Bard Graduate Center and assistant vice-president at Sotheby's New York). Under Nina Stritzler-Levine, the exhibition department contributed to the exhibition's organization and the catalogue. Ms. Stritzler-Levine served as Bard Graduate Center liaison with the Metropolitan Museum. I would like to thank Olga Valle Tetkowski for various administrative responsibilities she performed. Linda Stubbs coordinated the exhibition assembly at the Bard Graduate Center gallery. Han Vu helped with the computer and digital organization of the project and created a digital video for the exhibition. Ian Sullivan designed the installation and directed the installation crew with skill and professionalism. Yasmin Elshafei, Freyja Hartzell, Charlotte Nicklas, and Emma Wegner, four very able Graduate Center students, helped the exhibition department in various ways.

Tim Mulligan, director of external affairs, led the press campaign, assisted by John Chaich. In effect, Mr. Mulligan should be credited with the concept of collaborating with the Metropolitan Museum. I appreciate the support he provided early on in the development of this project. Susan Wall was in charge of the fundraising campaign, in collaboration with Brian Keliher. Lorraine Bacalles assisted with finance and administration. Bonnie Poon, director of programs, provided a fascinating roster of lectures and related programs, aided by Mykellann Ledden, public programs coordinator; Jill Gustafson, special projects coordinator; and Kristin Martin, public programs assistant. Jennie McCahey, gallery assistant, coordinated the exhibition tours and trained the gallery docents. Sandra Fell skillfully attended to many administrative details in the director's office. The day-to-day maintenance of the Bard Graduate Center is supervised by John Donovan and the facilities department. Chandler Small, coordinator of security, and his staff contribute substantially to the Graduate Center's daily operation.

Last but not least, I would like to thank those who worked with such diligence on the catalogue. I am grateful to Hans Ottomeyer, director of the Deutsches Historisches Museum, Berlin, for his enlightening essay on the vase in the broader context of neoclassicism; and Heather Jane McCormick, Ph.D. candidate at the Bard Graduate Center, for her illuminating history of neoclassicism at the Metropolitan Museum and her well-researched entries. I am proud of the Bard Graduate Center students who participated in the initial class and contributed catalogue entries to this volume. These include master's degree candidates Kei Chan, Katherine Danalakis, Marcella Ruble Harris, Jessica Lanier, Kristin Martin, Yasmine Mohseni, Irene Sunwoo, Jennifer Walton, and Emily Wheat Maynard; and doctoral candidates Sarina Forbes and Ezra Shales.

Peter Good and his colleagues at Cummings & Good, Chester, Connecticut, provided the elegant catalogue design. Susan F. Rossen was a skillful and sensitive editor. Cindy Kang, curatorial assistant, and Jennie L. McCahey, gallery assistant, at the Bard Graduate Center, and master's degree students Caroline Hannah and Michelle Hargrave authored several entries and provided additional editorial support. At Yale University Press in London and New Haven, I am grateful in particular to Sally Salvesen and John Nicoll. To all involved in *Vasemania*, my sincere thanks for the dedication and professionalism that contributed so much to its great success.

Susan Weber Soros
Director, Bard Graduate Center
for the Decorative Arts, Design and Culture

Preface

Opportunities for students of art history to examine and take the measure of works of art are critical for their education. While projected images in lecture halls and photographic reproductions in libraries remain the fundamental tools of instruction, the experience of handling an object in a museum storage room is essential to deepening one's understanding of art. I was delighted when Susan Soros, Director of the Bard Graduate Center for Studies in the Decorative Arts, proposed that The Metropolitan Museum of Art and the Bard Graduate Center collaborate in an exhibition that would promote students' close scrutiny of objects. It was only natural that such an alliance would focus first on the Department of European Sculpture and Decorative Arts: a generous grant from the Iris Foundation under the auspices of Susan Soros supported the computerization of the department's collection records. Furthermore, the richness of this department's holdings—so extensive that only a portion can be exhibited at any time—made it a prime candidate as a source for such an exhibition.

What the process of creating *Vasemania—Neoclassical Form and Ornament in Europe* demonstrated is that, while the Metropolitan's collection constituted its basis, the combination of curatorial expertise from the museum and the teaching skill at Bard became its vehicle. Under the leadership of Iris and B. Gerald Cantor curator-in-charge, Ian Wardropper, the department provided resources for this project. Curator William Rieder originated the concept and oversaw the organization of the exhibition. Associate curators Jessie McNab and Jeffrey Munger guided students in examining objects and reviewed their writing. Other departmental staff contributed substantially to the project, notably Perrin Stein, associate curator in the Department of Drawings and Prints. Associate Director Doralynn Pines was instrumental in implementing the agreement between the two institutions. Barbara Bridgers and the staff of the photography department took all new images of the objects selected for inclusion. Finally, the Registrar's office shouldered the burden of many logistical matters.

At the Bard Graduate Center, we are grateful to Stefanie Walker and Nina Stritzler-Levine, as well as other staff members, for their hard work in realizing this exhibition and the book it occasioned; hopefully, this is the first in a series of such initiatives. Yale University Press, with which both institutions enjoy long-time and fruitful collaborations, has produced this catalogue to its customary high standards.

Philippe de Montebello
Director
The Metropolitan Museum of Art

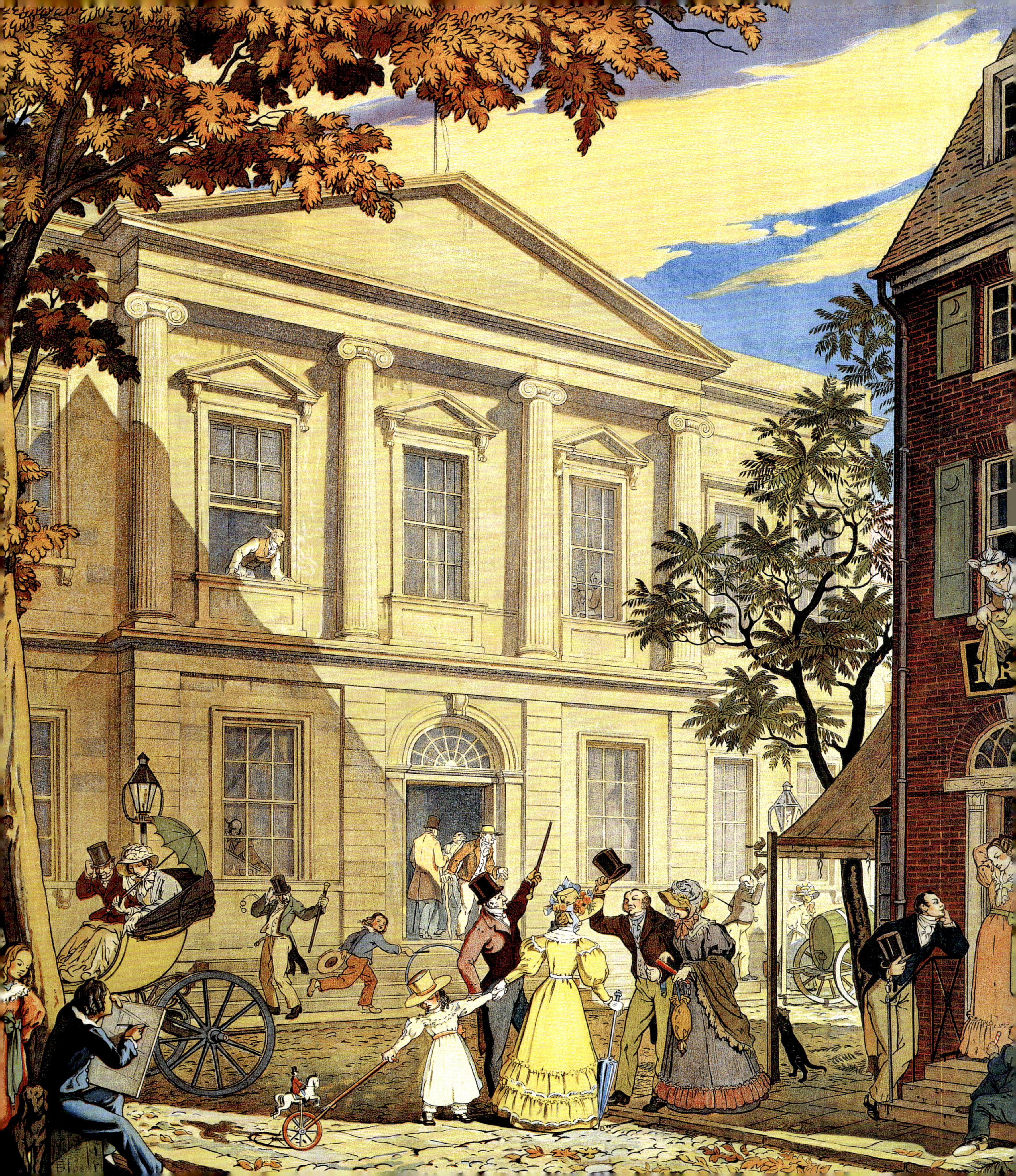

Collecting Neoclassical Decorative Art at the Metropolitan Museum of Art

Heather Jane McCormick

Among the thousands of works of art on display at the Metropolitan Museum, some of the most esteemed are those embodying the neoclassical spirit and style that pervaded Western art in the late eighteenth and early nineteenth centuries. Visitors who linger over interiors designed by Robert Adam, sculpture by Antonio Canova, paintings by Jacques-Louis David, furniture by Georges Jacob, and silver by Paul Revere see the work of the leading figures and foremost interpreters of neoclassicism, and experience the quality this collection offers.

With only a fraction of these holdings on view, however, what visitors seldom experience is their extent and depth. Comprising thousands of objects, paintings, sculptures, drawings, and prints, the Metropolitan Museum's collection of the art of this period is a rich starting point for the pursuit of almost limitless threads in the story of neoclassicism. Following one of them, *Vasemania* offers a glimpse of that depth as it draws on museum holdings in European decorative art, ancient art, prints, drawings, and books to comprehensively explore neoclassicism's most pervasive symbol—the antique vase—across media, decades, and national boundaries.

Extensive as the museum's collection of European neoclassical decorative art is today, it grew slowly, unevenly, and sometimes as a matter of chance, especially in the early years. One reason for this is that departments at the museum have predominantly represented individual media or cultures rather than styles; neoclassical decorative art, therefore, has never been viewed as a collection as such. In fact, the term "neoclassicism" does not seem to have been used to describe the museum's holdings of late eighteenth- and early nineteenth-century decorative art before 1926.[1] The history of this "collection," then, is in part the story of the museum's recognition of the movement as a basis for acquisition and interpretation. The earliest milestones in this development were the first displays of neoclassical objects as a group. Later, focused collecting, exhibitions of eighteenth-century decorative art, and publications signaled acknowledgment of this material as one of the Metropolitan's strengths.

While numerous histories of collecting at the Metropolitan Museum exist, this account revisits the museum's own record of its activities, as told in its annual reports, its *Journal*, and particularly its *Bulletin*.[2] Established in 1906, the *Bulletin of the Metropolitan Museum of Art*—especially the monthly issues published until 1971—beats with the pulse of the institution. With notices of new acquisitions and current and upcoming exhibitions, as well as articles on a compelling range of subjects, from individual artists and schools or styles to theories of art education and import tax laws, the informative and lively *Bulletin* tells the story of collecting neoclassical art between its lines, through its documentation of the important accessions, changing curatorial interests, and trends in collecting and exhibitions over the decades.

Today, the richness of the museum's collection of neoclassical decorative art is the result of over a century of acquisitions, and to a large extent, the generosity of some extraordinary benefactors. *Vasemania* pays tribute to a remarkable number of the most significant—including Alphonso T. Clearwater, Harris Brisbane Dick, the Samuel H. Kress Foundation, J. Pierpont Morgan, Jacob S. Rogers, Irwin Untermyer, Elisha Whittelsey, and Charles and Jayne Wrightsman, among others—through the objects that bear their names as donors and supporters.

* * *

The arts of late eighteenth-century Europe have had a place in the Metropolitan Museum's collections from its beginnings. In 1872, two years after its formal incorporation, the fledgling collection acquired its first piece of sculpture, Jean-Antoine Houdon's 1778 marble portrait bust of Benjamin Franklin (Fig. 1), the gift of John Bard, brother-in-law of the museum's first president, John Taylor Johnston, and a founder of Bard College. But it was the museum's mission—its guidelines for the acquisition not only of paintings and sculpture but objects illustrating styles and art movements of the past—that fostered its collection of neoclassical decorative art. Inspired by the example of London's South Kensington Museum (now the Victoria and Albert Museum), the Metropolitan's founders envisioned an institution that would not simply attract connoisseurs, but would educate an art-deprived American public, particularly its artists, designers, workmen, and manufacturers. To achieve their goal of "encouraging and developing...the application of arts to manufacture and practical life," the museum's first officers planned a collection that would comprise "not painting and statuary alone, but multiplied art, such as prints and bronzes, and industrial or decorative art of all kinds," as well as "representative examples of a school, a style or an epoch," two principles that promoted, from the museum's

Detail of Fig. 3

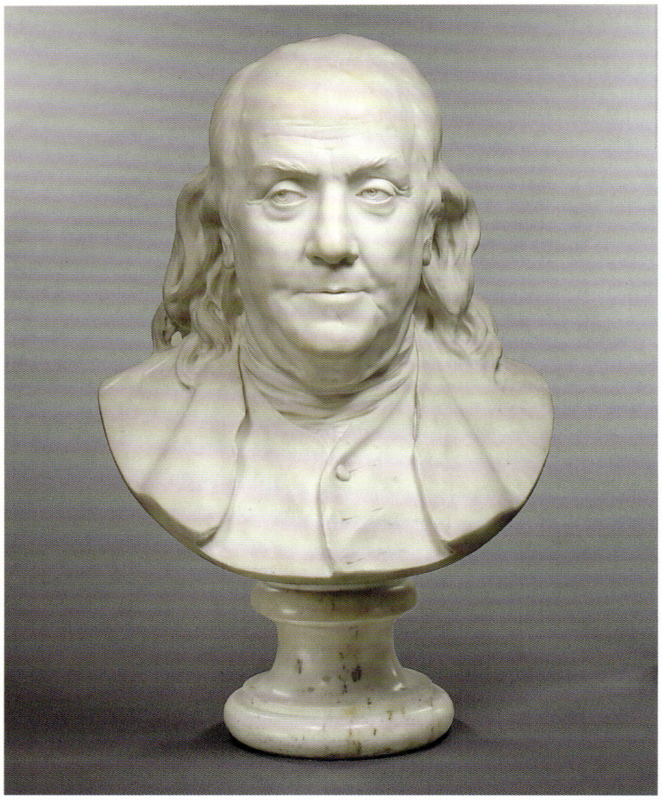

Fig. 1: *Benjamin Franklin*, 1778. Jean-Antoine Houdon. Marble; 22 ½ in. (57.2 cm) with base. Gift of John Bard (1872), 72.6.

earliest days, comprehensive period collections like those of neoclassical decorative art.[3]

While the circumstances were favorable for focused collecting, the initial development of the collection was shaped largely by opportunity and especially gifts. In the beginning, neoclassical objects entered the collection a few pieces at a time, singly or with larger gifts of diverse material. Perhaps the most storied object in *Vasemania*—Josiah Wedgwood's pottery replica (cat. no. 8) of the Roman cameo-glass *Portland Vase*—was one such early gift, donated in 1894 by New York financier Henry G. Marquand, then the museum's president. Best remembered for his donation of a valuable collection of seventeenth- and eighteenth-century paintings, Marquand gave the Metropolitan other notable examples of neoclassical decorative art, among them a five-piece Fürstenburg porcelain garniture (cat. no. 25).[4]

It was nearly forty years after it was founded, however, before the Metropolitan acquired its first substantial holdings of neoclassical decorative art, part of a characteristically grand gift from the legendary financier and benefactor J. Pierpont Morgan. In 1907, responding to reports in the press, the museum announced that Morgan had presented the institution with the Hoentschel collection of medieval and eighteenth-century French wood- and metalwork (Fig. 2). Morgan, who became well known for buying entire collections in single transactions (often for impressive sums) had acquired the material in 1906 from a reluctant Parisian architect, Georges Hoentschel, who finally relinquished it for one-and-one-half million dollars.[5] Hoentschel, who specialized in historically inspired interiors, had accumulated the carved ornaments, architectural woodwork, furniture, and gilt-bronze mounts and fittings over the course of his career. Placing the medieval woodwork on indefinite loan, Morgan gave the eighteenth-century pieces to the museum outright. Although Hoentschel's collection did not specifically focus on the neoclassical period, it included many examples of woodwork and gilded hardware in the style. Featured here, four wooden *Overdoor Panels*, probably from the Château de Bagatelle (cat. no. 74), are former Hoentschel pieces, as are a pair of tiny models for tripod-form *Perfume Burners* executed in gilded and painted wood (cat. no. 70), and five painted overdoors (cat. nos. 83–86).

Arriving at the museum in nearly four hundred packing cases, Hoentschel's collection could not be accommodated in the museum's galleries as they existed in 1907. Plans were immediately made for a new wing, also to be funded by Morgan. The scale of Morgan's benefactions prompted another event important to the development of the neoclassical collection: the formation in 1907 of the Department of Decorative Art, established to manage the growing holdings of post-classical European, Near Eastern, Asian, and American wood- and metalwork, furniture, textiles, ceramics, and glass. In 1910, coinciding with the museum's fortieth anniversary, the new department unveiled Morgan's Wing of the Decorative Arts with chronological displays of French, Italian, and English objects, as well as the American items the museum was then just beginning to acquire.

One strength of the Metropolitan's early representation of neoclassical decorative art was in fact its American holdings. Around the time the Hoentschel collection arrived, the museum began to experiment with the idea of exhibiting and acquiring the "industrial arts" of the American past, reflecting a growing popular interest that had already stimulated private collectors and a similar acquisitions program at the Museum of Fine Arts, Boston.[6] At first the Metropolitan's collection was dominated by early colonial furniture, but it did not take long for curators to begin acquiring Federal pieces in the neoclassical style. Indeed, historical and political associations made late eighteenth-century

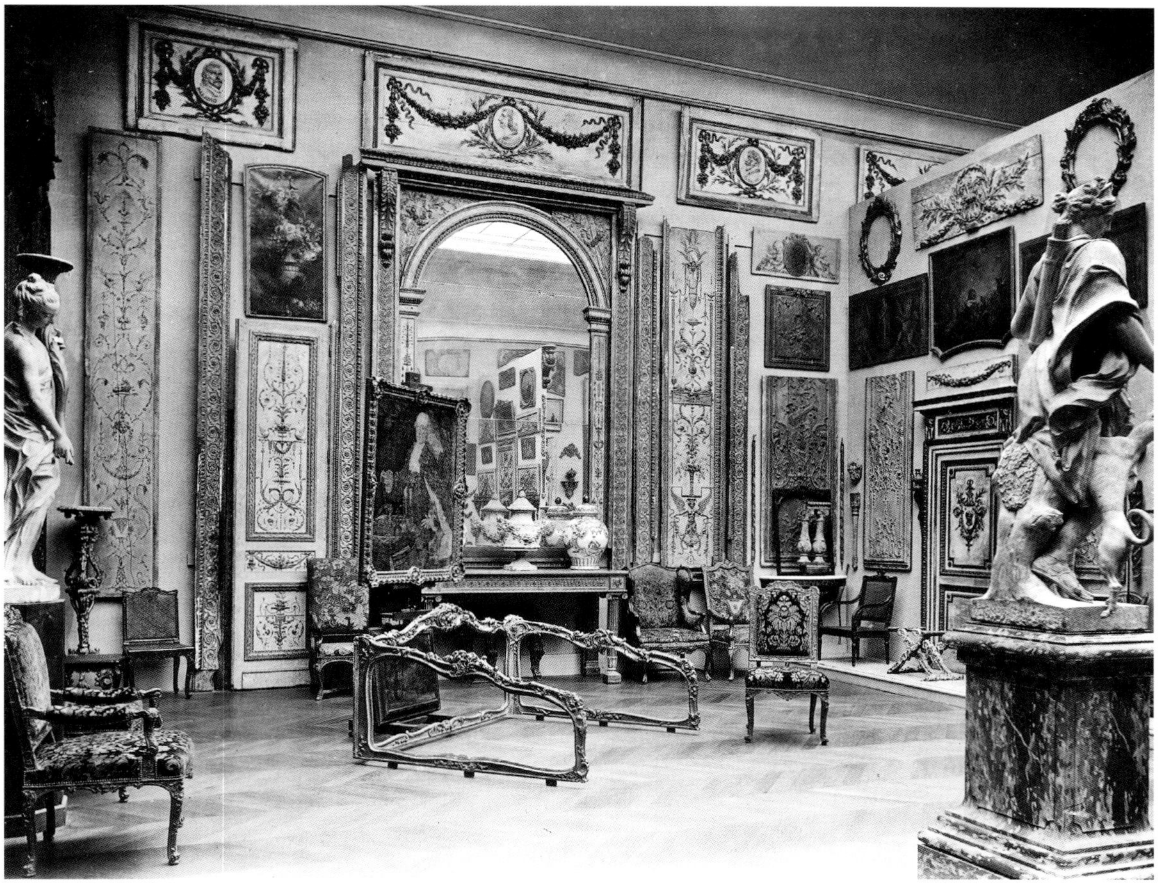

Fig. 2: View of the Georges Hoentschel collection at 58 Boulevard Flandrin, Paris, about 1904. Plate 10 from an undated album. Thomas J. Watson Library (139.1 H67 Q).

American decorative art the focus of more attention at the museum than the neoclassical decorative arts of other countries. Coinciding with the early nationhood of the United States, neoclassicism must have seemed the most American of idioms, its chaste lines an extension of republican simplicity at its most ideal. Moreover, the most famous American craftsmen—Paul Revere, Duncan Phyfe, and the Salem carver Samuel McIntire—worked in the Federal period. During the first half of the twentieth century, their work, along with that of their contemporary, the painter John Singleton Copley, was emphasized in museum publications, exhibitions, and collecting. In 1920 the Metropolitan celebrated its fiftieth anniversary with an exhibition dedicated to American decorative art of the Federal period. Two years later, curators mounted a loan exhibition of Phyfe's pre-1825 furniture. And when the American Wing opened in 1924, a neoclassical treasure was both its emblem and its façade: the elegant Greek Revival front of the United States Branch Bank, moved to the museum from its original location on Wall Street (Fig. 3).

In *Vasemania* two pieces of silver illustrate the Metropolitan Museum's strong emphasis on American decorative art and the allure of associative history. Although made abroad, both objects have historical relationships with prominent Federal-era figures in the United States. One, a presentation cup honoring Chief Justice William Murdoch (cat. no. 60), is among several pieces of American-owned English silver that Judge Alphonso T. Clearwater included in his large 1933 bequest of the American silver, which he had gradually deposited with the museum as loans during the preceding decade. Similarly, a splendid *goût grec* tureen by Jacques-Nicolas Roettiers was once owned by New York chancellor Robert R. Livingston. Given during the American bicentennial year by Dorothy C. Livingston, the tureen is part of a group of rare Roettiers silver which has been growing since 1933 when the first of seven pieces now in the collection was acquired (see cat. no. 50).

Another development of the early twentieth-century that enhanced the Metropolitan's holdings of neoclassical decorative art was the establishment of the Print Department in 1917,

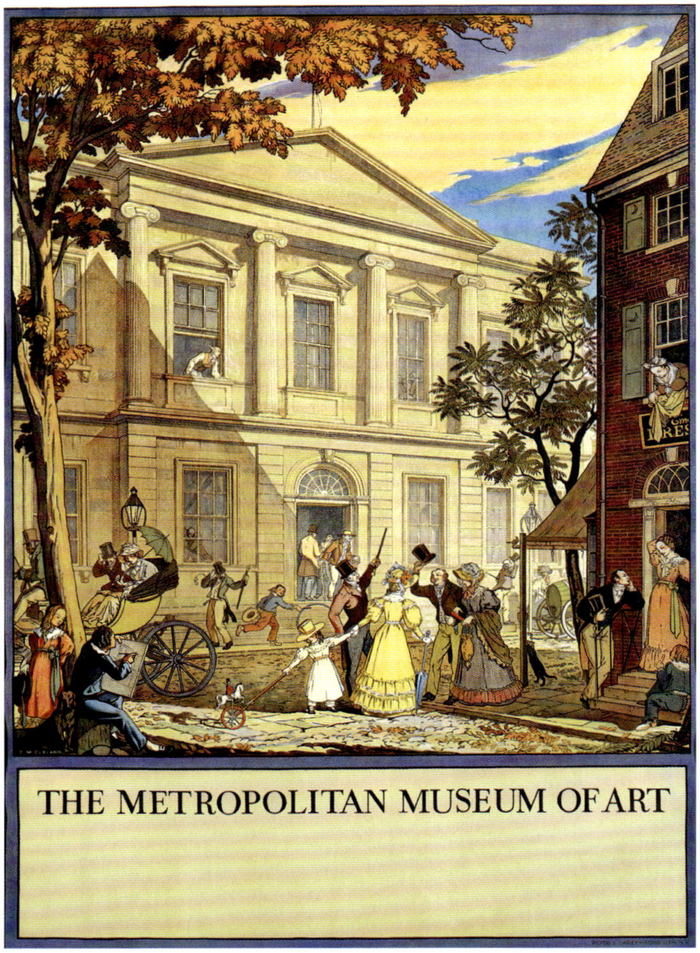

Fig. 3: Artist's proof for a poster advertising the American Wing of the Metropolitan Museum of Art, depicting the 1824 façade of the Wall Street United States Branch Bank, about 1926. Thomas Maitland Cleland. Commercial color lithograph.

inspired by New York publisher Harris Brisbane Dick's bequest of his collection of prints and more than one million dollars for acquisitions. With this gift, and under the direction of the Print Department's accomplished first curator, William M. Ivins, Jr., the collection flourished. One of Ivins's interests was the so-called "ornament print"— sheets, print series, and pattern books produced from the late fifteenth century onward that illustrate designs for objects and interior decoration, as well as pure ornament. Ivins developed this aspect of the museum's print collection by acquiring, exhibiting, and writing about ornament prints, including those of the late eighteenth century, when this material proliferated. The twenty-two engravings and etchings featured in *Vasemania*, supplemented by books and other works on paper, demonstrate the quality and breadth of the museum's neoclassical ornament collection. Another gift that has substantially supported the acquisition of both prints and drawings at the museum is that of Elisha Whittelsey, whose bequest came to the Museum as the Elisha Whittelsey Fund after his wife's death in 1941. *Vasemania* includes twenty-four drawings and prints acquired for the collection with partial or full support from the Whittelsey and Dick funds (see also Fig. 4).[7]

Endowments such as those of Whittelsey and Dick have played a key role in the formation of the Metropolitan's neoclassical collection because they have allowed curators to develop areas that the museum lacks or that need strengthening. Undoubtedly, the most important of these gifts was Jacob S. Rogers's 1901 bequest which, while not specifically directed toward the collection of neoclassical decorative art, has enhanced it enormously. Rogers's story, one of the most remarkable in the history of the Metropolitan, was crucial to its emergence as an institution of

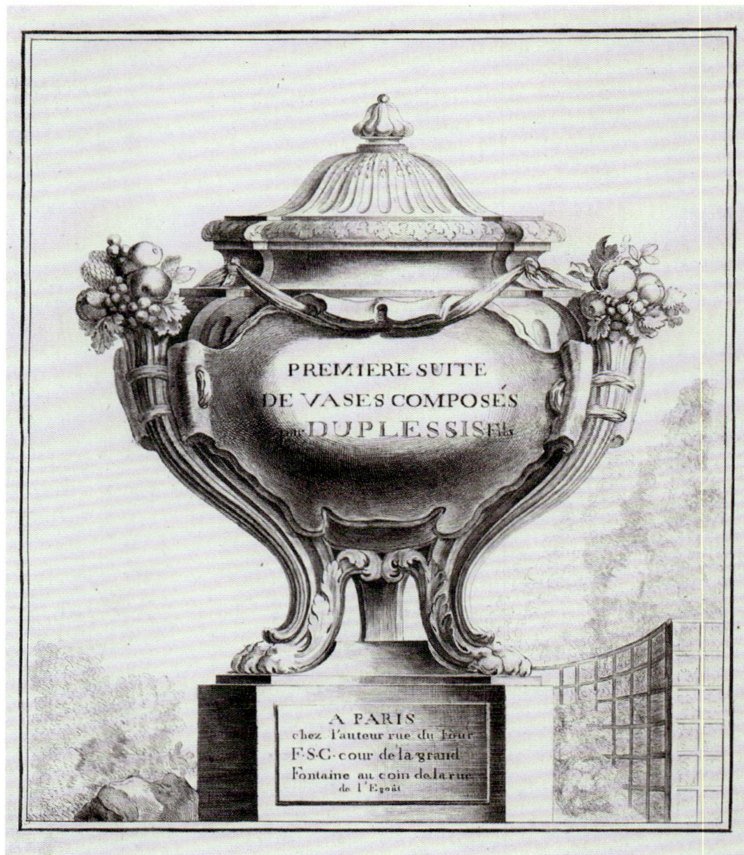

Fig. 4: *Title page from "Premier Suite de vases composés,"* 1775–80. Jean-Claude-Thomas Chambellan the Younger (called Duplessis the Younger). Etching; image: 10 ½ × 9 in. (268 × 228 mm). Harris Brisbane Dick Fund (1941), 41.135.8.(1).

consequence early in the twentieth century. A wealthy New Jersey locomotive manufacturer with no reputation as an art lover or philanthropist, Rogers had made periodic visits to the museum late in life and maintained a ten-dollar membership there, but he astonished museum trustees when, at his death in 1901, his will granted the institution more than five million dollars for the acquisition of books and works of art (see Figs. 5, 9). By far the largest gift the museum had received to that date—and not exceeded for years afterward—Rogers's bequest established an endowment that provides acquisition funds to this day, making his one of the most frequently occurring names in museum credit lines. In *Vasemania* an impressive eighteen objects and works on paper, collected between 1909 and 1978, were purchased with Rogers funds.[8]

* * *

As the history of the museum's American Wing illustrates, interest in neoclassicism, and the decorative arts in general, effloresced during the 1920s. Public enthusiasm for the neoclassical style—not just its American expression—was at a high point in the period. It was, after all, a neoclassical age itself in many ways. Classicism was the accepted idiom in public building of all kinds in early twentieth-century America. Two generations of Paris-trained American architects, aided by widespread admiration for the pristine White City of the 1894 Chicago World's Columbian Exposition, had encouraged ambitious Beaux-Arts building projects. This included renovations at the Metropolitan Museum itself, where classical additions were quickly obscuring the Gothic-inspired polychrome architecture of Calvert Vaux and Jacob Wrey Mould's first museum building in Central Park. In domestic architecture, wealthy clients sought serene, elegantly-proportioned Federal- and Adam-style interiors by architects like Charles Platt and the firm of McKim, Mead, and White. Meanwhile, as Americans increasingly explored their own history and arts, Colonial Revival—or, perhaps more accurately, Federal Revival—architecture was becoming the style of choice in neighborhoods all across the country. Reacting against the tastes of their nineteenth-century grandparents, many Americans of the 1920s found these neoclassical styles refreshingly modern.

In 1920 Ogden Codman's loan to the museum of his collection of books, prints, and photographs reflected the period's increased emphasis on eighteenth-century design. A society architect and interior decorator, Codman, with novelist Edith Wharton, had championed eighteenth-century European-style interiors in their book, *The Decoration of Houses* (Fig. 6). Reflecting his interests, Codman's collection focused on French, English, and American architecture and interiors of the late seventeenth through the early nineteenth centuries. Describing the collection, which was installed in a room Codman decorated, Ivins summed up prevailing attitudes, writing of the eighteenth century that it "happens, in addition to its great favor among those lucky people who have been building and redecorating their houses within the last score years, to be in many ways the most important period in the history of the dwelling house" when "social conditions first required solution of problems in the planning and proportion of private dwellings which are now dominant." He added, in the modernist rhetoric of his time, "Our contemporary house as a machine dates from that time.... The claim for the century rests, therefore, on a simple mechanical basis—the construction of a still utilized mechanism for living."[9]

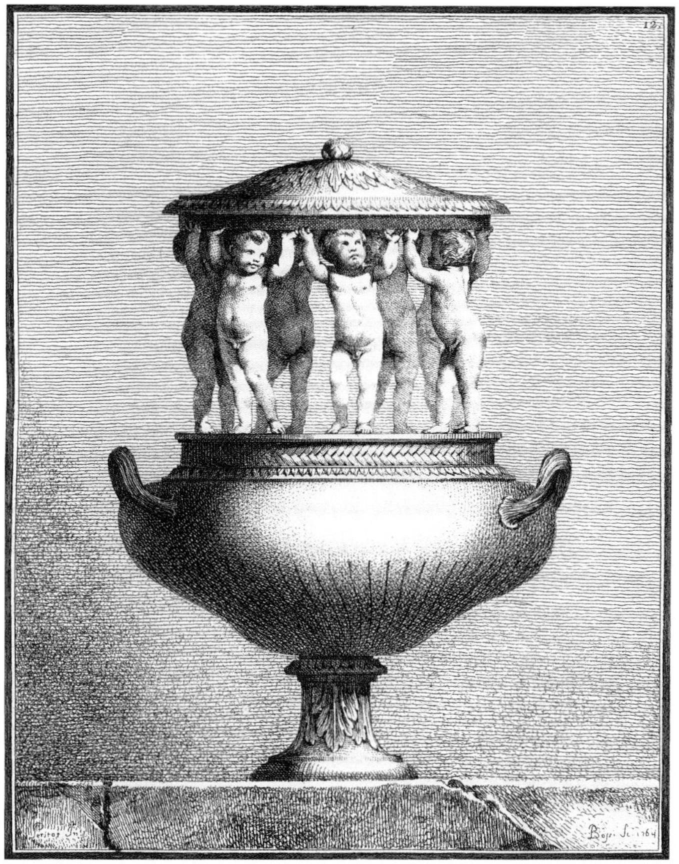

Fig. 5: *Vase with Putti*, 1764. Designed by Ennemond-Alexandre Petitot; etched by Benigno Bossi. Plate 12 from *Suite de vases*. Etching; plate: 9 3/16 × 6 5/8 in. (223 × 168 mm). Rogers Fund (1985), 1985.1024 (12).

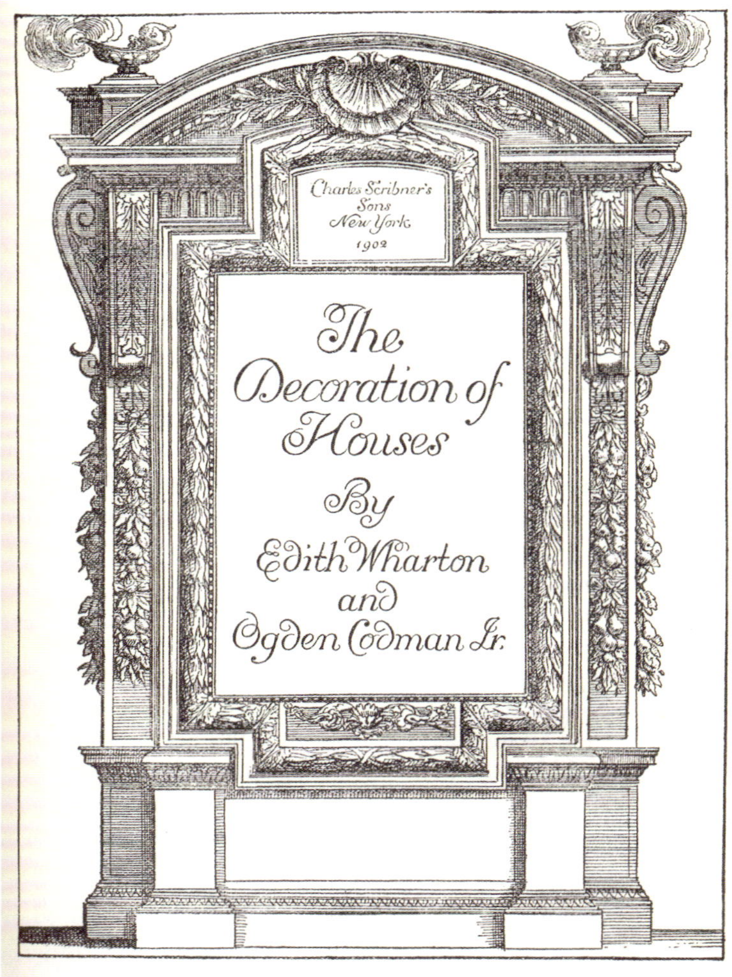

Fig. 6: Frontispiece from Edith Wharton and Ogden Codman, Jr., *The Decoration of Houses* (New York, 1902). Photograph © 2004 Bard Graduate Center/Han Vu.

In the same year, an exhibition of eighteenth-century French decorative art at the museum featured neoclassical objects loaned by a number of private collectors and benefactors, including Jules S. Bache, Mrs. Albert Blum, Mrs. W. P. Douglas, Mr. and Mrs. F. Gray Griswold, the Misses Hewitt, and Mortimer L. Schiff. A review of the exhibit by Royal Cortissoz echoes Ivins, and reflects how the museum's activities both embodied and influenced the tastes of the period: "There runs through [the exhibition], in countless mutations of form and decoration, the feeling for measure and balance, for grace and elegance, for fitness and for charm, which made the eighteenth-century French craftsman an unparalleled servitor of the art of living. How more pointedly could the museum have marked its function as a source of ideas not only in the art of America but in our social life?"[10]

Paralleling these points of view, the growth of the museum's holdings of French neoclassical decorative art increased substantially in the 1920s. Throughout the period, Pierpont Morgan's son Jack gave a series of eighteenth-century French architectural settings to the museum, in part to accommodate the Hoentschel collection, as well as his own recent gifts from his late father's estate. Of these, two represented the neoclassical style: in 1920 Morgan donated a wooden shop front from Paris's Quai Bourbon now installed in the museum's galleries of eighteenth-century French decorative art; two years later, he presented a suite of 1770s interiors from the Hôtel Gaulin, Dijon, including a salon, library, and a bedchamber probably designed by Jérôme Marlet. In 1923 the Wing of the Decorative Arts—which had been renamed the Morgan Wing in 1918 when Jack Morgan's gifts from his father's London collection were installed there—was rearranged, and several rooms devoted to neoclassical, "Louis XVI" objects opened, including the Hôtel Gaulin interiors. Displaying objects "[g]rouped rather by type of design than by actual date," according to a museum account, the French rooms emphasized the style of their contents. Architectural fragments like overdoors, painted panels, and carvings were shown with related furnishings, to bring this material "into close connection with the other objects...with which it was associated in actual use."[11]

The enthusiasm for historical interiors in the 1920s made it an extraordinarily busy decade for the museum's Department of Decorative Art, as staff installed room settings and rearranged galleries to accommodate the rapidly growing collection. The increasing stature of the decorative arts during that period is amply illustrated by the reinstallation of a series of wall panels painted by Hubert Robert for the Comte d'Artois's 1777 renovation of the Château de Bagatelle, and later used by Josephine Bonaparte at Malmaison. Donated by Jack Morgan in 1917, the panels were displayed as framed paintings at the museum until 1924, when they were reinstalled in an architectural setting designed for them. In 1926 the museum opened Wing K, dedicated to the collections of continental European and British decorative art. Making way for holdings that had never before been comprehensively displayed, like Benjamin Altman's large 1913 bequest of eighteenth-century French and other decorative art, the new wing featured the museum's first galleries of nineteenth-century objects—which included displays of later phases of neoclassicism—and a room devoted to the museum's collection of Wedgwood pottery, which by that date contained many examples of the firm's late eighteenth-century wares.[12] Years of vigorous growth for the museum's neoclassical holdings, particularly those of French decorative art, the 1920s and 1930s saw many major gifts—including Darius Ogden Mills's 1929 bequest of forty-six examples of French furniture and clocks, George and Florence Blumenthal's gift of a pair of *athéniennes* in 1933, and Kate Read Blaque's 1938 bequest of sixty-three snuffboxes and *étuis* in memory of her husband, Valentine Alexander Blaque—as well as large and popular exhibitions of eighteenth-century French painting, sculpture, and domestic silver.

* * *

The grim era of war in Europe and the Pacific in the late 1930s and early 1940s disrupted collecting and other normal museum functions, as the most valuable and vulnerable collections were placed in protective storage and heavily attended public concerts replaced much of the regular programming. With the war's end, the sense of renewed purpose and the museum's recognition of its role as a steward of humanity's cultural patrimony, especially in the wake of the destruction in Europe, pervade the early numbers of the post-war *Bulletin*. During the 1920s and 1930s, neoclassical decorative art had emerged as a distinctive area for collecting and interpretation at the museum; in the post-war period, from the 1940s through the 1960s, those collections became among the institution's strongest, as a series of outstanding gifts from Irwin Untermyer, the Kress Foundation, and Charles and Jayne Wrightsman established the Metropolitan Museum as a leading repository of eighteenth-century French and British decorative art.

Having increased steadily since the arrival in 1907 of the Hoentschel collection, the museum's holdings of eighteenth-century French decorative art continued to grow in the late 1940s and 1950s. Delayed by the war, installation of Mrs. Herbert N. Straus's 1942 gift of *boiseries* from the neoclassical reception room (*sale du dais*) of the Hôtel de Tessé, Paris, was complete in 1954 (Fig. 7). A committed benefactor, Mrs. Straus also donated an exquisite round room of the 1780s from a house on the Cours d'Albret, Bordeaux (Wrightsman Galleries), and gave more at her death, including a pair of rare Sèvres *vases à monter* featured in *Vasemania* (cat. no. 16).

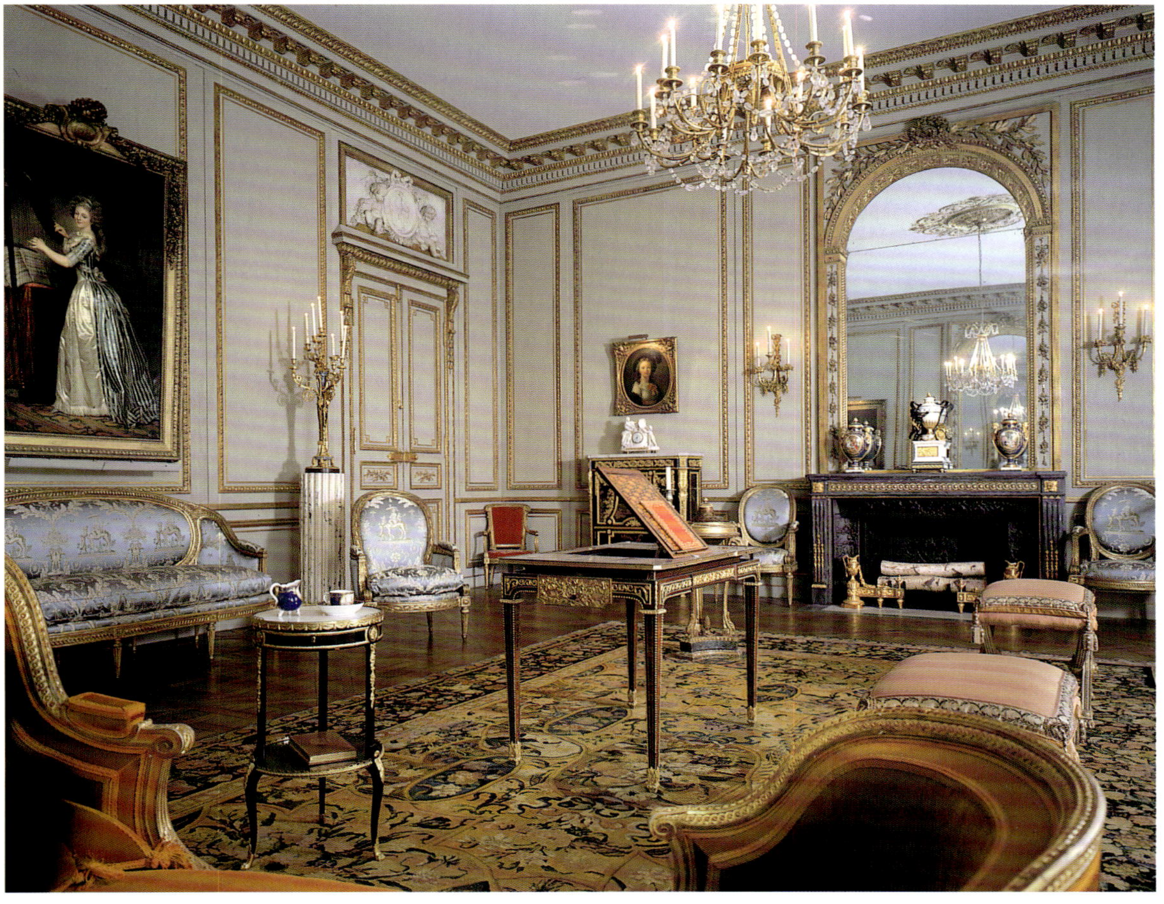

Fig. 7: Reception room from the Hôtel de Tessé, Quai Voltaire, Paris, 1768–72. Designed by Pierre-Noël Rousset, with painted and gilded oak woodwork by Nicolas Huyot, and stucco overdoors attributed to Pierre Fixon. Gift of Mrs. Herbert N. Straus (1942), 42.203.1.

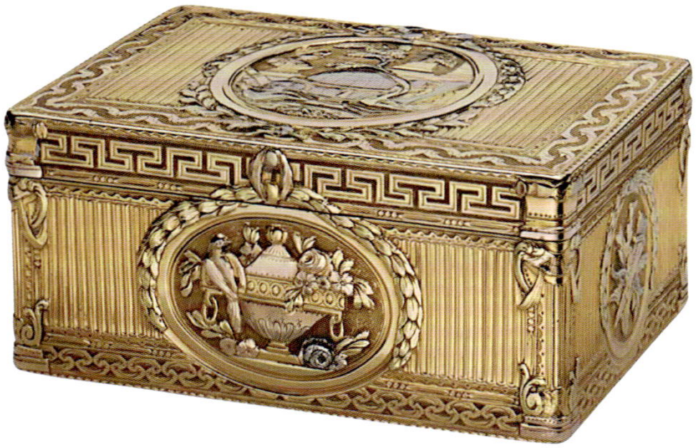

Fig. 8: *Snuffbox*, 1763–64. Jean Frémin. Gold; 1 ½ × 3 ⅛ × 2 ¼ in. (3.8 × 7.9 × 5.7 cm). Bequest of Catherine D. Wentworth (1948), 48.187.423. This *goût grec* snuffbox features a panel with a vase similar to the designs of Jean Charles Delafosse. In 1945, after World War II ended, the Wentworth snuffboxes—then on loan to the Metropolitan Museum—were brought from protective storage and exhibited in the Morgan Wing.

Another highlight of the post-war period was a spectacular gift of French eighteenth-century domestic silver and other objects from Santa Barbara native Catherine Wentworth in 1949 (Fig. 8). Exceeding four hundred pieces, the Wentworth collection was the largest of its kind in the United States. Like Wentworth, donors whose collecting focused on individual materials have played an important role in building the museum's neoclassical collection in specific areas. Others who made similar contributions include Judge Clearwater, whose bequest of American and English silver has already been mentioned; R. Thornton Wilson, who gave hundreds of pieces of porcelain and other ceramics in his wife's memory between the 1940s and the 1960s; Frank K. Sturgis, a collector of coaching prints and portraits of racehorses, who donated and sold many pieces from his collection of Wedgwood and other pottery to the museum in 1932 (see cat. nos. 6, 40, 42–44); as well as numerous donors of large gifts of laces and other textiles which included late eighteenth-century pieces.

While the museum's holdings of French decorative art grew in size and substance during the first part of the twentieth century, the British collections developed more slowly. Indeed, during the 1910s the museum's English decorative art was seen in part as a supplement to the American collection; exhibited with American objects, English furnishings were intended to provide instructive examples of the Old World antecedents of American forms and styles. Early gifts of English objects from donors like Marsden J. Perry and John L. Cadwalader included some neoclassical pieces, such as a soft-paste porcelain vase painted by Richard Askew, bequeathed to the museum by Cadwalader in 1914 (cat. no. 27). As late as 1929, however, curator Preston Remington wrote that British decorative art of the later eighteenth century had "never been adequately represented" at the museum, and noted that recent accessions of furniture, metalwork, and glass illustrating the neoclassical style were intended to redress the deficiency.[13] Over the next decade, the Metropolitan Museum continued to make purchases of late eighteenth-century English objects in a range of materials, including a pair of Staffordshire Church Works medallion vases featured here (cat. no. 37). In the same period, it received donations which supplemented these acquisitions, such as Sturgis's gifts of pottery and several delicate painted English tables from Louis J. Boury.

The late 1940s through the 1960s were the most significant years of growth for the museum's collection of eighteenth-century British decorative art. One highlight of the era was the installation of two English period rooms, including the museum's first British neoclassical interior, Robert Adam's elegant dining room for Lansdowne House, London, which opened in 1954 (Fig. 9). Throughout these years, however, one individual was responsible for what became—over the course of some thirty years—the most substantial single contribution to the museum's representation of British decorative art, including its neoclassical holdings. In a series of yearly donations and loans made from the 1940s, Judge Irwin Untermyer deposited a superb collection of English furniture and silver, as well as English and continental embroideries, porcelains, and early bronzes, at the museum (see Fig. 10). Dominated by the English pieces, Judge Untermyer's gift amounted to an astonishing 2,129 objects at his death in 1973.

Among the neoclassical objects in the Untermyer collection, some unusual examples of marquetry case furniture stand out, including a commode attributed to Messrs. Ince and Mayhew with a rare single front panel depicting a candelabrum and a pair of vasiform burners. Several examples of metalwork by Matthew Boulton and John Fothergill's Birmingham manufactory are also noteworthy Untermyer gifts, as is an unusual Roman-inspired ivory candelabrum probably made in India for the English market about 1800. The Untermyer collection is represented here by two sets of objects: a pair of mahogany pedestals with water and wine cisterns in the form of classical vases (cat. no. 69); and a pair of Boulton and Fothergill spar and gilt-bronze perfume burners

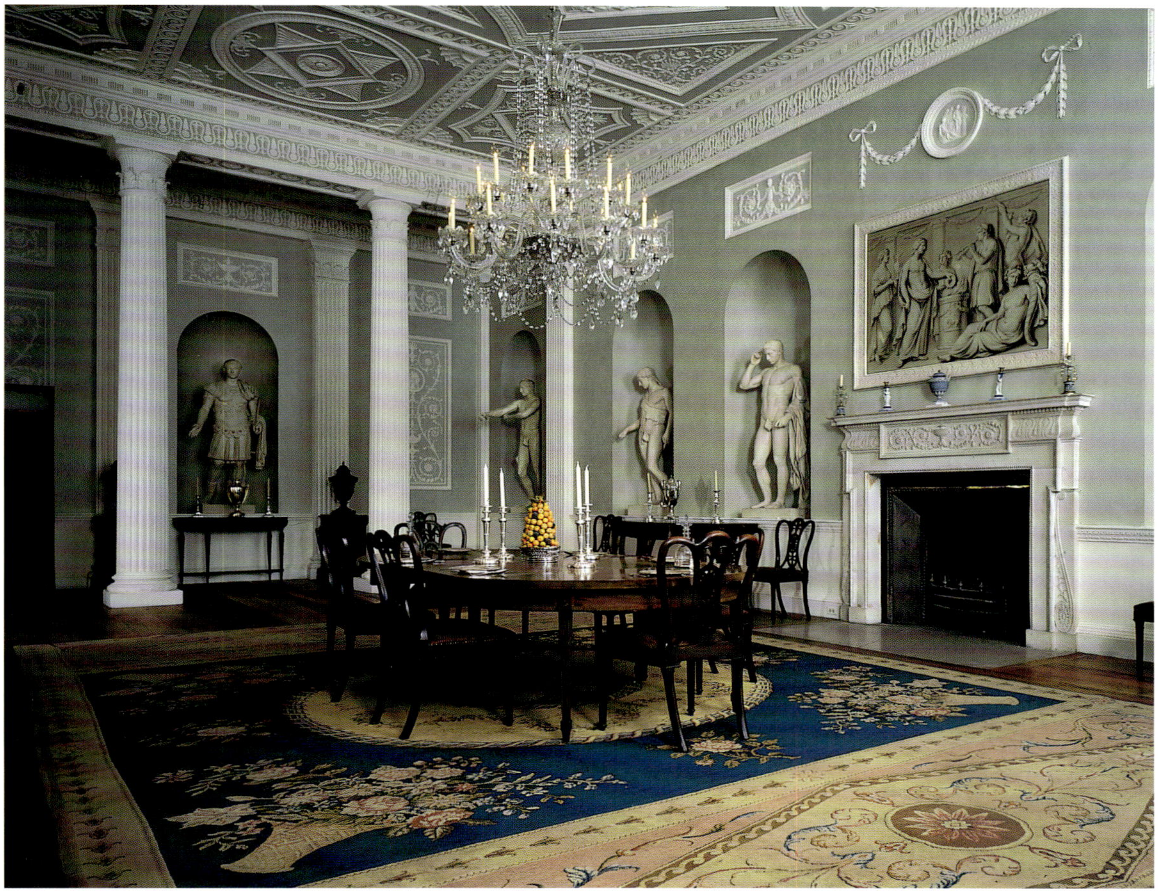

Fig. 9: Dining room from Lansdowne House, Berkeley Square, London, 1766–69. Designed by Robert Adam. Wood, plaster, and stone. Rogers Fund (1931), 32.12.

(cat. no. 45). Amassing his collection with the intention of leaving it to the museum, Judge Untermyer acquired objects with an eye to strengthening existing holdings and avoiding duplications. As a result, and because of his constant refinements, Untermyer's gift had maximum impact. By 1977, when a memorial exhibition highlighted selections from the Untermyer collection, the museum could boast that its representation of English decorative art was nearly without rival.[14]

The museum's pride in its British collection rested partly on another remarkable addition to its late eighteenth-century holdings which had arrived in 1958. That year the Kress Foundation presented the Metropolitan with nearly one hundred examples of European decorative art, unmatched for their quality and uniqueness in the museum's existing collections. Among the many extraordinary items in the Kress gift, the Earl of Coventry's drawing room from Croome Court, designed in the 1760s by Robert Adam and hung with the trompe l'oeil Beauvais tapestries that were made for it, became a jewel in the crown of the English collections (Fig. 11). A rare surviving example of a style of decoration

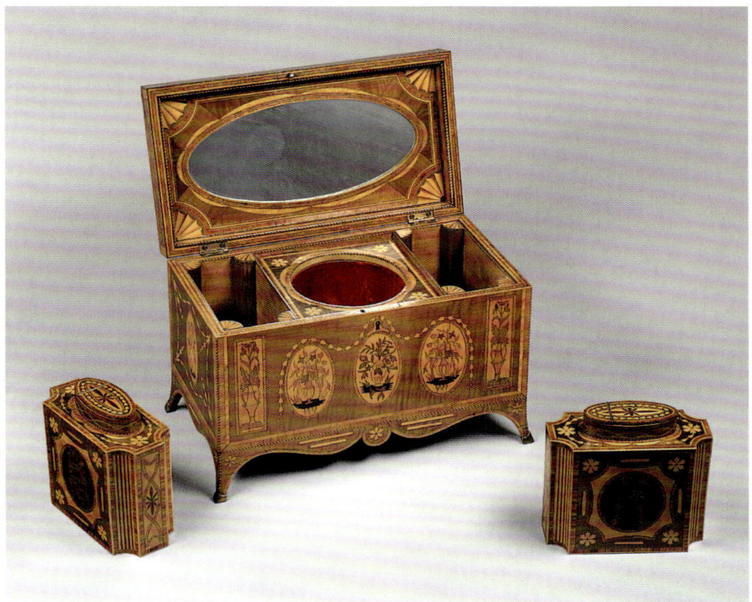

Fig. 10: *Tea Chest*, English, about 1790. English. Hardwood veneer, inlaid with tulip and other woods; 8 ¼ × 12 ¾ × 7 ½ in. (21 × 32.4 ×19 cm). Gift of Irwin Untermyer (1964), 64.101.1192–.1194.

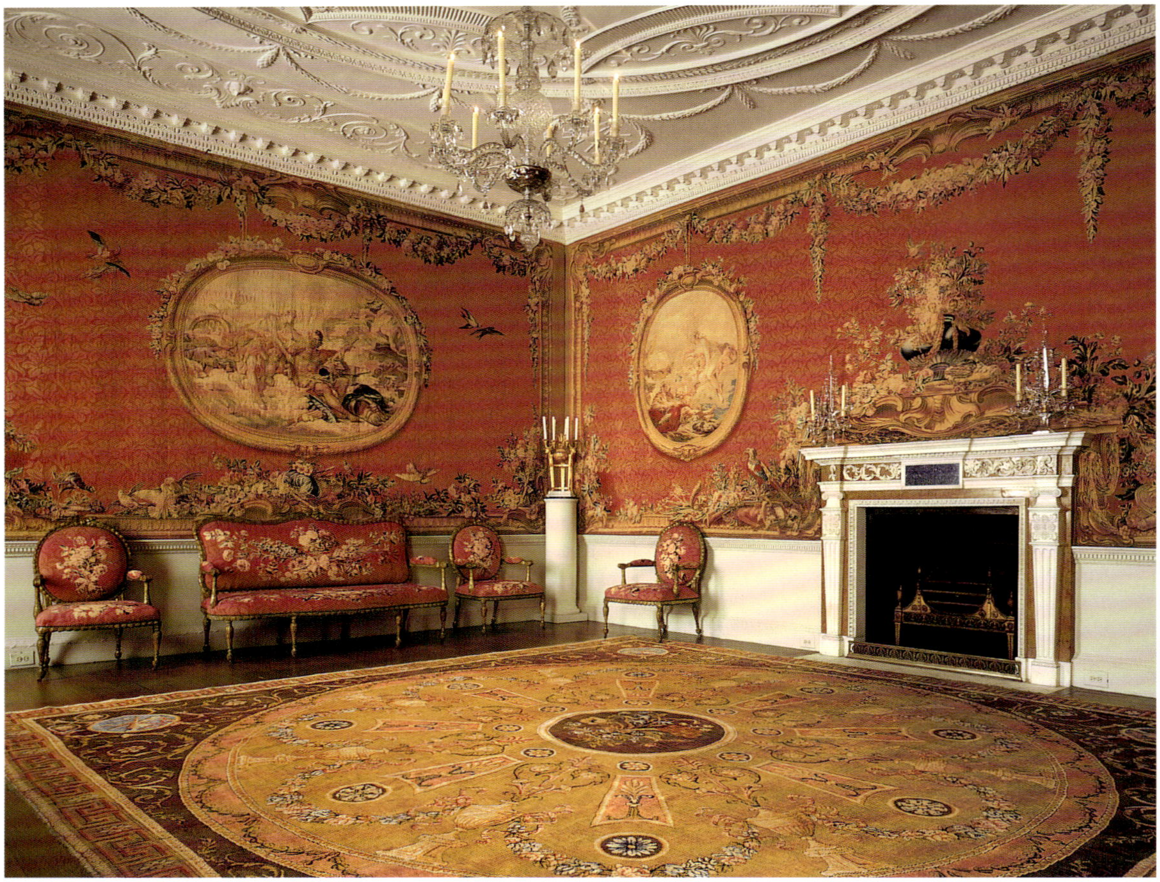

Fig. 11: Tapestry Room from Croome Court, Worcestershire, 1763–71. Plaster ceiling designed by Robert Adam and executed by Joseph Rose; wool and silk tapestries designed in 1758–67 by François Boucher, and woven in 1764–71 by the workshop of Jacques Neilson (1714–1788) at the Royal Gobelins Manufactory, Paris. Gift of the Samuel H. Kress Foundation, 1958. (58.75.1–.23).

that was unique to England, the Croome Court "tapestry room" arrived at the museum as a remarkably complete ensemble, brought together following extensive research which had included locating and reuniting the seating furniture with its original covers. Recently reinstalled, the Croome Court drawing room joins Adam's refurbished Lansdowne dining room and other interiors dedicated to British neoclassicism in the 1995 Annie Laurie Aitken Galleries of English Decorative Art, the most recent milestone in the development of the museum's collection of British neoclassical decoration.

Croome Court was only part of the Kress Foundation's unprecedented gift. Described by the museum as "one of the most important donations of decorative art ever made to any museum at any time,"[15] it also included the Hillingdon collection of French furniture and Sèvres porcelains, assembled by the British royal banker Charles Mills in the early nineteenth century (see Fig. 12). Comprising fourteen pieces of furniture embellished with porcelain plaques (seven signed by Martin Carlin, and others by Bernard van Risenburgh and Joseph Baumhauer), furniture by David Roentgen and Guillaume Kemp, and fifty-five pieces of Sèvres porcelain (see cat. nos. 15, 17–18, 20–23), the Hillingdon collection thrust the Metropolitan Museum into the first rank of international collections of French decorative art. In addition to the Hillingdon pieces, the Kress gift included a suite of furniture made for Marie-Antoinette, some of it stamped by Jacob, as well as other stamped pieces and *objets d'art*. Supplementing its donations, the Kress Foundation supported the 1964 publication of a catalogue of the objects, making an important contribution to the literature of European neoclassicism in the decorative arts.[16]

The splendid Kress gifts of 1958 all but overshadowed another important donation noted in the museum's annual report that year—one that would begin one of the most sustained and rewarding relationships the Metropolitan Museum has had with any benefactor. In 1958 Charles and Jayne Wrightsman provided funds to acquire the textiles needed for draperies and seating furniture in the Tessé salon, which the museum was then

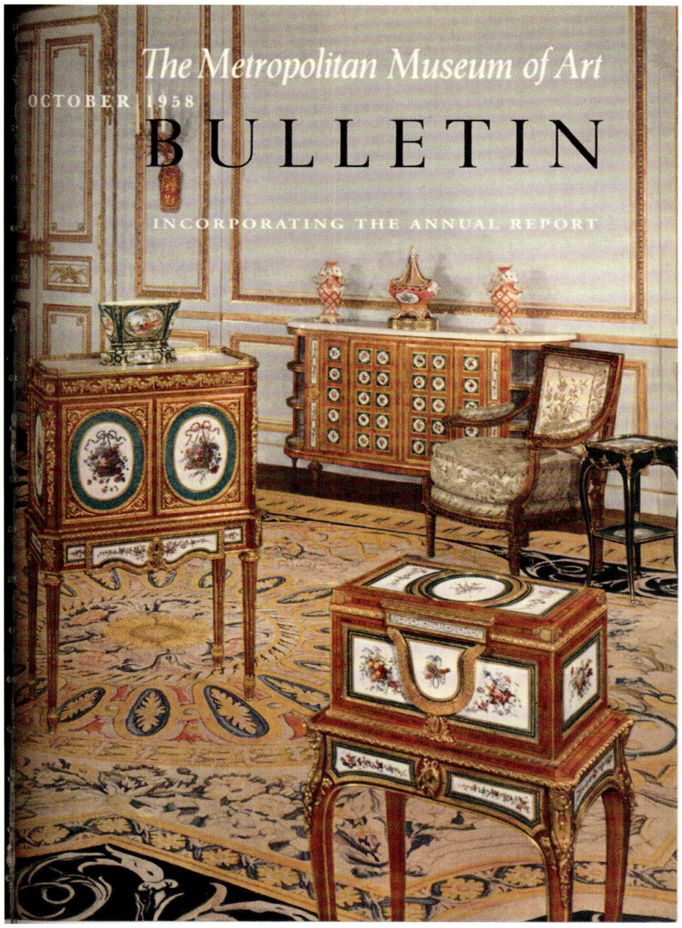

Fig. 12: Cover of *The Metropolitan Museum of Art Bulletin* (Oct. 1958). Vol. XVII, No. 2. Seen here in a temporary display in the museum's Tessé Room are some of the furnishings from the former Hillingdon collection, presented in 1958 to the Metropolitan Museum by the Samuel H. Kress Foundation.

reinstalling (Fig. 7). The Wrightsmans were in the process of amassing the foremost collection of eighteenth-century French furniture and related decorative arts in the United States; in the following years, through a series of gifts and loans, they deposited large parts of their collection and other objects at the museum. In 1962, as the Kress catalogue went to press, the Wrightsmans initiated plans with the museum to publish a catalogue of their own collection, a testament to the relationship that had already developed between the donors and the institution. That publication, ultimately amounting to five volumes published between 1966 and 1973, became a landmark in the study of French eighteenth-century furniture.[17]

As work on the catalogue progressed through the early 1960s, the Wrightsmans supported the 1963 purchase of two mid-eighteenth-century interiors, beginning a series of similar gifts that anchor the twelve Wrightsman Galleries of French decorative art at the Metropolitan Museum today. With their sponsorship of a new late eighteenth-century gallery in 1966, representation of French neoclassicism at the museum bloomed: through the 1970s, Wrightsman funds supported the completion of the Tessé room, the installation of a dining room from the Hôtel de Cabris, Grasse, given anonymously in 1972 (Fig. 13), and the purchase of an interior from collector Baron Pichon's Hôtel Lauzun, Paris. Throughout this period, the Wrightsmans supplemented these large-scale projects with ongoing gifts and loans of hundreds of objects to enhance the rooms. Today, the furnishings housed in the Wrightsman Galleries are a highlight of the museum's collection.

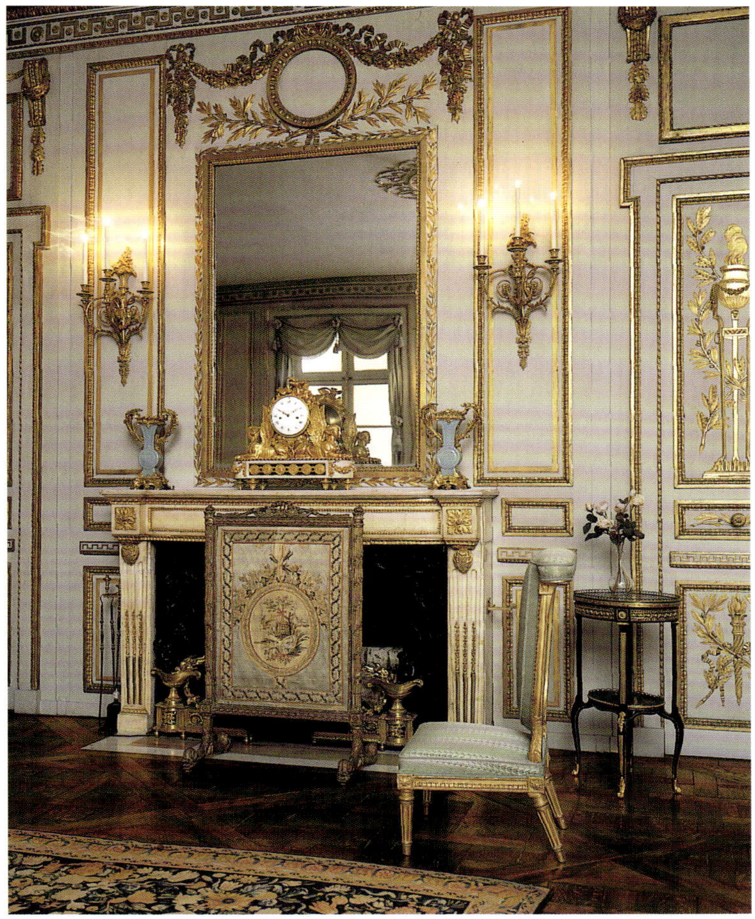

Fig. 13: Dining room from the Hôtel de Cabris, 13 Boulevard Carnot, Grasse, 1775–78. French. Carved oak and plaster, painted and gilded, with gilt-bronze, mirror glass, and oak flooring. Purchase, Mr. and Mrs. Charles Wrightsman Gift (1972), 1972.276.1–.2.

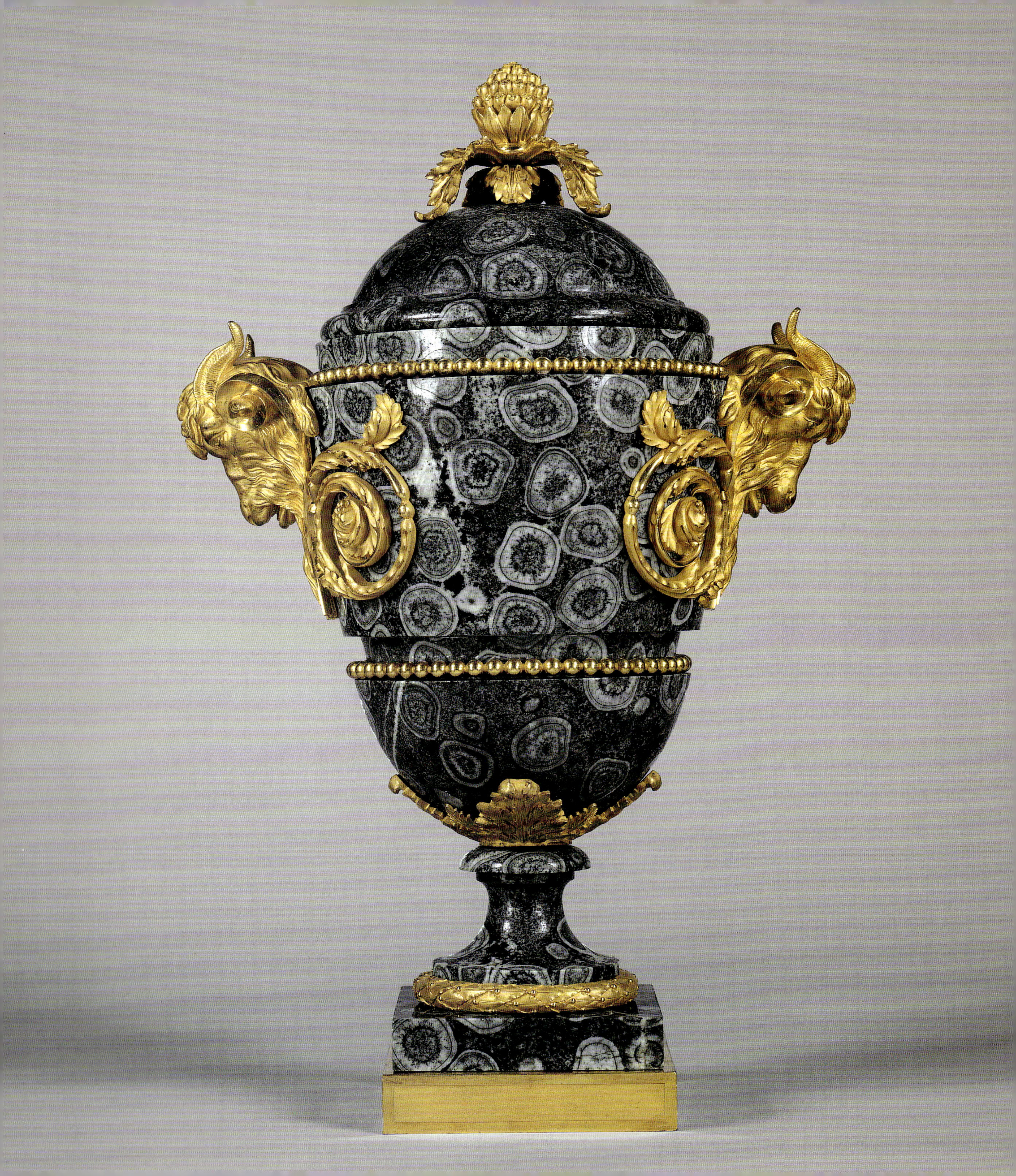

The Wrightsman legacy of giving continues to expand the museum's neoclassical holdings. In the 1990s, Mrs. Wrightsman presented the department of European Sculpture and Decorative Arts with a pair of very early *athéniennes*, close in style to the first known example of the form in France, as well as a pair of large granite and gilt-bronze vases which complement the objects featured here (Fig. 14). Recently, the Wrightsman Fund has supported acquisitions in areas of European neoclassicism that the museum's collection lacks. Purchased in the year 2000, an 1834 gilt-bronze pedestal table with a Berlin porcelain top—possibly based on the designs of visionary neoclassicist Karl Friedrich Schinkel—illustrates the later neoclassical style of a region not as well represented in a collection dominated by holdings of French and English decorative art.

The acquisition of the Schinkel table underscores the point that, rich as the Metropolitan Museum's holdings are, there are aspects of European neoclassicism yet to be explored. Eighteenth-century Italian neoclassical design, for instance, is represented by the prints of its foremost figure, Giovanni Battista Piranesi, but related collections of Italian as well as Eastern European and German decorative art remain largely undeveloped. Auspiciously, newer acquisitions—such as a pair of Luigi Valadier candelabra, an allegorical figure of the Nile in biscuit porcelain by Giovanni Volpato, a pair of late eighteenth-century Russian pierced bone vases by Nikolai Stepanovich Vereshchagin, and a pair of Viennese silver wine coolers made by Ignaz Josef Würth (cat. no. 51)—show that recent collecting is beginning to fill the gaps.[18] Nearly one hundred years after J. Pierpont Morgan's gifts from Georges Hoentschel's collection were announced, the work of rounding out Metropolitan Museum's neoclassical holdings represents a new stage in confirming their status as a collection. In telling their history, *Vasemania* is yet another.

Fig. 14: *Vase* (one of a pair) about 1760. French and Russian. Orbicular diorite and gilt-bronze mounts; 24 in. (61 cm), diam. 16 in. (40.6 cm). Gift of Mrs. Charles Wrightsman (1999), 1999.122.1.

1 In 1926 the Metropolitan Museum opened its first galleries devoted to nineteenth-century decorative art. Describing the installation, the museum's *Bulletin* noted that, along with displays representing "Romanticism" and "Antiquarianism," two rooms of furniture and other objects from the early nineteenth century illustrated "Neo-Classicism"; see "The Decorative Arts in the Nineteenth Century," *MMA Bulletin* 21 (Dec. 1926): 287–93.

2 Noteworthy books on the general development of the Metropolitan's collection include Winifred Howe's *History of the Metropolitan Museum of Art*, 2 vols. (New York, 1913–46); Leo Lerman, *The Museum: One Hundred Years and the Metropolitan Museum of Art* (New York, 1969); Calvin Tomkins, *Merchants and Masterpieces: The Story of the Metropolitan Museum of Art* (New York, 1970); John L. Hess, *The Grand Acquisitors* (Boston, 1974); and Howard Hibbard, *The Metropolitan Museum of Art* (New York, 1980). The collecting and philanthropic activities of a number of the donors noted in this essay have also been well documented; many are discussed in Germain Seligman, *Merchants of Art, 1880–1960: Eighty Years of Professional Collecting* (New York, 1961).

3 "An Appeal to the Public," by the Trustees of the Metropolitan Museum of Art, Mar. 3, 1871, repr. in *MMA Bulletin* 15 (May 1920): 98–101.

4 On Marquand and his collection, see Danielle Kisluk-Grosheide, "The Marquand Mansion," *MMA Bulletin* 29 (1994): 151–81; and more recently idem, "A Piano for a New York Millionaire," *CAI: Journal of the Clark Art Institute* (2000): 48–57.

5 A recent issue of the *Bulletin* is devoted to Morgan's gifts to the Metropolitan Museum; see Jean Strouse, "J. Pierpont Morgan, Financier and Collector," *MMA Bulletin* 62 (Winter 2000). Hoentschel is the subject of a recent full-length study; see Nicole Hoentschel, Vincent Bouvet, et al., *Georges Hoentschel* (Saint-Rémy-en-l'Eau, 1999).

6 A 1907 loan exhibition assembled by the Colonial Dames of New York State was perhaps the earliest effort, but the Metropolitan's experiment was rewarded with a successful display of American decorative arts at the museum's 1909 Hudson-Fulton Celebration Exhibition. Shortly after the exhibition closed, Mrs. Russell Sage's gift of the large Bolles collection of American furniture established the museum as a leading collector of decorative art made in the United States.

7 For acquisitions supported by the Harris Brisbane Dick Fund, see cat. nos. 9, 28, 34, 80–82, 90–92. For those acquired with the Elisha Whittelsey Fund, see cat. nos. 14, 46–49, 53, 63–67, 88–89, 93–94.

8 For acquisitions supported by the Rogers Fund, see cat. nos. 2–4, 7, 12, 31–33, 37, 41, 54, 57, 61–62, 64–65, 68, 76, 79.

9 William M. Ivins, Jr., "The Ogden Codman Collection," *MMA Bulletin* 15 (Nov. 1920): 251–52.

10 Royal Cortissoz, from a review published in the *New York Tribune*, May 9, 1920, quoted in "French Decorative Art and Sculpture of the Eighteenth Century," *MMA Bulletin* 15 (June 1920): 132.

11 "Rearrangement in the Morgan Wing," *MMA Bulletin* 18 (May 1923): 106.

12 See note 1.

13 Preston Remington, "Late Eighteenth-Century English Furniture: Recent Accessions," *MMA Bulletin* 24 (Nov. 1929): 295.

14 Olga Raggio, "Introduction," *Highlights of the Untermyer Collection of English and Continental Decorative Arts* (New York, 1977): 7.

15 "Eighty-Eighth Annual Report of the Trustees for the Fiscal Year 1957–58," *MMA Bulletin* 17 (Oct. 1958): 63.

16 Carl Christian Dauterman, James Parker, and Edith Appleton Standen, *Decorative Art from the Samuel H. Kress Foundation at the Metropolitan Museum of Art...* (Aylesbury, Buckinghamshire, 1964).

17 Francis J. B. Watson, et al., *The Wrightsman Collection*, 5 vols. (New York, 1966–73).

18 Valadier candelabra: Wrightsman Fund, 1994 (1994.14.1–.2); Volpato, "The River Nile": Purchase, The Isak and Rose Weinman Foundation Inc. Gift, 2001 (2001.456); Vereshchagin bone vases: Purchase, Friends of European Sculpture and Decorative Arts Gifts, 1998 (1988.13.1–.2).

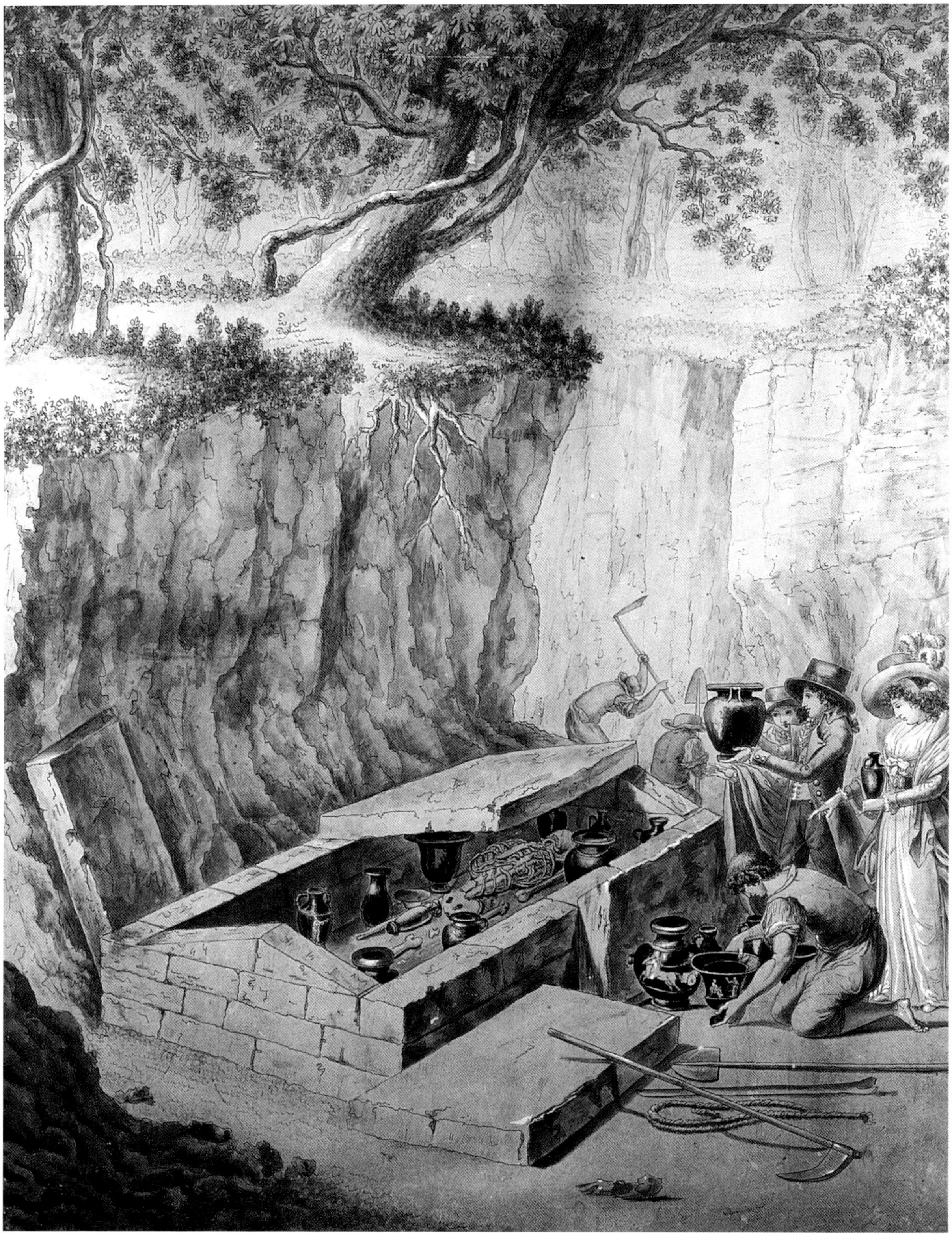

The Metamorphosis of the Neoclassical Vase

Hans Ottomeyer

The vase was a leitmotif during each of the three main phases of neoclassicism: the early, experimental *goût grec*; the mature, elegantly exuberant *goût étrusque*; and the late, more restrained *goût antique*, often called Empire. Designing and producing vases for a multitude of purposes became one of the main challenges for eighteenth-century designers. Why were vases so attractive to them? The reasons transcend issues of stylistic development or historical events and go back to the elemental properties of the form.

The vase is a vessel whose walls rise vertically from its foot. Depending on its shape, it can be described as a basin, bowl, ewer, goblet, pot, urn, and more. Vases are constituted of basic parts with names derived from the human form: foot, neck, and handles, the latter often set like arms onto the shoulders of the vessel's body. This reversible metaphor was well known since antiquity. Cicero, for example, compared the human figure to a vase: "for the body is like a vase and like the vessel of the soul."[1] Beyond its anthropomorphism, the vase possesses another quality that made it an obvious choice for neoclassical designers. Its body offers a continuous, ring-shaped image field that permits a friezelike composition to run around it like a panorama turned inside-out. Without the limits of a frame, sculptors and painters found that they could employ a vessel's continuous surface as a background for reliefs, bands of figures, and an endless variety of repeating patterns. Regarding their placement and display, vases are similar to sculptures. They stand on bases, socles, or plinths or they can serve as crowning features of both an interior and exterior setting. In this context, they become exclusively objects of aesthetic contemplation, removed from any utilitarian purpose.

The function of vases has changed considerably over time. In antiquity they were used as containers for wine, oil, and water, as well as for ceremonial purposes at feasts and sacrifices. They also served as presents, trophies, or temple offerings, and thereby conferred honor on their recipients or provided a source of aesthetic contemplation. This was the case, for example, for the Panathenean games, where the winner received a trophy vase as a prize. However, what secured the form's survival over millennia was its use as a burial gift and funerary urn.

It was not customary in ancient Greece to provide the dead with artistic vases to accompany them on their final journey. However, among the Etruscans and those Greeks who lived in the Roman Campania and in southern Italy, the bestowal of precious Greek vases upon the deceased became common; this was true as well as for the Thracians, Macedonians, and Celts. Later, for centuries in the areas dominated by Rome and its extended empire, marble- and ceramic-relief burial urns (called *cippi* in Italian, from the Latin *cippus*, or headstone) were produced to contain the ashes of the dead. Collected since the Renaissance, first in Italy and then throughout Europe, vases had an immense aesthetic impact, equal to that of monumental sculptures; together, they determined the image of the ancient world. In fact, because stone vases adorned with friezes were frequently excavated undamaged, their influence could exceed that of frequently fragmented sculpture and architectural examples. With their visual vocabulary intact, vases became guiding forms for studying and interpreting the ancient world. As Werner Oechslin, author of the most complete exhibition and monograph on the vase to date, stated, "The vase belongs [entirely] to antiquity."[2]

The Renaissance, mannerist, baroque, and neoclassical periods—each with its own view of antiquity—absorbed the designs of classical vases and made choices, variations, and new interpretations that entered their own artistic canon. Renaissance designers defined the vase as a clear, geometric body; bizarre and grotesque motifs on ancient vessels fascinated proponents of mannerism; baroque artists approached them as sites for lavish figural compositions; and devotees of neoclassicism admired their noble simplicity and serene contours. With its inexhaustible diversity, the classical vase provided each of these revival periods with models to imitate and elaborate upon. In that regard, the vase's principle function changed. From an object intended for use, it turned into a multilayered template for ornamental and iconographic invention. Thus, it seems safe to say that the concept of "art for art's sake" did not first occur with the aesthetic movement in the late nineteenth century, but rather at least one hundred years earlier. In the neoclassical period, the vase fascinated artists as a format unencumbered by functionality, one that offered infinite possibilities to express design and style in a play of abstract overall shape, geometric planes, and two- and three-dimensional images.[3]

In Europe the vase enjoyed its greatest popularity as a cultural icon between 1700 and 1825. Those conceiving, decorating, and making vases included architects, designers, sculptors, painters, bronze workers, porcelain makers, and iron founders—

See Fig. 7.

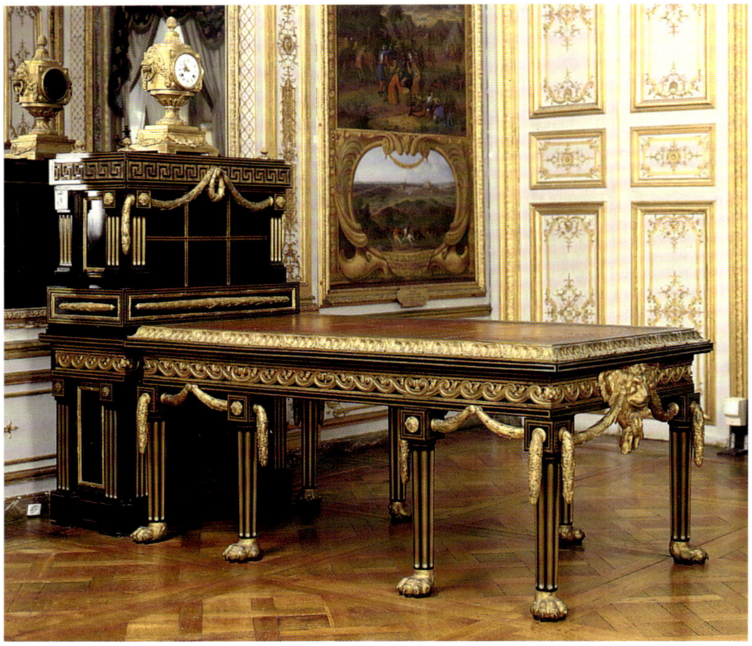

Fig. 2: *Writing Desk and Filing Cabinet*, 1756–57. Designed by Louis-Joseph Le Lorrain; bronze mounts by Philippe Caffiéri II. Wood, ebony veneer, bronze, and gilt bronze. Musée Condé, Chantilly. Photograph © Réunion des Musées Nationaux/ Art Resource, NY.

Parisians hurried to see the ultramodern furniture. Lalive de Jully wrote in 1764, "Since the execution of this cabinet, the taste in works 'à la grecque' has spread. Now one applies it in a ridiculous fashion to everything: to table china, jewelry, fabrics, haircuts, and even to shop signs, all of which are today 'à la grecque.'"[11] Critic Charles-Nicolas Cochin declared:

> [At the time,] all upstanding people used ancient ornaments entirely falsely. They distorted them, decorated their window seats with heavy decorative bands and committed a thousand other stupidities. The painter Le Lorrain designed in a heavy style the furnishings for the apartment of M. de la Live, a wealthy art lover, who also drew a little. The fame of the [ensemble] increased when the Comte de Caylus praised this undertaking enthusiastically. From that stemmed the garlands [that resemble] well ropes; vases that, according to ancient usage, contained liquids became clocks with hour-dials rotating around them, beautiful inventions that were imitated by all the ignorant; and Paris was flooded with every possible thing *à la grecque.*[12]

From such accounts and many surviving examples, it is clear that practically all types of decorative arts produced during the economic boom that followed the end of the Seven Years' War in 1763 were conceived in the *goût grec* idiom, but, because of overuse and exaggeration, the style was already beginning to be discredited by 1770 for being unnecessarily heavy and ornate. The preeminent artist-designer of the short-lived *goût grec*, Jean-Charles Delafosse, must be mentioned in this context, for he published his influential *Nouvelle Iconologie historique* in 1768, after a number of years of preparation (see Fig. 3). His detailed, bizarre designs, which almost always feature monumental vases, are complex. The designer's vessels consist of individual motifs interlaced or stacked within or atop one another, resulting in an overall abstract effect (see cat. no. 9). Delafosse's forceful and attractive compositions were adapted, directly and indirectly, to numerous vase and ornament designs.[13] But the vases actually produced in his style simplified the repertory of his forms, resulting in a lighter, finished appearance that helped keep their production cost down (see Fig. 4). Enhancing the significance of these vase designs for the decorative arts was the devotion of practitioners of early neoclassicism to monumentality and grandeur, which they achieved using a vocabulary of columns, pilasters, architraves, friezes, and vases. In this typology, the central and crowning elements were classical vases, which thus became the iconic motif of the *goût grec.* They appear frequently as finials, for example, on cartel clocks and silver vessels (see cat. no. 50), as well as on sconces atop a main structure shaped by pilasters.[14]

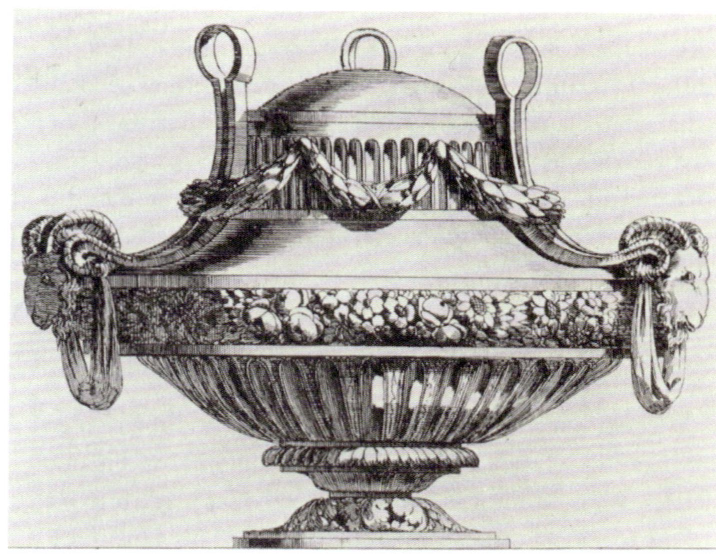

Fig. 3: *Vase with Ram's Heads*, 1768. Jean-Charles Delafosse. Detail of Vase 46 from *Nouvelle Iconologie historique.* Engraving. Photograph Hans Ottomeyer and Peter Pröschel, *Vergoldete Bronzen: Bronzearbeiten des Spätbarock und Klassizismus*, vol. 1 (Munich, 1986): 183, fig. 3.8.6.

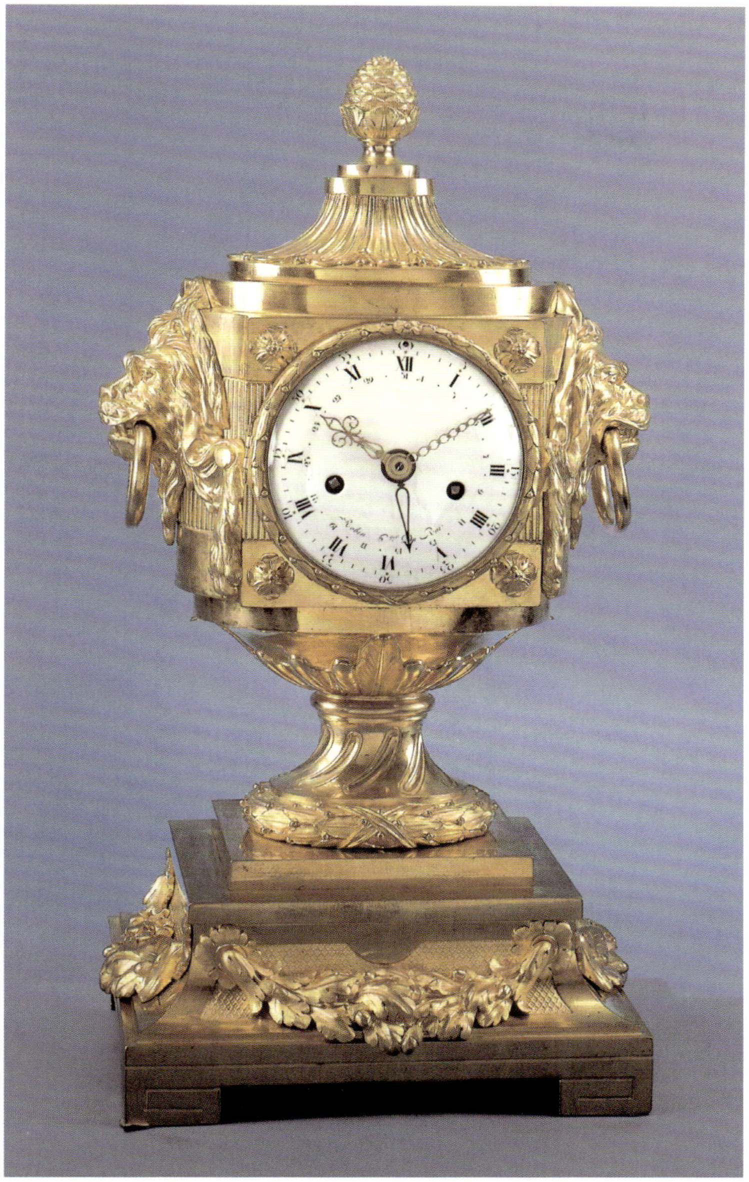

Fig. 4: *Mantle Clock with Lion Heads*, about 1770. Case by Robert Osmond; movement by Robert Robin. Gilt bronze, enamel; 24 3/8 × 11 13/16 × 10 in. (62 × 30 × 25.5 cm). Cleveland Museum of Art.

Nonetheless, between 1755 and 1770, when the *goût grec* dominated, a lingering taste for the rococo is discernible in many transitional designs. Not everyone was prepared to fully embrace the *goût grec*'s strict and logical approach, one that broke with familiar and friendly forms. Commenting on the ubiquitous presence of the acanthus-leaf motif in *goût grec* designs, the Marquis de Marigny, brother of Madame de Pompadour, declared: "I want neither the modern cabbage [style], nor the austere ancient [one]. Half of one, half of the other."[15] This expression of ambivalence toward the new, neoclassical manner—which, after the spirited asymmetry of the rococo, seemed overly formal, even rigid—reflects the attitude of the court and aristocracy, who would only consider such a break with the long-lasting, previous development as a fashion alternative, not a binding choice. Appearing first in this archaizing and dogmatic form, *goût grec* neoclassicism attempted to legitimize itself by evoking a revered and distant era, considered more moral and sound than the present. One is tempted to call it the first historicizing style, except for the fact that its proponents lacked accurate knowledge of actual historical sources. It remained for the subsequent generation to dedicate itself to a more scientific study of antiquity and to forge a revised image of it.

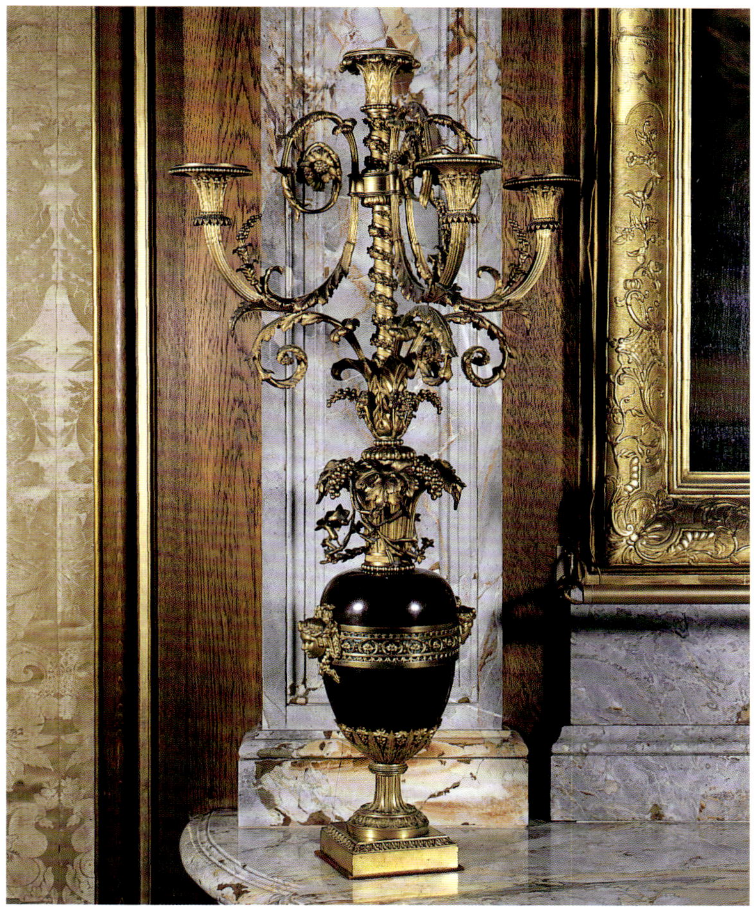

Fig. 5: *Girandole*, one of a pair, about 1780. French. Bronze, gilt and patinated; 38 × 20 in. (96.5 × 50.7 cm). Waddesdon Manor (2581). Photograph Waddesdon, The Rothchild Collection (The National Trust), Mike Fear.

The grotesque style, or the "goût étrusque"

The next phase of neoclassicism differed from the *goût grec* by incorporating other views and new models. Evolving uninterrupted from about 1775 to the end of the ancien régime, in 1789, the style was encumbered with a confusing series of names that contemporaries used to identify it: *goût étrusque* ("Etruscan taste"), *genre arabesque* ("arabesque type"), or a combination of the two.[16] While neither Etruscans nor Arabs had anything to do with the style's forms or ornamental vocabulary, one can determine what was intended with these consistently and long-employed designations. In descriptive terms, the *goût étrusque* should more correctly be called a grotesque style. Grotesques (*grotteschi*) refer to the stucco and painted ornaments created by Renaissance artists, such as Raphael and his followers, in the early 1500s. They based such designs on the findings yielded by contemporary excavations of Roman palaces, for example the Domus Aurea, where wall and ceiling decorations were found underground, in what were thought to be grottos. Derived from both new archeological discoveries and Renaissance prototypes, the visual language of this eighteenth-century revival relied upon naturalistic ornamental motifs and chimerical figural inventions, arranged in mirror image around a central axis, almost always in the form of a plant or fountain rising from a vase (see cat. no. 79). The complex ornamental motifs of this manner were well suited for vase decoration and quickly assimilated into new designs (see Fig. 5). The compositional principle of the style is embodied in the acanthus volute: axio-symmetric, curvilinear, terminating in a spiral. Reflecting the new manner, vases for example were embellished with curved fluting and curling designs on the friezes of volute handles. Designers inserted grotesque forms into these foliated compositions: sphinxes, chimeras, griffins, harpies, herms, and frequently humans (see Fig. 6). On panels designed for interiors, for instance, one typically finds male or female torsos emerging from a frond of acanthus that itself rises from a vase. Accompanying these fantastical figures is the entire Bacchic retinue—satyrs, maenads, billy-goats, rams, and Pan's children—as well as their attributes: grapevines, grapes, musical instruments, overflowing cornucopias, garlands, festoons, and fruit baskets. This naturalistic and hedonistic imagery, evoking the world of Bacchus and exuding a pronounced *joie de vivre*, contrasts emphatically the heavy architectural forms, sober abstract friezes, and rigorous intellectual imagery of the earlier *goût grec* manner.

This infusion of vitality and playfulness was the result of a new understanding of everyday life in the ancient world. Behind

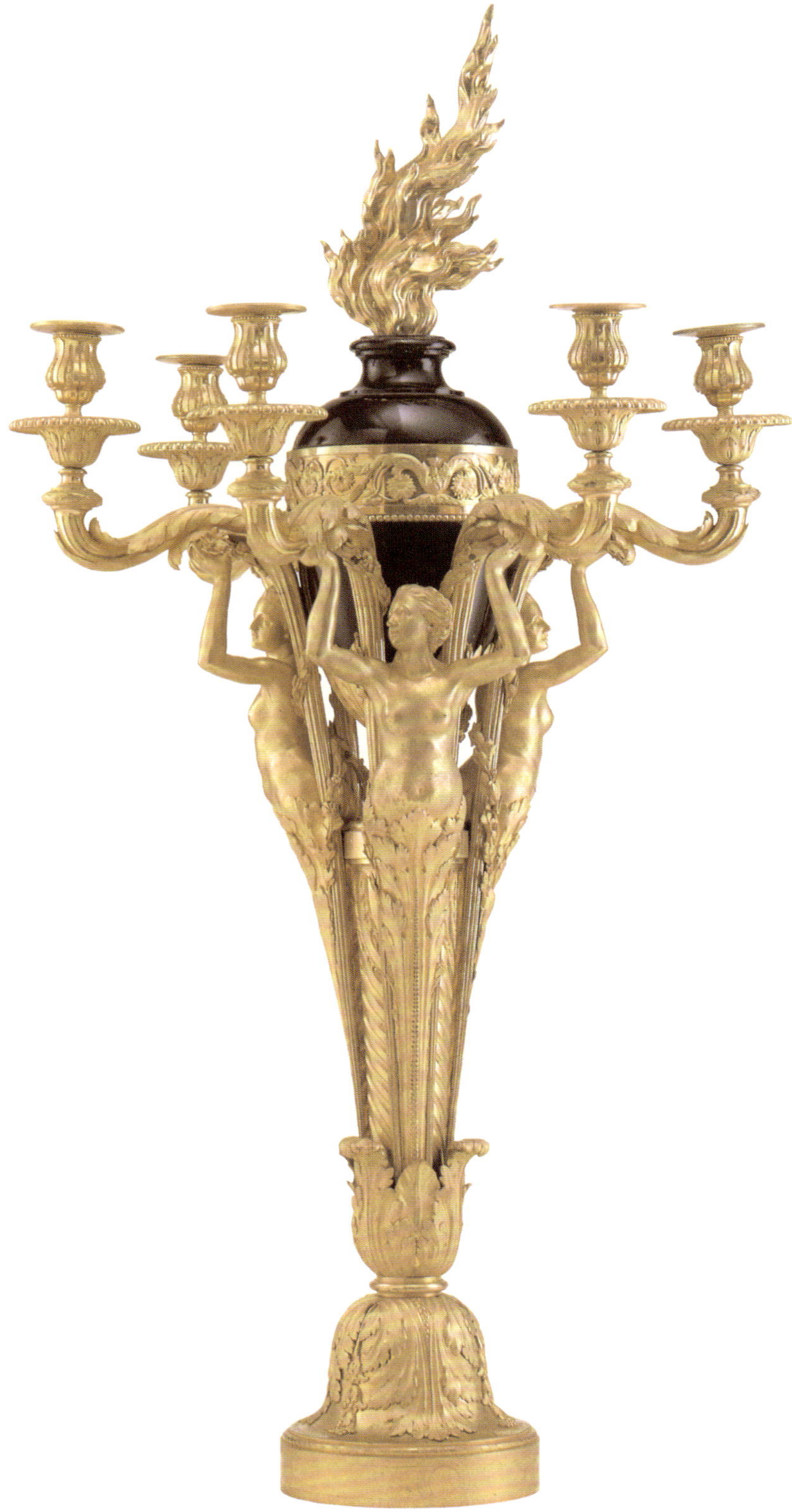

Fig. 6: *Girandole*, one of a pair, about 1775. Jean-Claude-Thomas Chambellan the Younger (called Duplessis the Younger). Gilt and blue enameled bronze; 59 × 30 in. (149.9 × 76.2 cm). Detroit Institute of Arts, Founders Society Purchase, Mr. and Mrs. Horace E. Dodge Memorial Fund (71.395).

or beyond the august temples and other columnar structures, archeologists found, to their amazement, remains of Roman palaces and houses with lively and colorful decorations that often included vase motifs. Such discoveries occurred in or near Rome, mainly in the imperial palaces in the Palatine Hill area, and in Tivoli, where Hadrian's Villa was investigated. Designers adopted motifs from these painted and stuccoed wall decorations and from sculptural elements of buildings in the imperial fora. High Renaissance architects and decorators had been inspired by these models, but also by other ancient sites that had since been destroyed or buried. This made sixteenth-century works an obvious source for neoclassical artists.

The appellation "Etruscan style" for objects and interiors of the mid-1770s to the 1780s requires further explanation. For a long time, "Etruscan" was a catchall term, encompassing the entire archaic period of ancient art. This included vases actually of Greek origin but found in Campania in southern Italy, as well as archaizing works dating from Roman imperial times.[17] The term was used as well to describe work that was considered ancient, but neither Greek nor Roman. Caylus entitled the second chapter of each volume of his *Recueil d'antiquités* "Des Étrusques"; here, he attempted to define the character of "Etruscan" art. He could not resolve the contradiction between the elemental forms of genuine Etruscan bronzes and the compositional complexity of Attic vases; thus, from the beginning, *étrusque* was a vague concept. Caylus, who changed his mind about Etruscan art from volume to volume, admitted: "One cannot say anything definite about the Etruscans."[18]

A related issue was the prominent debate about the nature of Greek versus Roman art, and which of the two was superior. The most vocal proponent for the Etruscans and Romans against the views of Caylus and Winckelmann, who, as we have seen, supported Greece, was Giovanni Battista Piranesi (see cat. no. 2), the famous printmaker and designer, who was also a respected antiquarian scholar and archeologist in his own right. Piranesi's *Diverse maniere d'adornare i cammini*, published in 1769, figured in the discussion. In his preface, Piranesi acknowledged a partial influence of Greek on Roman art, but championed the latter as having successfully combined elements from different periods into a new, superior style. This then became the justification and model for his own eclectic designs. For the most part, the sixty-five pages of *Diverse maniere* contain designs for fireplaces, which Piranesi chose as a subject precisely because they did not exist in antiquity and he could exhibit his ability in adapting ancient ornamentation to modern needs. The volume also includes ideas for andirons, tripods, chairs, clocks, candle holders, console tables, sedan chairs, commodes, and of course vase designs that combine Greek and Roman models. Piranesi drew from his extensive knowledge of antiquity to rethink the embellishment of furniture and other objects, but only in terms of motifs he applied to his forms; he did not adopt any single style entirely.

Although in his foreword Piranesi emphasized the role of Etruscan art, the volume contains no design that could truly be described as such. Avoiding a clear definition, Piranesi explained that, like the Romans, he mixed this style with that of the Greeks, leaving it to the connoisseur to distinguish one from the other.[19] For Plate I, in the preface, he created 114 small images with all the objects and motifs he considered Etruscan. He argued that they should be employed for wall decorations, as he believed they had been in antiquity, an opinion echoed by Sir William Hamilton in the foreword of the first volume of his *Collection of Etruscan, Greek and Roman Antiquities...*, published in 1767.

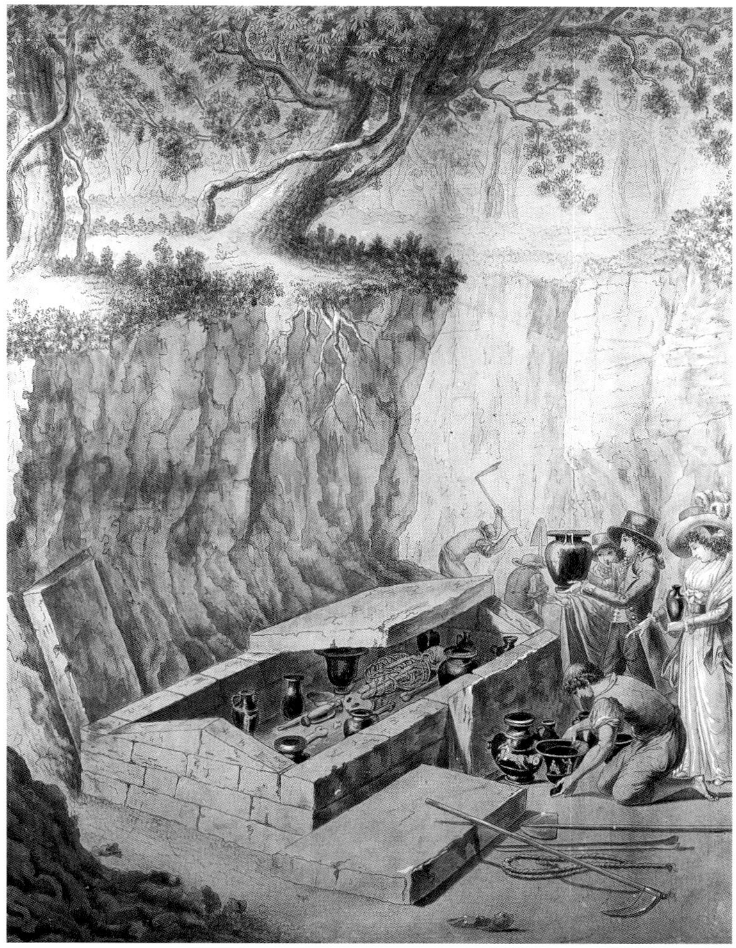

Fig. 7: *Opening of a Tomb at Nola in the Presence of Sir William Hamilton and Emma Hamilton*, about 1790. Christoph Heinrich Kniep. Pen and wash. Photograph Deutsches Archäologisches Institut, Rome; Neg. 38.149.

Hamilton began the first of his influential collections of ancient vases—some of which he designated as "Etruscan"—while serving as British ambassador to the Kingdom of Naples (see Fig. 7). He sold this collection to the British Museum, London, in 1772. With the profit, he created a second collection, whose origin he correctly recognized as Greek; his publication of this group (1791–95) features line drawings—derived from the original vases—by the German artist Johann Heinrich Wilhelm Tischbein. Hamilton's collections and catalogues prompted a taste for Attic red- and black-figured vases and quickly set the standard as sources for contemporary artist-designers. The illustrations also encouraged a vogue for the linear rendering of forms (*gravure au trait*).[20] This soon became a particularly pervasive method of Rome-based artists working in all media, a direction reinforced by Winckelmann's influential aesthetic pronouncement on the primacy of the contour in ancient sculpture. In addition to this emphasis, frieze decorations restricted to two colors became an essential design principle for vases and other arts toward the end of the century.

Among the canonical items in this refined "Etruscan" style were the internationally distributed vases of the English ceramist Josiah Wedgwood. His work, inspired by ancient and Renaissance models, as well as the publication of Hamilton's first vase collection, constitutes a major achievement in the field of neoclassical vase design and manufacture. In the 1760s, Wedgwood produced cream ware (see cat. nos. 31–33) and black basalt vases in classical shapes (see cat. nos. 36, 38–39), the latter often decorated in orange-red encaustics, in imitation of Attic prototypes (see cat. no. 5). At his manufactory, which he established in 1769 and called Etruria, he employed neoclassical designers, most notable among them the sculptor John Flaxman (see cat. nos. 7, 42). Working assiduously to create a line of vessels and tableware that were closely allied to antique styles, Wedgwood based his jasper ware, which involved applying a light-colored relief over a darker ceramic background, on ancient cameos and cameo glass (see cat. nos. 6–8, 40–44). Flaxman took this two-dimensional approach to another level, especially in his illustrations for the *Iliad* (1793) and the *Odyssey* (1795). Wedgwood's most famous ceramic copy of antique cameo glass is certainly that of the Portland Vase, an extraordinary tour de force and elaborate homage to antiquity (cat. no. 8).

Like Hamilton and Wedgwood, the Scottish architect Robert Adam played a decisive role in the development of the mature, refined phase of neoclassicism by translating his experience of the art of antiquity into concepts and forms suitable for contemporary interiors and vase design. On his extended Grand Tour, Adam hired the influential neoclassicist Charles-Louis Clérisseau in 1757 as his drawing teacher, and they traveled in Italy together.[21] By 1778, when Adam published the first volume of his *Works in Architecture*, he noted the confusing terminology used by contemporaries, adding a decided swipe at French rivals:

> The French, who till of late never adopted the ornaments of the ancients, and jealous as all mankind are of the reputation of their national taste, have blended together those ornaments with the vague and fantastical appellation of arabesque, a style which, though entirely distinct from the grotesque, has notwithstanding been most absurdly and universally confounded with it by the ignorant.[22]

Indeed, the floral and vegetal components of the grotesque manner are shared by the arabesque, a type of ornament thought to have derived from Near East. The term was used from the late sixteenth century on; by the seventeenth, it was [mis]applied to any type of scrolling design, causing confusion in terminology.

Despite his desire for distinct nomenclature, however, Adam, like Piranesi, recommended that one not follow antique sources slavishly, for they represented a style developed over time based on numerous historical precedents. Instead, he believed that architects and designers should grasp the spirit of antiquity, which Adam expressed as "...all gaiety, grace and beauty."[23] In his case, it is clear that "Etruscan" was not to be understood literally or historically, but rather as a new, synthetic style based upon Roman art and its rediscovery and interpretation in the High Renaissance. Adam described his method as follows: "We have introduced a great diversity of ceilings, friezes, and decorated pilasters, and have added grace and beauty to the whole, by a mixture of grotesque, stucco, and painted ornaments together with the flowing *rinçeau* with its fanciful figures and winding foliage."[24]

Parallel and very much in competition to developments in England, a similar grotesque style came to characterize eighteenth-century interior decoration in France, exhibiting a lexicon of forms that frequently included vases. An early such example is provided by a series of eight panels originally from the Hôtel Grimod de la Reynière, Paris, and datable to 1773–75 (they are now in the Victoria and Albert Museum, London).[25] These panels were set above a wainscoting decorated with garlands between three windows which corresponded to mirrors on the opposite wall. They display vases in the center of axially symmetrical grotesque ornamentation. The drawings of the original disposition of the interior by Jan Chrystian Kamnsetzer show that four candelabra, in the form of large vases on tripod supports, complemented the painted decoration (see Fig. 8). In this instance, the two-dimensional, painted vase motif moved from the wall into the room, as it were, as a volumetric, independent form.

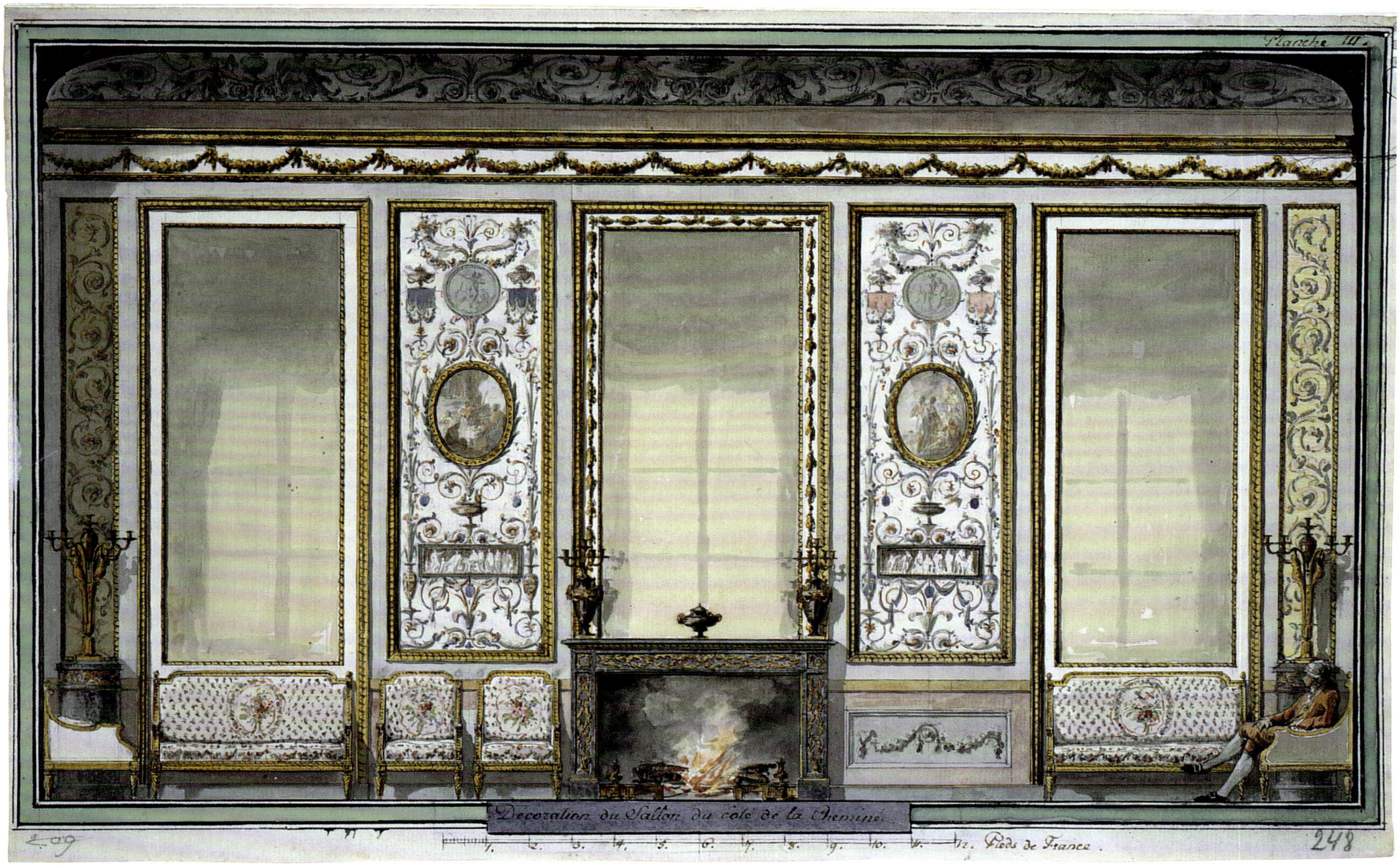

Fig. 8: Elevation of the Salon of Hôtel Grimod de la Reynière, Paris, 1773–75. Jan Chrystian Kamnsetzer. Pen, wash, and watercolor. University Library, Warsaw (Kat. res. Nr. 995). Photograph T. Zoltowska-Huszcza, Warsaw.

Some of the most beautiful neoclassical vase designs of the 1770s came from the architect-designer François-Joseph Bélanger.[26] He interpreted the ancient grotesque style in a new way, drawing upon hand-colored publications of the Vatican Loggie,[27] which inspired his 1777–81 decoration of the salon at the Château de Bagatelle in the Bois du Bologne, near Paris. This interior featured pink-ground panels with low-relief figures by Nicholas-François-Daniel Lhuillier. Set against the pastel ground were medallions containing grisailles or cameos resembling Wedgwood designs, as well as beautifully carved, gilded, and painted overdoors with vases as their central motif (see cat. no. 74).[28] Bélanger's greatest contribution to neoclassical vase design, however, was a group of vessels he created for the Duc d'Aumont.

Around 1772 the duke, inspired by ancient marble and porphyry vases, began to search for valuable French hardstone deposits that could be made into a variety of vessels, which would be sanded, polished, and fastened together with gilt-bronze settings. He hired Bélanger to design the shapes and mounts; the latter were executed by the most famous bronze caster and chaser of the era, Pierre Gouthière (see Fig. 9).[29] This eminent collection was well known, because it was dispersed at auction soon after the duke's death in 1782. For the auction, a catalogue was printed that describes and illustrates the precious vessels.[30] This publication was particularly important for the dissemination of neoclassical vase forms and their decoration; its engraved line drawings reveal the influence of ancient Greek vase painting. Most of Aumont's collection was purchased by Louis XVI and

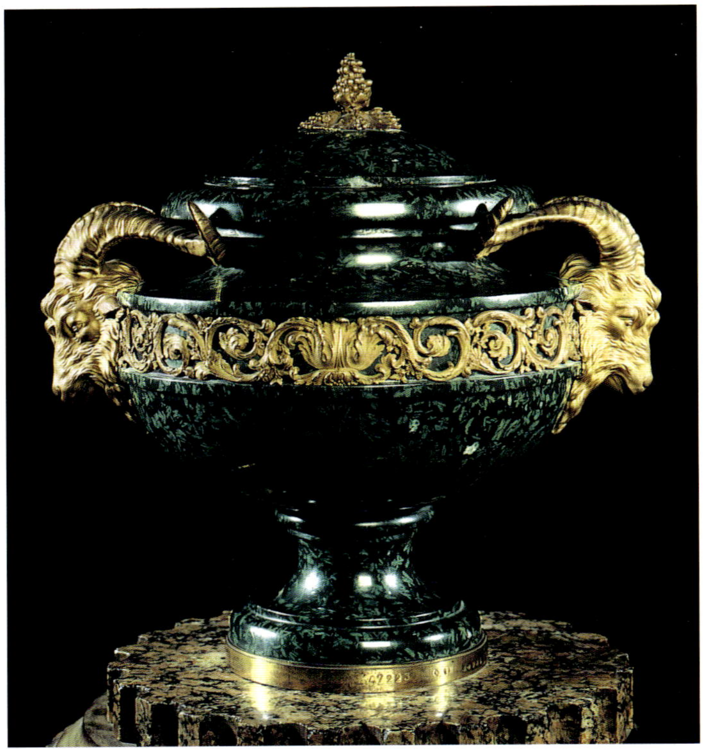

Fig. 9: *Vase with Ram's Heads*, about 1772. Pierre Gouthière. Serpentine and gilt bronze; 15 × 14 ⅟₁₆ in. (38 × 36 cm). Musée du Louvre, Paris (OA 5179). Photograph © Réunion des Musées Nationaux/ Art Resource, NY.

Marie-Antoinette; after the French Revolution, these objects entered the collections of the Louvre. The purchase prices, between 3,000 and 5,000 *livres*, were some of the highest ever paid in this period for documented eighteenth-century mounted vases and demonstrate the enormous value ascribed to finely wrought pieces in the antique style. Moreover, the Aumont vases inspired a continuing effort to promote the production of hardstone manufactories.[31]

Only after 1785 did decorators employed by the French royal family begin to work in the so-called "genre arabesque," the French version of the "Etruscan" style, whose name, as we have seen, Adam decried as a misnomer. For example, in 1787, under the direction of Richard Mique, the artist Jean-Siméon Rousseau de la Rottière decorated a gambling room, boudoir, and Turkish-style cabinet at the Château de Fontainebleau for Marie-Antoinette. In these rooms, the French grotesque manner attained its greatest expression. Vases remain central to these designs: they appear in the middle of the wall panels, form the bodies of chandeliers and candelabra, and stand on fireplace mantels as candlesticks, perfume burners, clocks, and ornamental ceramic objects.

While discussions of the origin of neoclassicism usually cite Herculaneum and Pompeii as critical sources, French wall designs or motifs that specifically reflect the excavations of these cities, buried by Mount Vesuvius in 79 A.D., appear only toward the mid-1780s.[32] One must remember that most eighteenth-century architects and designers did not know about these ruins because of the lack of reliable illustrated publications. Despite the excitement news of these recently discovered classical sources engendered, serious books on the findings did not appear immediately. Instead, numerous titles were published that falsely promised first-hand information. These volumes may feature line drawings of individual figures or groups from the walls of excavated buildings but hardly ever the comprehensive decorative layout of a room.[33] Decorative wall painting from Roman sites, however, was well known from a variety of publications and was soon combined with the related grotesque ornamentation of the Renaissance.

As outlined above, the resulting eighteenth-century grotesque style informed the conception, structure, and motifs of all decorative arts. Since the grotesque had its origin in wall decoration, objects created in this manner were meant to harmonize, even merge with the overall composition of an interior. During this luxury-loving phase of neoclassicism, such motifs became increasingly subordinate to the devices that divided wall compositions. In addition to Clérisseau and Bélanger, other Paris-based designers, such as Gilles-Paul Cauvet (see cat. no. 79), Jean-Demosthère Dugourc (see cat. no. 78), and Mique, popularized the style. Their work evolved seamlessly into the early furniture designs of architect-designer Charles Percier. The appeal of the *goût étrusque* persisted through the French Revolution into the Directory. But a shift in taste was evident by 1797 with Percier's works, which signaled a new movement, the so-called Empire style. Yet outside France, in Austria, England, Italy, and Russia, the grotesque style and its vase motifs remained in vogue well into the 1820s.

Late neoclassicism, or "Empire"

There are critical discrepancies regarding the chronology of what is often called Empire style and its division into early, main, and late phases. This new manner was in fact called the *goût antique* by contemporaries. It emerged in France after the economic chaos caused by the Revolution subsided, around 1796. It started, in other words, before Napoleon established his empire and remained popular into the mid-1820s, well after its fall.[34]

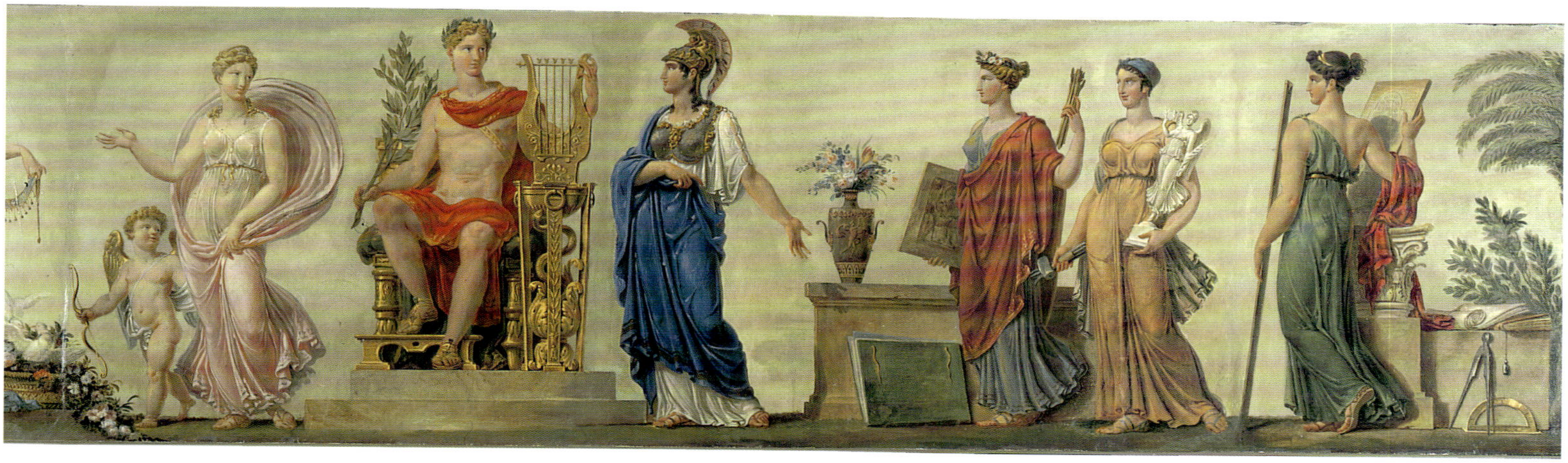

Fig. 10: *Frieze with Venus, Apollo, Minerva, Painting, and Sculpture,* originally from the Hôtel de la Victoire, *Paris,* 1797. French. Oil on plaster. Châteaux de Malmaison et Bois-Préau. Photograph © Réunion des Musées Nationaux/Art Resource, NY.

From 1796 on, Percier and his partner, Pierre-François-Léonard Fontaine redefined furnishings and interiors for fashionable Parisian society. Both trained as architects at the Académie Royale in Paris, and both concluded their studies with prolonged stays in Rome. In the Italian capital from 1786 to 1791, Percier examined buildings with archeological meticulousness and absorbed the grotesque decorations of antiquity and the Renaissance.[35] With thousands of sheets in his portfolio, he returned to Paris via Florence, Mantua, and Genoa. He initially worked as a stage designer, adapting what he had learned to opera and theater; his efforts in this vein included productions of moralistic plays set in the ancient world. Through his success in this area, by 1796 Percier began designing interiors for French financiers, merchants of military supplies, and generals who had amassed fortunes quickly and constituted the new leadership of the Directorate. In 1797 the partnership of Percier and Fontaine received its first commission from its most famous clients, General Napoleon Bonaparte and his wife, Josephine: the Hôtel de la Victoire in Paris (see Fig. 10).[36] Two years later, Napoleon, now Consul, named the pair his official architects; when he became emperor in 1804, he appointed Fontaine First Court Architect. As such, Fontaine determined the direction of architecture and furnishings of Napoleon's royal residences in France and beyond. This remarkable artist managed to retain his position throughout the political turbulence that beset France after Napoleon's downfall, serving Louis XVIII, Charles X, and Louis-Philippe, and even the Second Republic.

Percier's role was that of designer, while Fontaine supervised the planning and actual execution of their projects. Even when Napoleon dismissed Percier as court architect on December 29, 1804, because the emperor had never laid eyes on this shy man, the two colleagues remained faithful partners. Until his death in 1838, Percier supplied Fontaine with projects and executed design drawings. The irony here is that Percier, the true inventor of the Empire style, is not documented as such, because his name does not appear in the official records. In addition to his position as the emperor's architect, and his many independent commissions, Percier exercised his greatest influence on late neoclassicism through his publication *Recueil des décorations intérieures* (Fig. 11).[37] It has been assumed incorrectly that the publication's outline engravings of interiors, furniture, and utensils were created during the Empire period, that is, after 1804, but the first twelve sheets of this "bible of the Empire" were issued in 1801, reproducing designs Percier actually created in 1797–98. By 1805 the book had grown to fifty-four pages. The *Recueil,* along with the

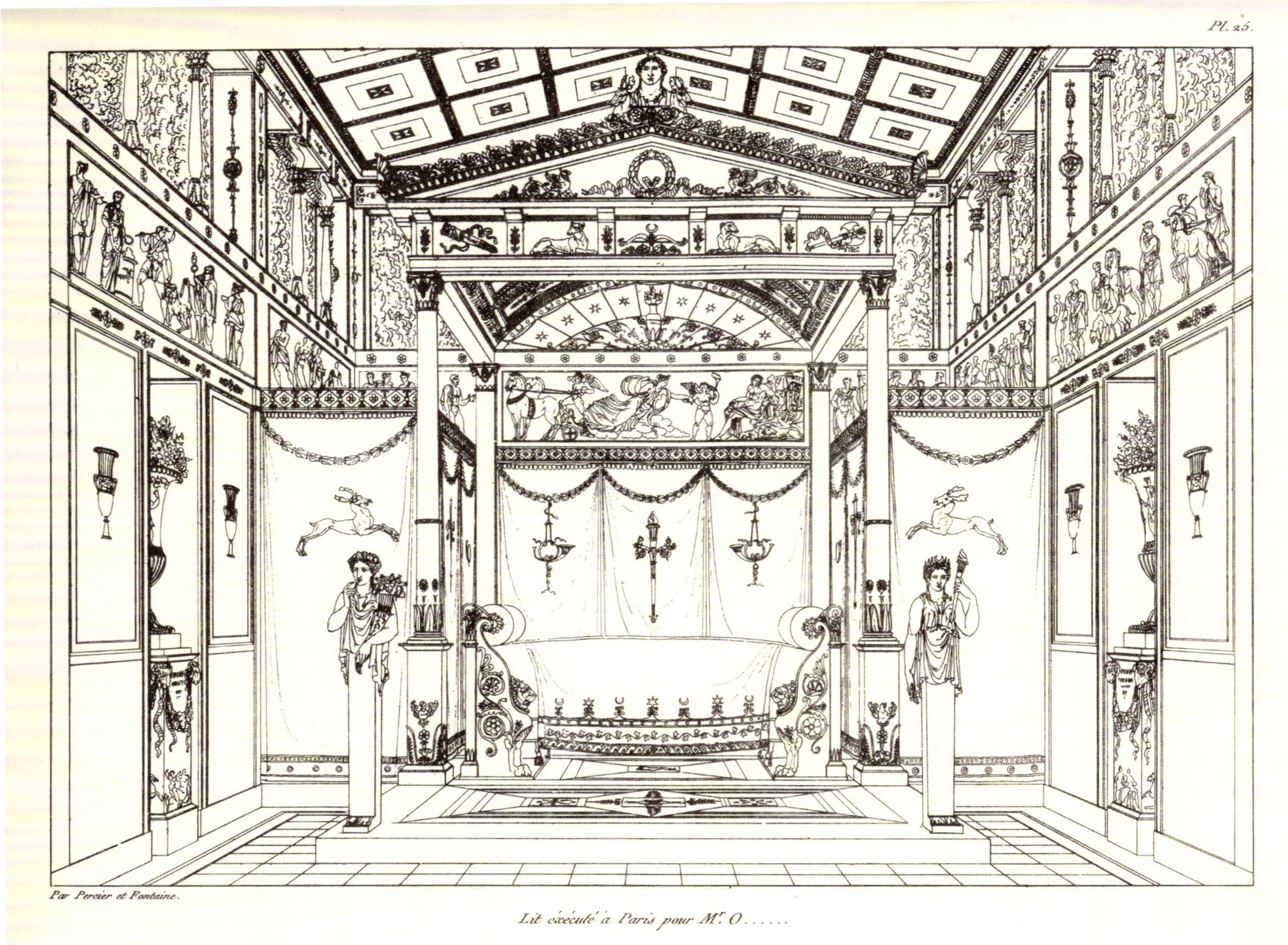

Fig. 11: *Bedroom for Madame Ouvrard,* designed 1797, in the form of a temple of Diana. Charles Percier. Plate 25 from *Recueil des décorations intérieures.* Engraving. Photograph Bard Graduate Center, Han Vu.

activities of the partners' followers, helped disseminate Percier's ideas for complete neoclassical interiors throughout Europe.

In 1812 Percier wrote that it was most important to recognize and preserve the basic historical model of a given motif, to maintain the original, and to follow the universal rules of classical design. A new invention, he believed, could succeed only if it were part of the continuing history of a type.[38] Not surprisingly, the *Recueil* includes numerous vase designs referring to Greek models that Percier and Fontaine had studied in Roman and Florentine collections (Fig. 12). Percier turned the two-dimensional, painted decoration of these ancient pieces into low reliefs, covering the bodies of the silver vases and gilt utensils he designed for silversmiths Martin-Guillaume Biennais (see cat. no. 62) and Henri Auguste (see Fig. 13). Percier's linear manner in the *Recueil* is also linked to his veneration of Greek vase painting. Previously, in reproductions of ancient sculpture, engravers employed contour lines of varying breadth to distinguish the side in light (thinner lines) and that in shadow (thicker lines). Percier, however, treated

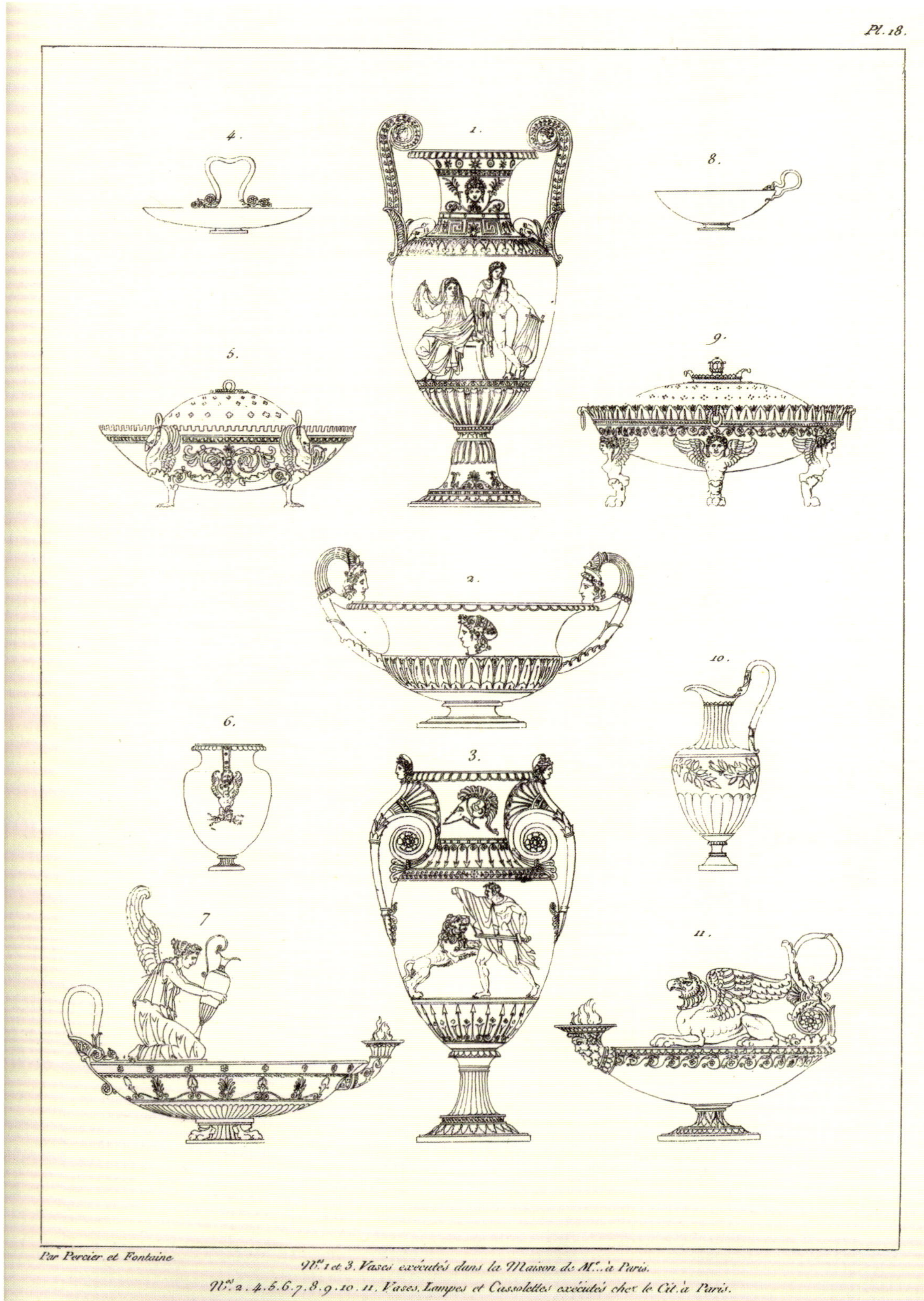

Fig. 12: *Vases for Madame Moreau and other vessels*, designed 1801–2.
Charles Percier. Plate 18 from *Recueil des décorations intérieures*. Engraving.
Photograph Bard Graduate Center, Han Vu.

Frequently cited classical vase types:

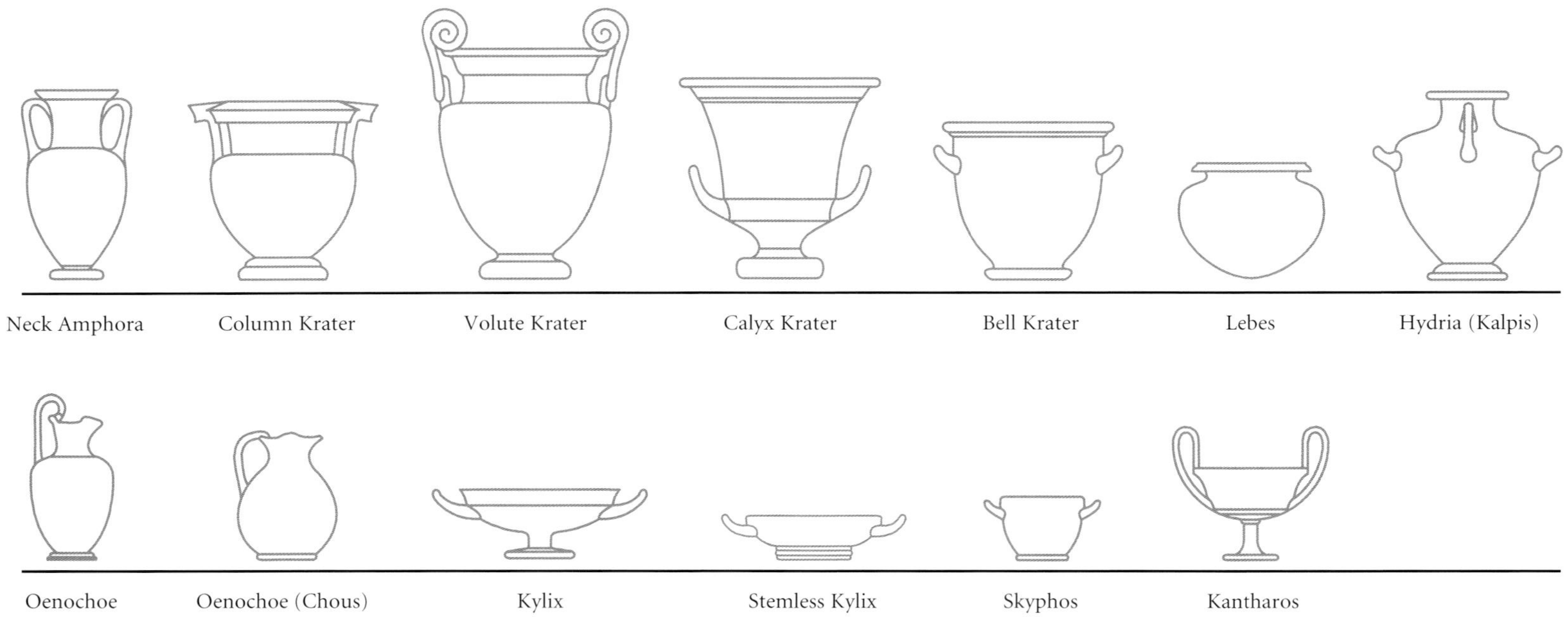

Drawings based on images from Gisela M.A. Richter, *Attic Red-Figured Vases* (New Haven/London, 1946).

Editor's Note:

Unless otherwise noted, only the measurements for height are provided. In those instances in which more dimensions are required, height precedes width precedes depth.

The location of marks is listed where this information was available.

All objects illustrated are in The Metropolitan Museum of Art, New York, unless otherwise noted.

Technical and foreign terms are explained in the Glossary.

Frequently cited references are listed in a short form in the notes to entries; full references are provided on pages 187–88.

The pottery founded by Josiah Wedgwood was known under a variety of names during the eighteenth and early nineteenth centuries; to avoid confusion, this publication uses a single generic identification for all Wedgwood wares.

Contributors and their initials:

K.C.	Kei Chan
K.D.	Katherine Danalakis
C.H.	Caroline Hannah
M.H.	Michelle Hargrave
W.K.	Wolfram Koeppe
J.L.	Jessica Lanier
K.M.	Kristin Martin
J.L.M.	Jennie L. McCahey
H.J.M.	Heather Jane McCormick
P.M.	Paul Micio
E.S.	Ezra Shales
E.W.M.	Emily Wheat Maynard

The Hamilton Collection and Its Influence

Stefanie Walker

By the time Sir William Hamilton arrived in the Kingdom of Naples as British envoy in 1764, the Grand Tour to Italy had become a requirement for men of culture and many educated women as well. They were attracted by the glories of the Roman past, especially the rediscoveries of the Vesuvian cities of Herculaneum (1738) and Pompeii (1748). Visiting sites where excavations were ongoing, many of these well-to-do travelers were thrilled by a sense of adventure and discovery (see Ottomeyer, Fig. 7). Acquiring classical art and objects allowed a direct identification with this much-admired ancient civilization. Its remains—architectural ruins, statuary, wall paintings, vases—were regarded as ideals of beauty in form and ornamentation.

Not only were vases more plentiful and affordable than sculpture, for example, but their painted decoration also presented an extensive, new pictorial repertory of ancient art that could be related to contemporary traditions of drawing and painting. Initial confusion about the origin of the vases found in Italian burial sites, concerning whether they were Etruscan or Greek, spurred lively debate among scholars and dilettantes throughout Europe. The printmaker-scholar Giovanni Battista Piranesi championed Etruscan and Roman superiority; he was opposed by the antiquarian Comte de Caylus and historian Johann Joachim Winckelmann, among others, who considered Greek art the highest form because of what they perceived to be its naturalism and simplicity. Vases and other excavated artifacts were seen as the best, most direct documents with which to reconstruct the daily life of antiquity. The intense study and discussion of vases—the need to explain, order, and catalogue them—indicate not only the eighteenth century's general fascination with taxonomy but its passionate connection and identification with the ancients.

Hamilton was an exemplar of the Enlightenment, a polymath interested in geology, garden design, and music. The career diplomat was also an art connoisseur and passionate collector, though not as wealthy as many of his contemporary competitors. Motivated by his historical and aesthetic interests, he assembled, during his long tenure in Naples, two extensive and outstanding collections of antiquities. He sold the first in 1772 to the British Museum, London, and the second in 1801 to Thomas Hope. The collections comprised mainly vases: over seven hundred in one, almost a thousand in the other. Doubtless, their beauty, number, and variety impressed the many international visitors to Naples who saw them at his residence.

Above all, Hamilton's decision to publish his collections in multivolume folio editions (see cat. no. 1) exerted a lasting impact on scholars, collectors, artists, and, most specifically, vase manufacturers. By publicizing his first collection of vases in elaborate hand-colored plates, he and his collaborator, the publisher Pierre d'Hancarville helped the vase form, whether they intended to do so or not, to become a fashionable commodity. By the time Hamilton's second collection was published (1791–95), under the supervision of Johann Heinrich Wilhelm Tischbein, the vase was firmly entrenched as an icon of neoclassicism. Vases were avidly collected and admired: original pieces served as valuable testimonials of antiquity, and modern versions were adapted to current use and taste. The relationship of English ceramist Josiah Wedgwood with Hamilton and his collections proved critical to his evolution as one of Europe's foremost designers and manufacturers of wares inspired by ancient prototypes. His interpretations range from direct copies, such as the "Hydria" and Portland Vase (cat. nos. 5, 8), to adaptations, such as his jasper ware version of the famous Borghese Vase (cat. no. 6), and, finally, freer interpretations of ancient prototypes, such as his Pegasus Vase (cat. no. 7).

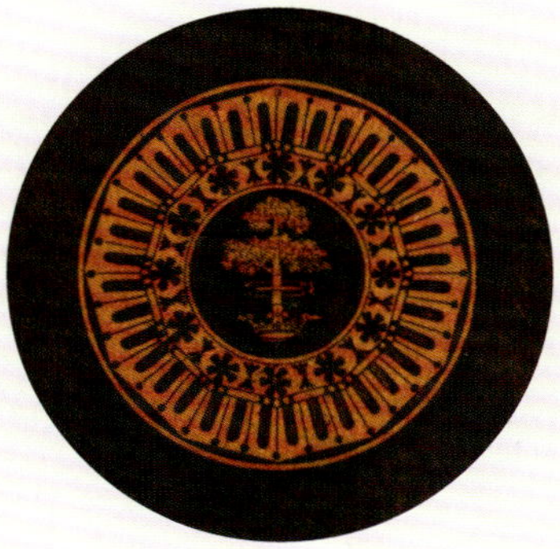

Detail of cat. no. 1

1. FRONTISPIECE

William Hamilton (1730–1803) and Pierre d'Hancarville (1719–1805), publisher
1767
Hand-colored etching, from *Collection of Etruscan, Greek, and Roman Antiquities from the Cabinet of the Hon. W. Hamilton His Britannick Maiesty's Envoy Extraordinary and Plenipotentiary at the Court of Naples* (hereafter Hamilton/d'Hancarville 1767–76), vol. 2
Sheet: $19\frac{11}{16} \times 14\frac{15}{16}$ in. (500 × 380 mm)
Thomas J. Watson Library, Gift of Christian A. Zabriskie, 1954
579.2 H186 F

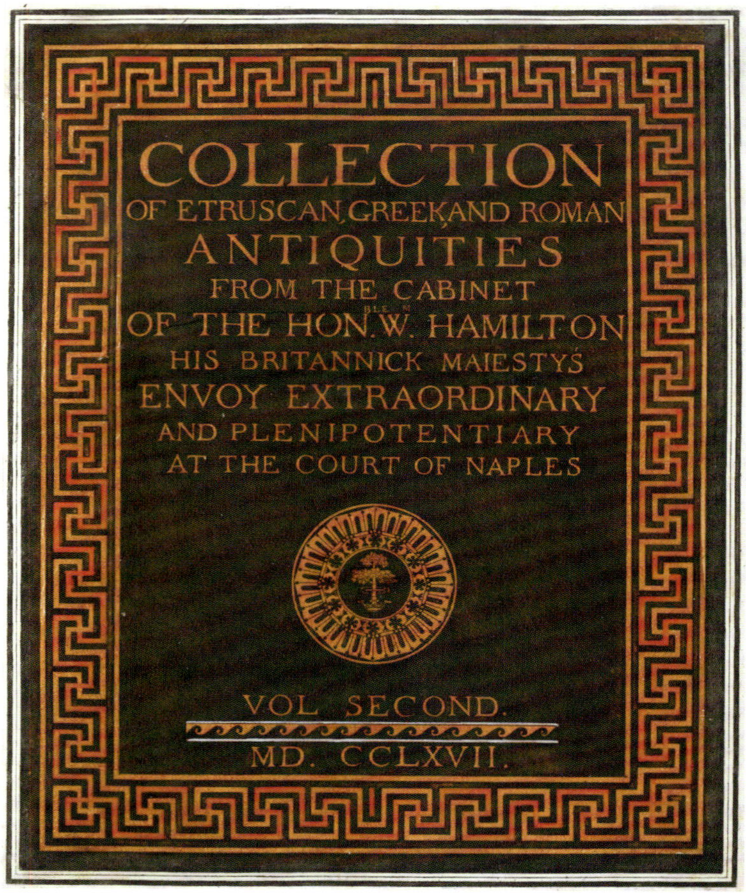

Cat. no. 1

Sir William Hamilton commissioned from publisher Pierre d'Harcanville an elaborate four-volume book of his first collection of ancient vases with the aim of showcasing his objects and giving them added cachet and value. Hamilton and d'Hancarville's publication was a significant factor in building appreciation for Greek vases, and increasing their value from the late eighteenth century on.[1]

The decorative borders of the frontispiece were copied from a set of illustrations in Giovanni Battista Piranesi's *Della introduzione e del progresso delle belle arti*, published in 1765.[2] About his etchings of decorative classical friezes stacked one above the other, Piranesi stated that he adapted them from Etruscan tomb paintings. However, he probably only copied antique patterns such as the meander, and supplemented them with others of his own design.

One of the great novelties of the Hamilton/d'Hancarville publication are the plates showing detailed scenes painted on the vases. They were "unrolled" from the round shape of the vase onto the flat surface of the paper, and the figures were often adjusted to fit the surface and rectangular surround. Frequently, frames were embellished with decorative running borders—palmettes, meanders, or the like—similar to the one on this frontispiece. In many cases, the borders on the figurative plates do not correspond to those of the vases; rather, they were invented purely for decorative effect.

The production of the volumes was a lavish affair. D'Hancarville employed fifteen people for this project alone, including etchers, draftsmen, and printers; however, the author of this frontispiece remains unknown. The publication's etchings were hand-colored in ocher to imitate red-figure vases. Certain details were emphasized and scenes enhanced with a thin layer of additional colors.[3] The meander border of the frontispiece, for example, contains two shades of red. Since the work was done entirely by hand, each colored etching is an original work of art.

The process d'Hancarville employed was extremely laborious and expensive, putting a severe strain on Hamilton's finances. By the end of production, it cost him over £8,000; in contrast, he purchased the Portland Vase (see cat. no. 8) for £1,000![4] This prompted Hamilton to underwrite a much simpler publication for his second vase collection, published in 1791–95 under the supervision of German painter Johann Heinrich Wilhelm Tischbein.[5] K.D./J.L.M.

1 Hamilton/d'Hancarville 1767–76. Michael Vickers and David Gill reevaluated the current importance of ancient vases and suggested that vases probably had little value in antiquity. See Michael Vickers and David Gill, "Value and Simplicity: A Historical Case Study," in idem, *Artful Crafts: Ancient Greek Silverware and Pottery* (Oxford, 1994): 1–32.
2 Nancy H. Ramage, "Piranesi's Decorative Friezes: A Source for Neoclassical Border Patterns," *Ars Ceramica* 8 (1991): 14–15, fig. 3.
3 Pascal Greiner, *Le Antichità Etrusche Greche e Romane 1766–1776 Di Pierre Hughes D'Hancarville* (Rome, 1992): 49–50.
4 Ibid.: 50.

5 William Hamilton, *Collection of engravings from ancient vases mostly of pure Greek workmanship discovered in sepulchres in the kingdom of the Two Sicilies but chiefly in the neighbourhood of Naples during the course of the years MCDDLXXXIX and MDCCLXXXX now in the possession of Sir Wm. Hamilton His Britannic Maiesty's envoy extry. and plenipotentiary at the Court of Naples with remarks on each vase by the collector,* 4 vols. (1791–95).

2. THE WARWICK VASE

Giovanni Battista Piranesi (1720–1778)
1778
Etching, plate 602 from *Vasi, candelabri, cippi, sarcofagi, tripodi, lucerne, ed ornamenti antichi disegnati ed incisi dal Cav. Gio. Batt. Piranesi*, vol. 1
Plate: 27 ¾ × 18 ½ in. (705 × 470 mm)
Rogers Fund, Transferred from the Library, 1941
41.71.1 (12)

Giovanni Battista Piranesi produced this series of etchings of antiquities toward the end of his career. It comprises 110 plates, which were gathered into two folio volumes in 1778. By the 1760s, the artist had become increasingly involved in the restoration and sale of classical antiquities to people on the Grand Tour, an activity that supplemented his printselling business.[1] He was at that point a renowned draftsman, printmaker, antiquarian, restorer, and architect. Through his often controversial writings on antiquities and his highly successful prints, Piranesi became one of the main instigators of eighteenth-century neoclassicism. He was a friend of the Scottish architect Robert Adam and mentor to several vase-designing artists, such as Jean-Laurent Legeay and Louis-Joseph Le Lorrain (see cat. nos. 88–90).

The Warwick Vase, illustrated here, takes its name from Hamilton's nephew, George Greville, the Earl of Warwick, who was its penultimate owner. The English painter Gavin Hamilton discovered fragments of the marble vase during his 1769–70 excavation at Lago Pantanello, near Hadrian's Villa at Tivoli. Piranesi assisted with a major restoration of the vase. Traditionally, he has been credited with its actual reconstruction; however, a recently published letter, dated February 26, 1772, from James Byres to Sir William Hamilton indicates that the Italian artist was responsible only for the vase's overall design.

The same letter also identified Hamilton (no relation to Gavin Hamilton) as the owner of the fragments and the financier of the restoration: "The model of your great vase is at last finished in stucco and all the ancient pieces inserted." Hamilton hoped that the British Museum, London, would buy the vase, making it a centerpiece of the collection it had purchased from him in 1772.

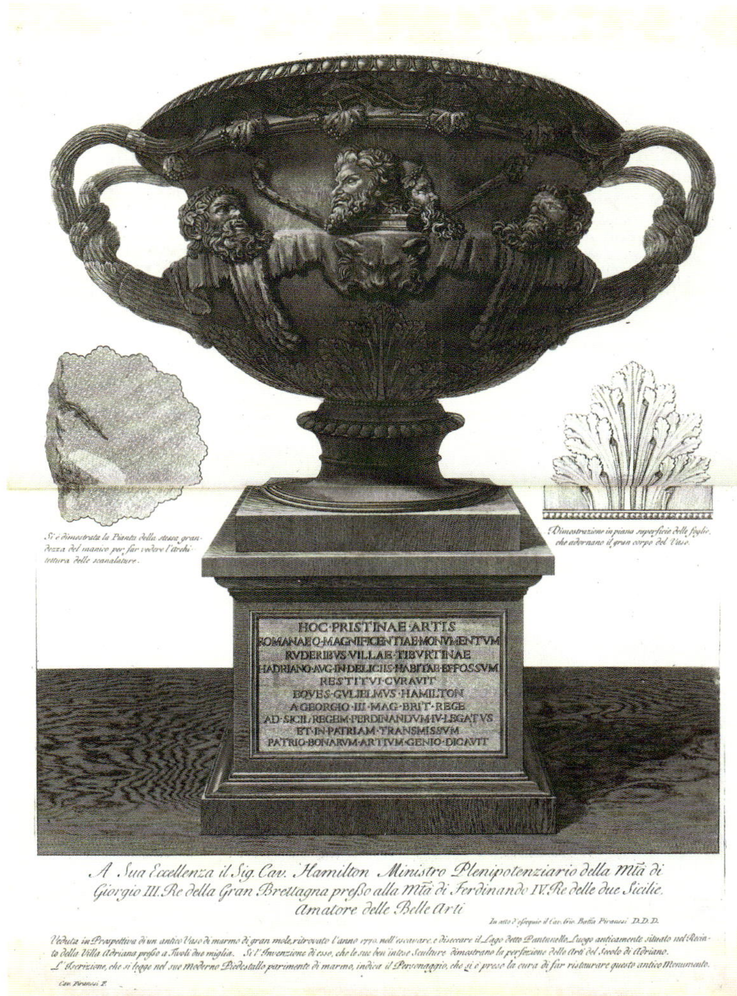

Cat. no. 2

Several attempts to effect the sale failed, and the vase eventually went to Hamilton's nephew. It remained in the family until 1979, when the Trustees of the Burrell Collection, Glasgow, acquired it.[2]

The vase is carved with the heads of Bacchus, Silenus, Pan, and a female, perhaps Ariadne.[3] The bowl, stem, and base were reconstructed in six sections, and the vase placed on a pedestal with a commemorative Latin inscription which translates as follows:

> Sir William Hamilton, ambassador from King George III
> of Great Britain to King Ferdinand IV of the Sicilies,
> had this monument of ancient art and Roman majesty,
> which was excavated among the ruins of the Tiburtine villa—
> delightful home of the emperor Hadrian Augustus—
> restored and, having sent it to his country, dedicated it to
> the native genius of fine arts 1774.[4]

5. WEDGWOOD "HYDRIA"

Josiah Wedgwood (1730–1795), Etruria, Staffordshire
About 1780
Black basalt ware with matte enamel decoration
18 in. (45.7 cm)
The Charles E. Sampson Memorial Fund, 1966
66.17

Sir William Hamilton's first collection of Greek vases and its catalogue (see cat. no. 1) provided a surfeit of classical designs and forms from which eighteenth-century decorative artists such as Josiah Wedgwood drew inspiration.[1] One of Wedgwood's copies from Hamilton's first collection of Greek vases was this "hydria," the original of which is in the British Museum, London (Fig. 1). The potter reproduced it in black basalt stoneware with red-enamel painting.

Wedgwood's arrangement of the figures on the Metropolitan Museum's example prove that he based his work on the etching in the first Hamilton catalogue rather than on the original vessel. For example, while the figural scene encircles the entire body of the ancient hydria, Wedgwood arranged his composition only on

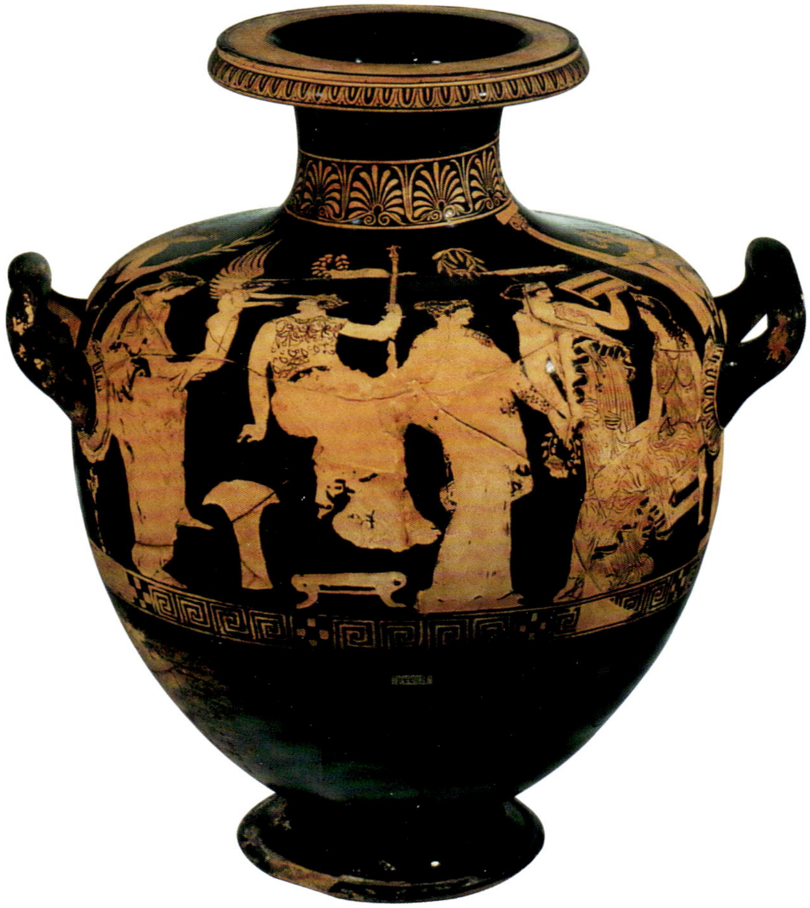

Fig. 1: *Hydria*, Attributed to the Nikias Painter. Attic red-figure terracotta. About 400 B.C. British Museum, London, E 225.

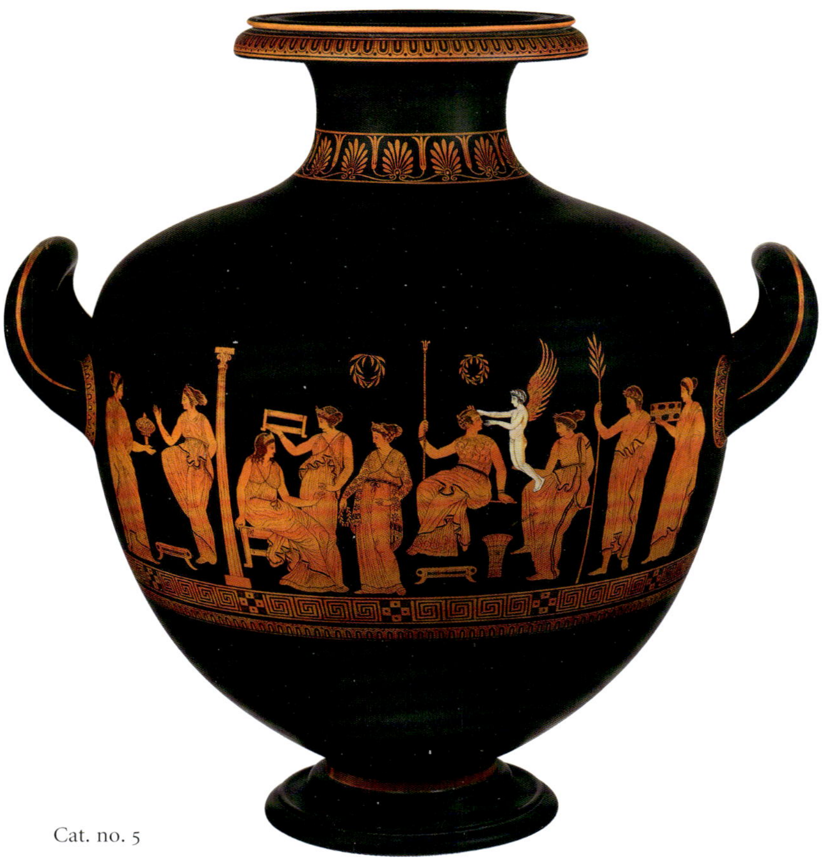

Cat. no. 5

one side of his basalt version, flanking the main scene on the original with pairs of figures from the sides (Fig. 2). He also disregarded or was unaware of the hydria's function as a water jug and eliminated the third handle on the rear of the vase, which was used to pour liquids (see cat. no. 3).

Apart from the placement of the figures and the lack of a third handle, Wedgwood's hydria shape and figurative scene are extremely faithful to Hamilton's ancient vase. The ceramist's border decorations, including the unique meander baseline under the frieze, the egg and dart motif around the mouth, and the palmettes encircling the neck follow those on the ancient model.

The Wedgwood vase, like its classical prototype, is decorated with ten figures in an interior setting; however, because they are not labeled, and do not hold any specific attributes, we cannot interpret the scene's meaning with certainty. While Pierre d'Hancarville, Hamilton's publisher, called the main figures Paris and Helen, more recent scholars have identified them as Apollo, who holds a tall sheaf of laurel, surrounded by eight Muses, and a small, white-bodied Eros floating in the center.[2]

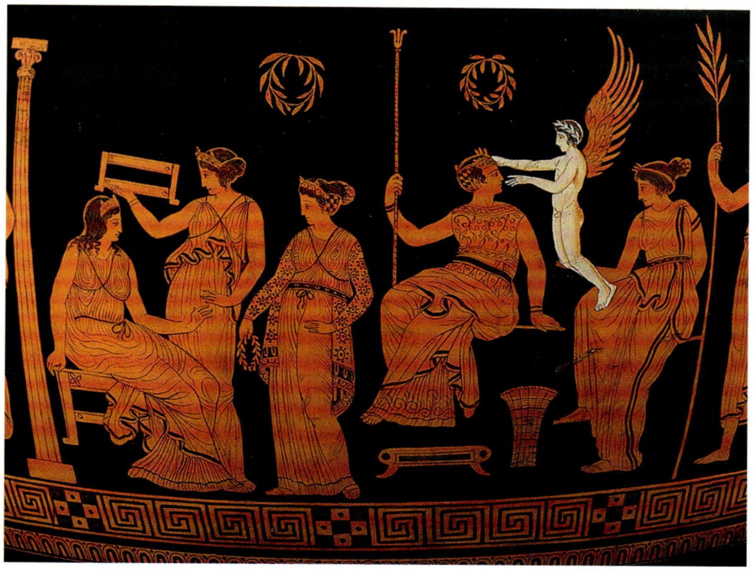

Fig. 2: Detail of cat. no. 5.

Wedgwood intended his decorated black basalt wares of this type, with hand-painted, so-called red "encaustic" pigments, to imitate the red-figure style of Greek vases. He described them, "not only... completely to imitate the Paintings upon the Etruscan Vases, but to do *much more*... to give to the Beauty of Design, the Advantages of Light and Shade in various Colours; and to render Paintings durable without the Defect of a varnished or glassy Surface."[3] Although Hamilton and Wedgwood eventually realized that these ancient vessels were Greek rather than Etruscan, the latter continued to call them by the false moniker in order to promote his wares and increase sales. In September 1768, Wedgwood admonished the company clerk not to forget to call his vases "Etruscans."[4] K.D.

[1] Hamilton/d'Hancarville 1767–76. The original vase was found in Capua in southern Italy and has been attributed to the Nikias Painter. Wedgwood first observed red-figure style decoration on Greek vases in the Comte de Caylus's influential multivolume catalogue of Egyptian, Greek, and Roman art (Caylus 1752–67). See Adams 1992: 73; and Reilly 1989: vol. 1, 414–15.

[2] Hamilton/d'Hancarville 1767–76: pls. 30–32; and Beazley 1984: 1334.27.

[3] Reilly 1989: vol. 1, 415: 1779 Catalogue, Class XIX 'Vases, Ewers, &c Ornamented with Encaustic Paintings.' Wedgwood may have named his painting technique "encaustic" after the process described by Pliny, which involved using pigment mixed with warm wax. Wedgwood added kaolinic clay, which he called "Ayoree," to his enamel colors in lieu of wax.

[4] Young 1995: 77.

6. URN ADAPTING THE "BORGHESE VASE"

Josiah Wedgwood (1730–1795), Etruria, Staffordshire; modeled by Jean de Vaere, also called John Devaere or Jean de Vere (1754–1830)
About 1785–1800
Vitrified stoneware (jasper ware)
18 ⅝ in. (47.3 cm)
Gift of Frank Sturgis, 1932
32.95.10

The colossal marble Borghese Vase, now in the Musée du Louvre, Paris, served as the source for the ornamental figures engaged in the bacchanalian scene on this solid jasper ware vase (see Ottomeyer essay, Fig. 1). The modeling has been attributed to Jean de Vaere, a member of Wedgwood's Rome-based modelers working under the guidance of the neoclassical sculptor John Flaxman. The circle in Rome, which also included Giuseppe Angelini and Camillo Pacetti, "improved" on classical designs, making them suitable for Wedgwood's use on vases and plaques.[1]

The Flemish-born Jean de Vaere trained as a sculptor with his uncle in Ghent, and by the age of twenty was working in Paris, primarily as a goldsmith. He subsequently moved to London, attended the Royal Academy in 1786, and thereafter spent seven years in Rome, where he assisted Flaxman, before returning to England to become Wedgwood's chief modeler at Etruria in 1794.[2]

During his second year in Rome, de Vaere adapted the famous Borghese Vase for jasper ware production. Flaxman, who praised de Vaere's designs as "full of the sentiment of the fine antique," had already supplied Wedgwood with an interpretation of the vase.[3] In a letter to the ceramist, the modeler suggested that he planned to refine de Vaere's version once it was finished.[4] Because the design was created in the late 1780s and this vase was made in solid blue jasper, it can be dated between 1785 and 1800. Following Wedgwood's death in 1795, the production of solid jasper for larger pieces declined and was replaced by jasper dip, a process of covering the white body with a thin coating of tinted clay.

The design of this urn represents Wedgwood's neoclassical eclecticism, in other words, his adaptation of historical models to contemporary decorative preferences. Both the urn's prototype—a Greek krater—and its decoration evoke the function of the ancient piece as a vessel for mixing wine and water. The urn features the wine god, Bacchus, and his retinue, including his drunken companion Silenus, dancing musicians, and satyrs. Yet, unlike specific imitations by Wedgwood such as that for the Portland Vase (cat. no. 8), this urn "corrects" the

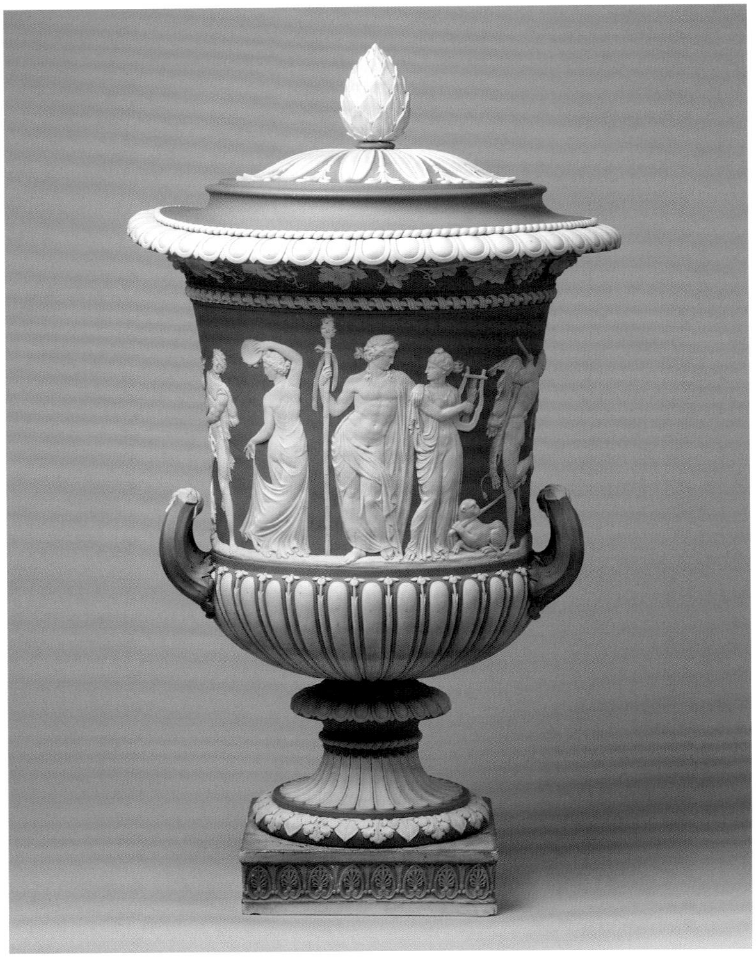

Cat. no. 6

original by adding the handles one would expect to find on a krater; adjusting the proportions of the shape, moldings, and decorations to an eighteenth-century aesthetic; and employing the newly invented jasper for the body. The base is a later replacement; other versions exist with grander pedestals. Perfectly finished on a lathe, this object was the product of an assembly line of several highly trained modelers and repairers. The warping on its lip is a flaw that occurred during firing, likely from a hot spot in the kiln.

The lid added to the piece identifies it as an urn, a form derived from Roman funerary vessels, rather than as a vase. Not sure about the usage of the terms "urn" and "vase," Wedgwood wrote to the architect William Chambers in December 1768 for clarification. Chambers declared that there was "a real difference between *Urns & Vases*," but that he did not have the time to explain the differences at length, as he was about "to wait upon the Queen."[5] Thereafter, Wedgwood abandoned the use of the term "urn," generally listing both types as vases. E.S.

1 See G. Blake Roberts, "The Wedgwood School in Rome," *Proceedings of the Wedgwood Society* 12 (1986): 215–27.

2 Reilly 1994: 155. *Allgemeines Lexikon der Bildenden Künstler... von Ulrich Thieme und Felix Becker*, vol. 34 (Leipzig, 1992): 33.

3 Reilly and Savage 1980: 131.

4 Samuel Smiles, *Josiah Wedgwood* (London, 1894): 222; and C. Louise Avery, "A Gift of Ornamental Wedgwood," *MMA Bulletin* 27 (Nov. 1932): 238.

5 Reilly and Savage 1980: 438.

7. PEGASUS VASE

Josiah Wedgwood (1730–1795), Etruria, Staffordshire; designed by John Flaxman (1755–1826)
About 1786–90
Vitrified stoneware (jasper ware)
17 ⅞ in. (45.4 cm)
Rogers Fund, 1910
10.101.1

Shortly after attempting to copy the Portland Vase in jasper ware, Wedgwood completed the model for this piece, adapting the scenes from a Greek vase, from the famed first collection of Sir William Hamilton, now in the British Museum, London. The original is a fifth-century B.C. red-figure calyx krater with a wide mouth and low handles, while Wedgwood's egg-shaped body appears to be derived from a Greek stamnos.[1]

The neoclassical sculptor John Flaxman designed the main figural frieze for the vessel in 1778 by copying the corresponding plate in the third volume of Pierre d'Hancarville's publication of Hamilton's first collection (see cat. no. 1).[2] Wedgwood's partner, Thomas Bentley, described the scene as "some Honor paid to the Genius of Homer," leading to the nineteenth-century name for the design as the "Apotheosis of Homer."[3] Yet this is certainly an incorrect interpretation, since there are no examples of ancient vases depicting the legendary poet. The scene, showing a *kithara* player mounting a platform and watched by a standing Nike (winged figure of victory), other seated figures, and attendants, is now usually referred to as the "Crowning of a Kitharist."[4]

On the reverse of the Wedgwood vase is a less elaborate scene, with the goddess of wisdom standing inside a temple, modeled after the Minerva Giustiniani, a Roman statue now in the Vatican Museum; next to her is a large palm tree, which was added as a decorative ornament.[5] Each handle, of white jasper, takes the shape of two serpents, terminating in a medusa mask. The finial, also of white jasper, assumes the form of the mythical Pegasus emerging from a cloud, which gives the piece its current name.

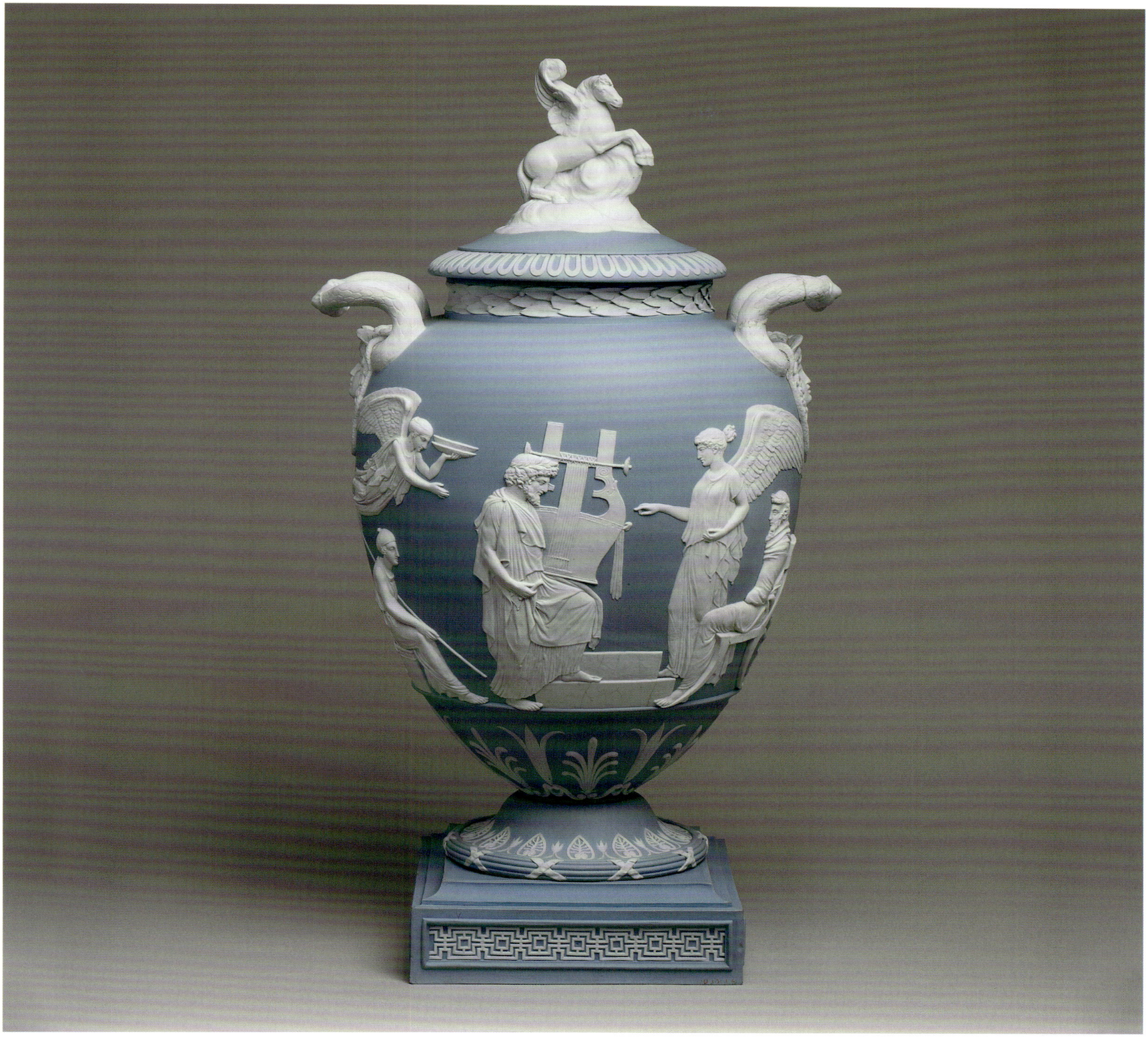

Cat. no. 7

Flaxman's "Apotheosis of Homer" design was used originally for plaques on chimney pieces in 1778 and was first adapted to this vase shape in 1786.[6] The artist followed the etching in the Hamilton/d'Harcanville catalogue fairly closely, although he provided the seated female figure with a stool, unlike the etching in the book, which depicts her seated in mid-air.[7] Flaxman also made some subtle additions, including the feathered wings, the instrument's strings, and the figures' musculature and facial features.

Proud of this design, Wedgwood quickly sent a plaque version to Hamilton in February 1779. The collector responded, "I was sure that your industry would produce in time something excellent in the way of Basreliefs from the specimens I saw before I left England but I realy am surprised & delighted to the highest degree with this proof of the hasty strides you have made towards perfection in your art."[8]

In 1786 Wedgwood presented one of the finer copies of the vase, in pale blue solid jasper, to the British Museum, London. Remarkably, the museum took it. Until then, it had been dedicated to collecting antiquities, and not contemporary works of art. Furthermore, by then it owned the ancient vase that inspired the copy. The acceptance of this gift, therefore, demonstrates the degree to which Wedgwood's neoclassical interpretation was admired. Because his Pegasus Vase was produced in relatively small numbers, it is more rare today even than his version of the Portland Vase (cat. no. 5), which was reissued regularly during the nineteenth and twentieth centuries.[9] K.D.

1 1772.3–20.26. The British Museum lists it as being in the "manner of the Peleus Painter." See J. D. Beazley, 1984: no. 1041.2. It is also depicted in Jenkins and Sloan 1996: cat. 61. For the reference to the stamnos form, see Aileen Dawson, *Masterpieces of Wedgwood in the British Museum* (London, 1984): 106.

2 A letter from Thomas Bentley to Hamilton in February 1779 suggests that, when Flaxman modeled the relief subject in 1778, he relied upon a drawing d'Hancarville had sent to Wedgwood: "Having modelled a large tablet from one of the unpublished designs in your Excellency's collection at the British Museum, which we copied from a drawing lent us by Mr d'Hancarville..."; see Jenkins and Sloan 1996: 184, cat. 60. However, a 1770 catalogue of books used by the factory as source designs, prepared by Bentley, demonstrates that it had at least the first two volumes of the Hamilton/d'Harcanville catalogue and undoubtedly received the remaining two soon after. Thus, the potters were aware of the engraving well before the 1778 date of Flaxman's design; see Johnson 1990: 15–16. Possibly, the drawing referred to in the letter was an auxillary source.

3 Reilly 1989: vol. 1, 583.

4 Jenkins and Sloan 1996; and Dawson (note 1): 102.

5 Reilly 1989: vol. 1, 646.

6 David Irwin, *John Flaxman 1755–1826: Sculptor Illustrator Designer* (New York, 1979): 22. However, Jenkins and Sloan 1996: 184, cat. 60, noted that the relief design for such a plaque was initially modeled by Flaxman in 1778. Reilly 1989: vol. 1, 644, stated that the first example of this vase in blue and white solid jasper ware, was fired on February 18 and 25, 1786; two dipped versions were created in March, according to Wedgwood's Oven Book.

7 Most likely the ancient scene would have depicted the figure seated a chair; however, since many such details were painted over black slip, they might have worn off over time.

8 Meteyard 1970: vol. 2, 380–81. In the letter that accompanied the plaque, Wedgwood asked for verification of the identity of the seated figure. Hamilton replied in June 1779 that it was the "Apotheose of Homer, or some celebrated Poet..."; see Dawson (note 1): 107.

9 Dawson (note 1): 109.

8. WEDGWOOD PORTLAND VASE

Josiah Wedgwood (1730–1795), Etruria, Staffordshire
About 1840–60
Vitrified stoneware (jasper ware)
10 3/8 in. (26.4 cm)
Gift of Henry G. Marquand, 1894
94.4.172

One of the most renowned surviving examples of ancient art, the original cameo-glass Portland Vase (British Museum, London), is a Roman work dating to the end of the first century B.C. or the beginning of the first century A.D. For Josiah Wedgwood, replicating it in jasper ware became a test of his artistic mastery and technical abilities. His ultimate success represents one of the most elaborate efforts to emulate antiquity during this period.

What the scene on the vase depicts exactly has been disputed ever since its discovery. Earlier scholars believed that the vase referred to the birth of Alexander the Great or symbolized the reign of Emperor Alexander Severus, in whose sarcophagus it reportedly was found.[1] A more recent interpretation identifies the main figure on one side as Atia, mother of the emperor Augustus, coupling with Apollo in the form of a snake and thereby alluding to the divine birth of the emperor. The scene may also refer to the marriage of Julia, daughter of Augustus, to his nephew Marcus Claudius Marcellus in 25 B.C. The scene on the other side of the vase has been interpreted as Theseus leaving Ariadne on Naxos while Aphrodite looks on.[2]

Before Wedgwood undertook his copy, the original Portland Vase had already acquired an illustrious provenance. The vase was first recorded as part of the collection of Cardinal Francesco Maria del Monte in 1600. After the cardinal's death in 1627, it was purchased by Cardinal Francesco Barberini and remained in his family's possession for 150 years. In 1780 Donna Cornelia Barberini-Colonna sold the piece to James Byres, a Scottish architect, to settle her gambling debts. Byres made sixty plaster copies of the vase before selling it three years later in 1783, to William Hamilton, who kept it for only one year, after which he sold it to the Duchess of Portland. Wedgwood first learned about the vase when it was in Hamilton's possession. Upon the duchess's death in 1785, her entire collection was sold at auction by Mr. Skinner and Co. When the vase went up for bid on the last day of the sale, it was acquired by a Mr. Tomlinson on behalf of the duchess's son, the third Duke of Portland. Three days later, the duke lent the Portland Vase to Josiah Wedgwood to be copied in jasper ware.[3]

Cat. no. 8

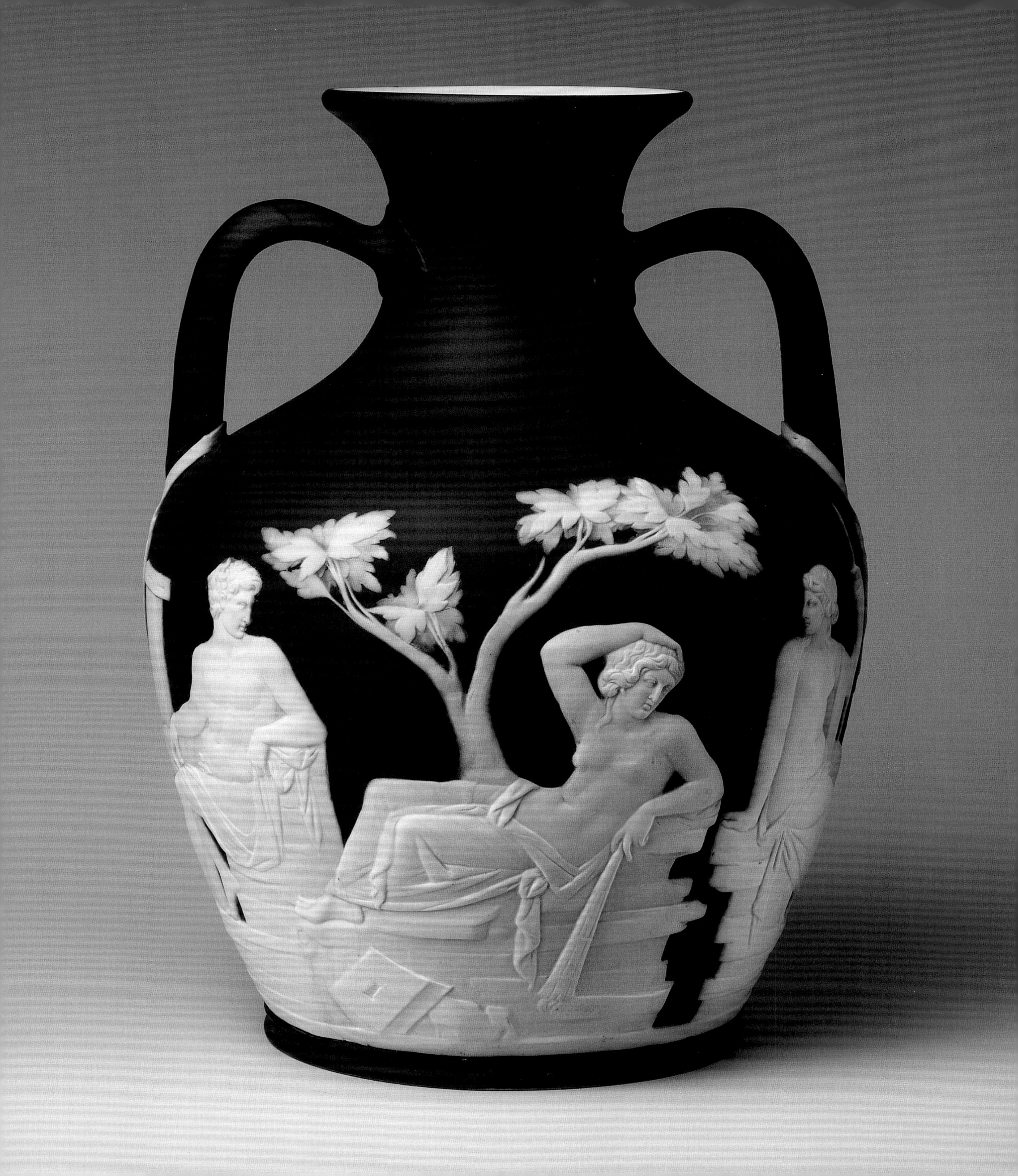

Wedgwood set out not only to reproduce the vase, but in fact to improve what he saw as flaws in the original. He realized that this undertaking would be very difficult technically, and in the end the project consumed almost four years. He first wished to adjust the ancient vessel's shape, which he considered squat and uneven. He also determined to correct what he felt were visible errors in the execution of the figures, but did so only with Hamilton's approval. The latter commented, "You are very right in there being some little defects in the drawing; it would however, be dangerous to touch that, but I should highly approve of your restoring in your copies what has been damaged by the hands of time."[4]

For the material, Wedgwood chose jasper ware, which he described as "much harder than glass, nearly as hard as agate...and like the agate, it will bear to be cut, and take a polish...."[5] However, his original jasper ware produced only a brown-black color, which led him to further experimentation with the material until he achieved the perfect shade of black that he sought.

In the first year of production, Wedgwood devoted himself to the figures, undertaking much of the modeling himself, assisted by Henry Webber and several others. The figures on the ancient cameo glass were carved in varying degrees of thickness, which enhanced the depth of the relief. Wedgwood's figures, on the other hand, were molded individually and attached to the body. He was troubled that he could not achieve a convincing cameo effect, and in fact he never succeeded in applying the figures thinly enough to produce the type of transparency visible in the original vase. To correct this, he applied a thin black slip over some areas of the white body.[6] The process was extremely labor intensive and risky, in that the vessel could be ruined during any one of the numerous steps of production. When the piece was finally assembled, it still faced the danger of being destroyed during firing.

Finally, on May 1, 1790, Wedgwood felt confident enough to show his vase to a select few. The viewing was recorded in the *General Evening Post*, which generated public interest, although the slow and tedious production of the vase made the cost much too high for a wide market. It is uncertain how many first-edition Portland Vases Wedgwood produced. The company's record books indicate that forty-three were fired, eleven of which are known to have broken; still others may have been defective.[7] Many years after Wedgwood's death, his son Josiah Wedgwood II wrote, "The number of copies was restricted by the difficulties and risks attending the execution of each copy, which were so great that I believe my father never sold ten copies...."[8]

The technique was eventually perfected, and several limited editions of the Portland Vase were reissued in the nineteenth and twentieth centuries; the most recent, in 1977, honored the Silver Jubilee of Queen Elizabeth II.[9] Like the Portland Vase featured here, the later copies could be produced on a mass scale in varying colors and sizes because of improvements in technology and manufacture.[10]

In 1810 Wedgwood returned the Portland Vase to the Duke of Portland, who lent it to the British Museum, where it has remained ever since (the museum purchased the vase in 1945). On February 7, 1845, a disgruntled museum visitor smashed it into hundreds of pieces. The vase was subsequently restored several times, most recently in 1989.[11] As a result, Wedgwood's first copies became important documents. In the nineteenth century, they were declared to be the only ones that faithfully reproduced the original.[12] K.D.

1 The vase was reportedly found full of ashes in a marble sarcophagus in the Monte del Grano, a funeral monument in the southeast section of ancient Rome. However, this is doubtful, since a 1594 account of the sarcophagus's discovery (published in 1771) does not mention the vase. The vase first appears in the literature in 1697, more than one hundred years after the sarcophagus was unearthed; see Donald B. Harden, *Glass of the Caesars* (Milan, 1987): 60.

2 Ibid.: 62.

3 D. E. L. Haynes, *The Portland Vase* (London: 1975): 7–13.

4 Cited in Wolf Mankowitz, *The Portland Vase and the Wedgwood Copies* (London, 1952): 29.

5 Ibid.: 23.

6 Reilly 1989: vol. 1, 667.

7 Reilly 1989: vol. 1, 678–79.

8 Mankowitz (note 4): 39.

9 Reilly 1989: vol. 2, 568.

10 The Metropolitan Museum owns a second, later edition, in light blue jasper ware (08.259).

11 Nigel Williams, *The Breaking and Remaking of the Portland Vase* (London, 1989).

12 Mankowitz (note 4): 18.

The Vase as a Template for Artists and Early Neoclassicism

Stefanie Walker

The major artists featured in this section, Jean-Charles Delafosse, Ennemond-Alexandre Petitot, and Joseph-Marie Vien, created vase designs that followed a centuries-old tradition. Since the Renaissance, designers, draftsmen, and printmakers were fascinated by the two basic challenges that the vase form offered: the proportions and curves of the shape or contour, and the ornamentation of the vessel. The possibilities were endless, but the goal was always a harmonious and inventive whole, usually one in which decorative qualities and artistic virtuosity were more prized than the feasibility of actually realizing such an object in stone, metal, or ceramics. The vase had become a template for artists to display their creativity, be they painters, sculptors, designers, or architects.

As European culture became engaged in a more rigorously scientific investigation of antiquity, as well as with expressing the Enlightenment ideals of rationality and simplicity, there was increasing dissatisfaction with the rococo, criticized as frivolous and exaggerated. By around 1760, this resulted in a return to symmetry, simpler geometric shapes, and ornament based largely on Roman architectural decoration, although the trend became known as the *goût grec*, or "Greek taste."

While experimental, the earliest neoclassical vase forms relate to the art of preceding decades. Among them can be mentioned the 1737 publication of vase designs by sculptor Edmé Bouchardon and those of Jacques-François-Joseph Saly from the 1740s. The dynamic and expressive figures on vases designed by Delafosse (see cat. nos. 9–10) and Petitot (see cat. nos. 12–13) are rooted in the sculptural vases of seventeenth-century baroque artists, while Vien (see cat. no. 11) drew on even earlier models, Raphael's decorative work in the Vatican, and sixteenth-century mannerist prints. But a new ornamental vocabulary, at once familiar and fresh, soon emerged. Objects displaying *goût grec* decoration feature ropelike garlands, medallions, masks, repetitive convex and concave shapes (fluting and gadrooning), and flowing meanders and borders. Powerful and popular symbols of antiquity, vases in this early form of neoclassicism can be both ponderous and whimsical.

Two examples made at the French royal porcelain manufactury at Sèvres attempt to translate these new impulses into three dimensions: The earlier object (cat. no. 15), from 1764–65, displays a striking geometric meander and distinctive curved and angled handles; the later one (cat. no. 16), actually a pair dating to the mid-1770s, exemplifies a particularly successful combination of glittering *goût grec* ornament in gilt bronze attached to a simple, soaring shape glazed in a single, strong hue.

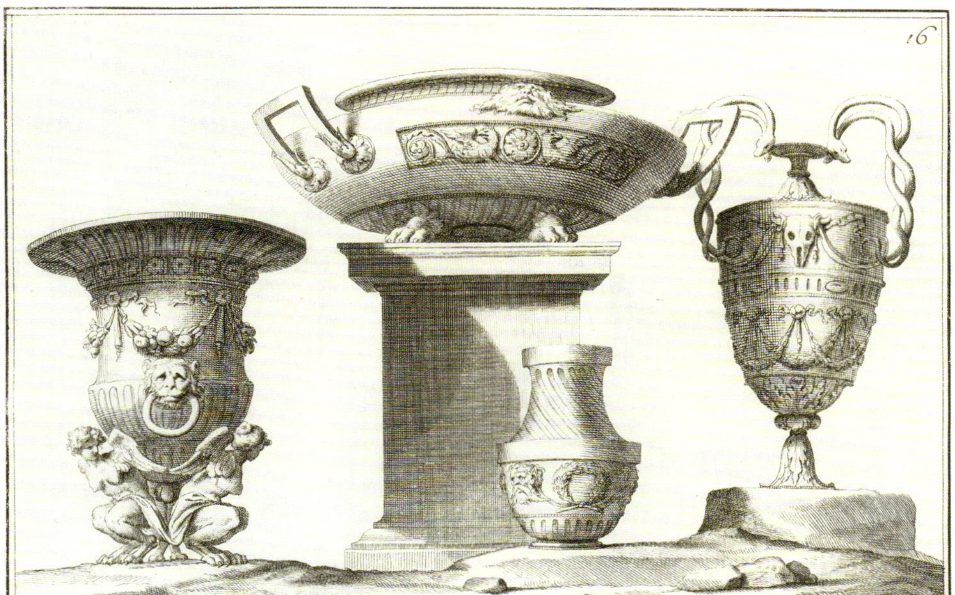

Four Vases, about 1778. Designed by Carlo Bianconi; etched by Angelo Ferri. Etching, plate 16 from *Esemplare di alcuni ornati*...; 6 ⅝ × 9 ⅞ in. (168 × 250 mm). The Elisha Whittelsey Collection, The Elisha Whittelsey Fund (1976), 1976.510.3 (16). See cat. no. 14.

9. DESIGN FOR A DORIC-INSPIRED VASE

Designed by Jean-Charles Delafosse (1734–1789), printed by Jean-Charles Baquoy (1721–1777)
Possibly before 1755
Engraving, plate 6 of *Vases antiques*
Plate: 9 15/16 × 6 1/2 in. (253 × 165 mm)
Inscribed in plate: [top right] 6; [bottom left] *J. C. de la Fosse inv. del*; [bottom right] *C. Baquoy excudit*
Harris Brisbane Dick Fund, 1929
29.58.31 (69)

French architect and designer Jean-Charles Delafosse treated the theme of vases in many of the hundreds of ornament engravings that comprise his prodigious output. Although his designs are easily recognizable for their bold architectonic character, few are more starkly geometric than this one, from a series of six plates depicting vessels of startling, almost uncompromising severity.

Known as the *Vases antiques*, these highly original designs bear little resemblance to vessels used in the ancient world. Unlike most of his contemporaries, Delafosse all but abandoned classical vases as his model. He drew instead from the forms of Greek and Roman architecture and sculpture to create vases that evoked antiquity but were entirely new. Like this one, several designs in the series are imaginative and unorthodox deconstructions of canonical elements of the fluted Doric order, playfully truncated, pierced, or inverted.

Delafosse's use of large-scale architectural forms and relationships make these vases appear colossal. Manipulating perspective and tone, the artist emphasized a sense of monumentality; a low horizon line and deep shadows convey the disorienting feeling of looking up at a large and unfamiliar structure, an impression that is enhanced by the suggestion of foliage, roughly rendered at the base of several vases in the suite.

The radical austerity of the *Vases antiques* series is not sustained in all of Delafosse's vase designs. For instance, while many vessels illustrated in the artist's famous 1768 *Nouvelle Iconologie historique* (see cat. nos. 10, 34) incorporate features familiar from *Vases antiques*, most are more dependent on traditional, historical prototypes, their severity mediated by twisting, grotesque figures typical of sixteenth-century mannerist designs. In a body of work that is already remarkable for its distinctiveness and originality, the plates in Delafosse's *Vases antiques* stand out for their innovations.

Details of its publication history make it possible to date *Vases antiques* about 1755.[1] Such a date would be remarkably early for these designs, placing them, along with Louis-Joseph Le Lorrain's writing table for Ange-Laurent de Lalive de Jully of 1756, among the earliest manifestations of the *goût grec* in the decorative arts (see Ottomeyer essay, Fig. 2). Yet, seen as architectural fantasies, Delafosse's imposing vases are not unlike Le Lorrain's and Jean-Laurent Legeay's visionary images of the same period (see cat. nos. 88–90). Although its date remains unresolved, the series raises interesting questions about the chronology of Delafosse's designs and his contributions to the earliest developments of neoclassicism. H.J.M.

1 *Vases antiques* was issued on at least two occasions, the first perhaps as early as the 1750s. Their title page bears the name of the publisher, F. Chereau, a firm founded by François Chereau and carried on under that name by his wife, Marguerite Caillou. It is possible that *Vases antiques* was issued before 1755,

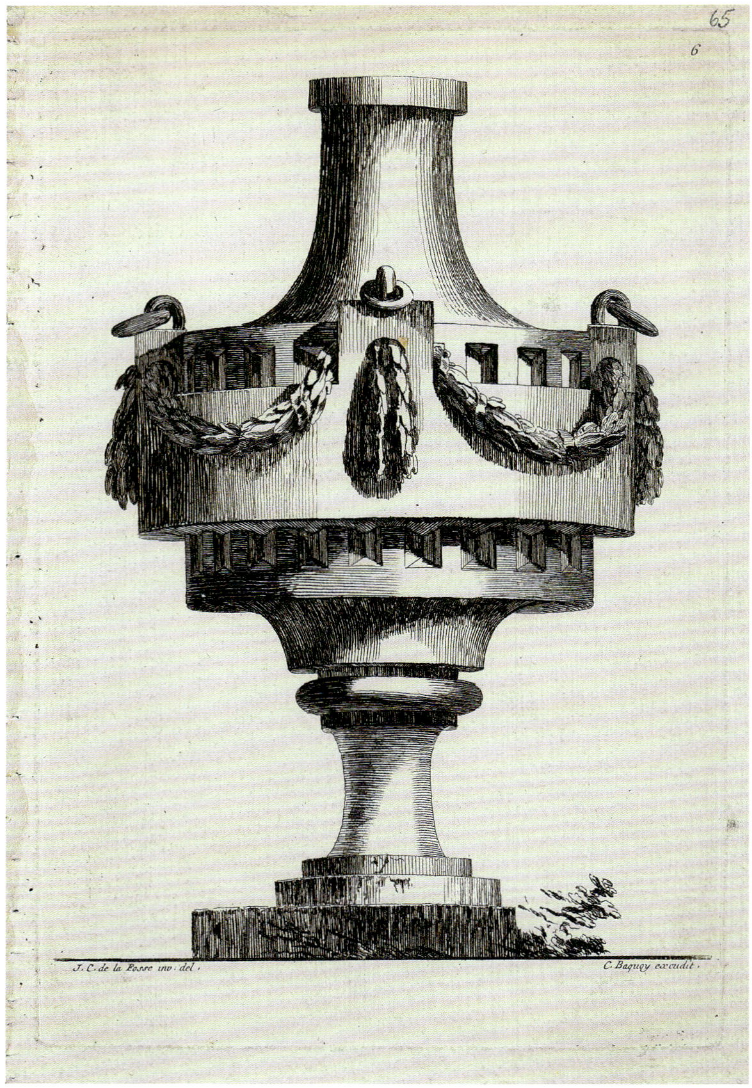

Cat. no. 9

when the business passed to their daughter-in-law, Geneviève-Marguerite, who adopted the imprint "la V[euv]e de F. Chereau"; see Maxime Préaud et al., *Dictionnaire des éditeurs d'estampes à Paris sous l'Ancien Régime* (Nantes, 1987): 79–84. It may be, however, that Geneviève-Marguerite Chereau used her father-in-law's imprint for several years before signing her publications as his son's widow. It is not clear, therefore, whether the present sheet represents an impression from the 1750s or the later one, issued after 1775 by Jacques-François Chereau, François's grandson.

10. DESIGN FOR A EWER WITH WATER DWELLERS

Jean-Charles Delafosse (1734–1789)
Mid-1760s
Pen and black ink, and gray wash
9 $\frac{9}{16}$ × 5 $\frac{9}{16}$ in. (244 × 141 mm)
Inscribed: [lower left, in brown ink] *J. C. Delafosse*
Gift of Cornelius Vanderbilt, 1880
80.3.664

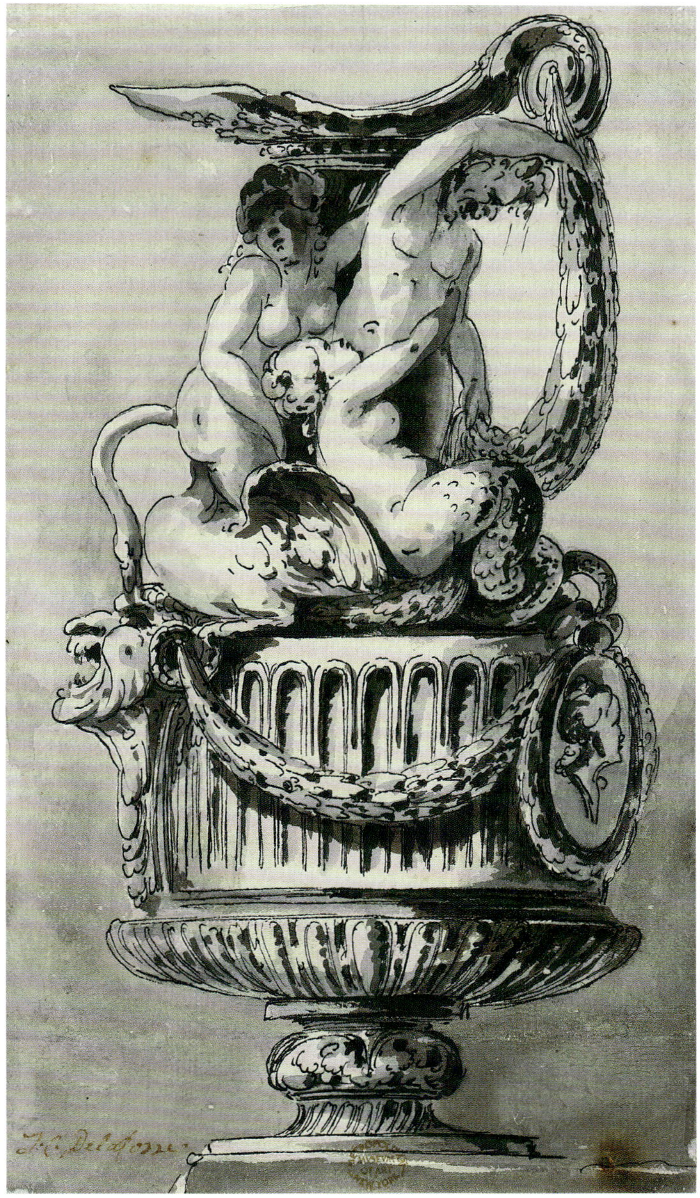

Cat. no. 10

Closely related to a design published in his best-known work, *Nouvelle Iconologie historique* (1768), Jean-Charles Delafosse's lush ink and wash drawing of a ewer probably dates to the mid-1760s, when the artist was preparing the designs and executing the engravings for this ambitious volume, which he announced in the *Avant-Coureur* in July 1767.[1] While different in form than its engraved counterpart (Fig. 1), Delafosse's drawing shares several ornamental details with the print, including the robust laurel garlands, applied medallion, and leafy pouring lip. Both designs—undoubtedly intended as water ewers—also feature mermaid motifs, although Delafosse treated the theme in a more complex, narrative manner in his drawing.

In the drawn design, a mermaid and two other figures are entangled in urgent gestures of embrace and flight. Twisting away from a swan poised for action, the group may allude to the story of Leda, who was ravished and impregnated by Zeus, in a swan's guise. The comely female head depicted on the medallion may suggest the beauty of Helen of Troy, who was said to be their child. Alternatively, it might represent Aphrodite, who sprang from the sea and who is often represented by a swan, her attribute. In this design, the bird arcs it graceful neck downward, pecking at the eyes of a grimacing mask in what perhaps is meant to be a further allusion to lustful feelings.[2]

Stylistically and conceptually, the ewer depicted in this drawing differs from the severity of some of Delafosse's vase designs, like those in his *Vases antiques* (see cat. no. 9), which are characterized by strict geometry and an architectonic structure unrelieved by the organic forms and figures that dominate this image. Conforming more closely to traditional models, the present drawing is consistent with a larger group of vase designs that Delafosse prepared for his *Iconologie*. Undoubtedly inspired by mannerist examples, he populated his designs with fantastic, composite figures typical of the work of sixteenth-century Italian artists such as Polidoro da Caravaggio and Enea Vico or the prolific French print publisher Jacques Androuet du Cerceau.

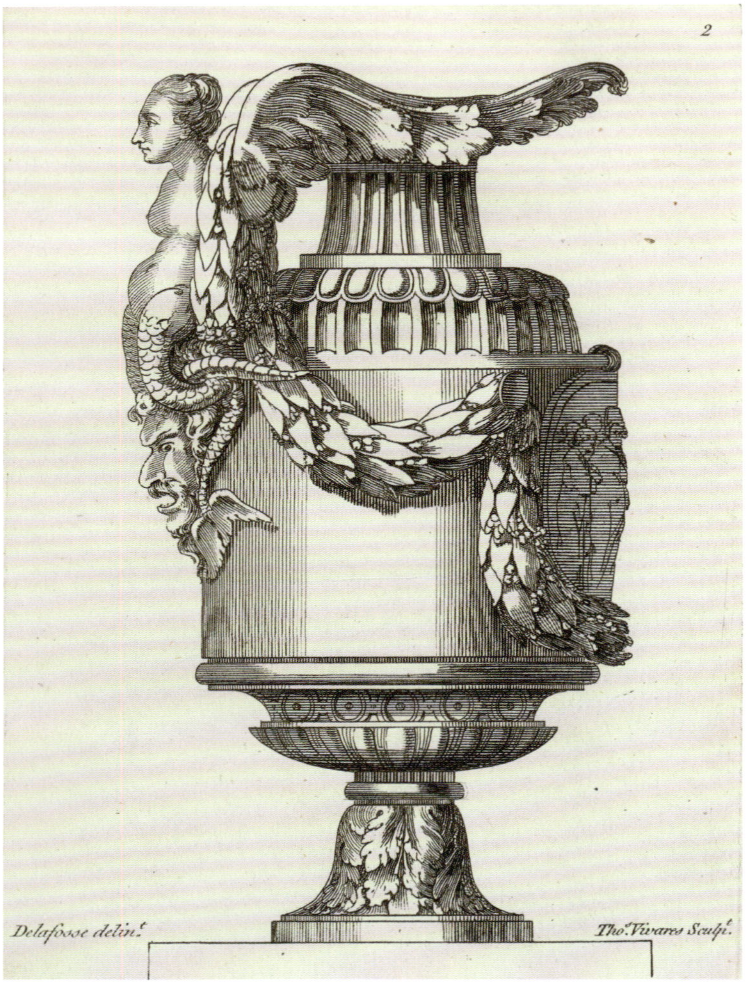

Fig. 1: Plate 2 from *A New Book of Vases* (London: Francis Vivares, 1770). Engraving; plate: 7 ⅝ × 5 ⅝ in. (194 × 142 mm). Harris Brisbane Dick Fund (1928), 28.94.36. In this engraving, Thomas Vivares closely copied, in reverse, a vase illustrated in plate 37 of Delafosse's *Nouvelle Iconologie historique* (1768). In Vivares's version, the engraver changed Delafosse's ram mask to one of a satyr. This engraving is one of six plates in a volume of designs for vases published by Vivares's father; for another plate from the suite, see cat. no. 34.

While the direct influence of mannerist prints is obvious in many later eighteenth-century designs for vases (cat. no. 36, for example), the learned and inventive Delafosse eschewed copying, and absorbed and synthesized his sources instead. In fact, although the present drawing is in keeping with the spirit of Delafosse's mannerist-inspired designs, it seems to draw on the compositional techniques of baroque sculpture as well.

Owing to the nature of his medium, Delafosse's confident ink and wash drawing achieves a freedom and richness that is very different from the crisp discipline of his engravings. Certain characteristics of its execution, however, suggest that this work might have been a preliminary model for a print. The artist's attention to volume and tone make the ewer's form readily understood, and would have rendered the image easy to translate into the medium of the copperplate, its deep contrasts being consistent with the style of Delafosse's other engravings. Related in content, execution—and even in size—to other images in *Nouvelle Iconologie historique*, it is possible that this design was prepared for but not included in that publication. H.J.M.

1 Marcel Roux, *Inventaire du fonds français: Graveurs de XVIIIe siècle* (Paris, 1949): vol. 6, 215.

2 A closely related drawing (80.3.663), signed by Delafosse, was acquired by the Metropolitan Museum at the same time as the present example. It depicts a similar ewer; see Myers 1991: cats. 31-32. Its design—featuring putti, garlands of grapes and other fruit, and ram's-head mask—suggests that it was intended for wine. Similar in their format, vigorous execution, and detail, the drawings may be ideas for pendant wine and water vessels, much like Wedgwood's ewers of Bacchus and Neptune (see cat. no. 39).

11. DESIGN FOR A TRIPOD PERFUME BURNER

Designed by Joseph-Marie Vien (1716–1809); etched by Marie-Thérèse Reboul (1738–1805)

1760

Etching; plate 12 from *Suite de vases composée dans le Goût de l'Antique* (Paris)

Plate: 7 ½ × 5 $\frac{5}{16}$ in. (190 × 136 mm)

Inscribed: [lower left] *.Vien.in.*; [lower right] *R. Vien. S.*

Harris Brisbane Dick Fund, 1947

47.108.22 (12)

An admirer of the classical tradition in seventeenth-century French painting, Joseph-Marie Vien was a firm believer in the superiority of classical models and a committed advocate of the study of antiquity by artists. Influenced by his friendship with the Comte de Caylus, Vien brought the inspiration of both contemporary archeology and the burgeoning *goût grec* to the carefully considered settings and details of his large compositions, such as his celebrated *Cupid Seller* (1763; Musée National de Fontainebleau), which helped to establish the norms of neoclassicism in painting.[1] In this and other scenes depicting Greek and Roman history and domestic life, Vien conveyed an impression of the ancient world with fanciful reconstructions of vases, censers, tripods, and other furnishings. It is not surprising, therefore, that the artist also produced

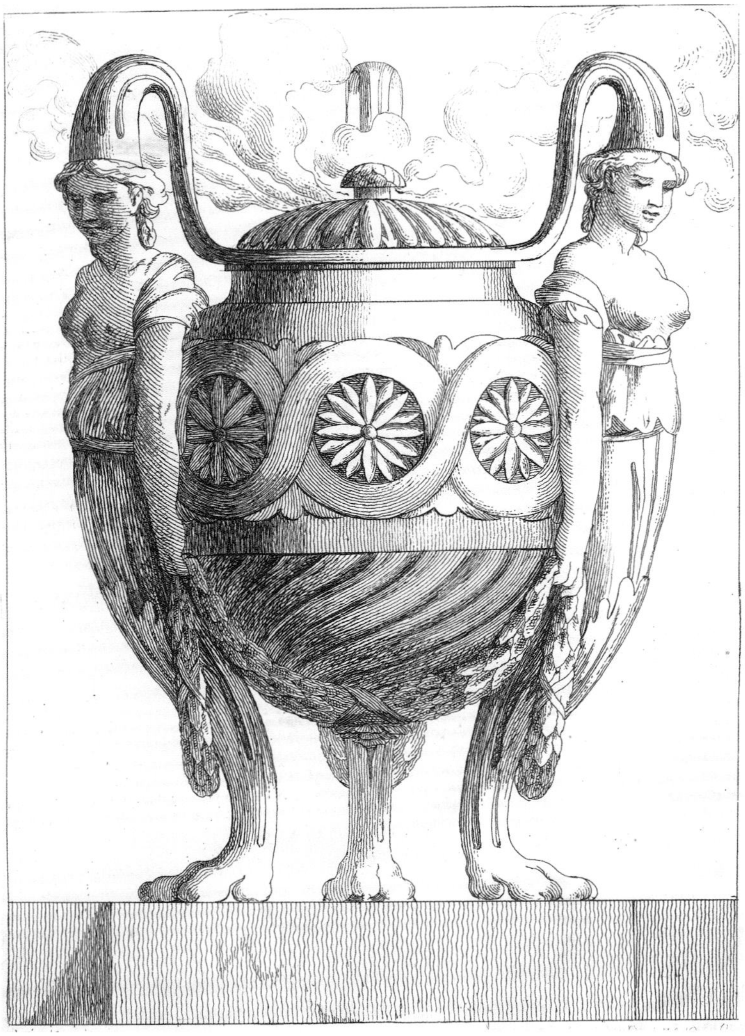

Cat. no. 11

designs for objects in a classical mode. One of them, this etching was published in Paris in 1760 as part of a suite of thirteen plates of vases "in the antique taste" designed by Vien and etched by his wife and fellow academician, Marie-Thérèse Reboul (see Fig. 1).

Depicted with billowing vapors escaping from its lid, Vien's distinctive globular vase was intended as a perfume burner. Its tripod form and motifs—three female monopodes with lion feet, supporting the vessel on a cradle of leafy swags—reinforce that conclusion. By the later eighteenth-century, scent burners usually incorporated iconography alluding to Greek and Roman sacrificial rites (see cat. no. 70). The tripod form alludes to the shape of the three-legged stands and three-sided marble bases used in antiquity to support altars.

In Vien's design, the partly clad female figures may refer to the celebrated Roman shrine of the goddess Vesta, a circular temple supported by caryatids where virgins tended an eternal fire for the divine patroness of the hearth. Although rare in eighteenth-century perfume burners, Vien's symbolism does have art-historical antecedents; in the sixteenth century, Raphael produced a design for a tripod-form incense burner that features three bare-breasted caryatids supporting the vessel's scent reservoir.[2] Other neoclassical objects associated with flames and burning, like candlesticks and fuel burners for hot-water vessels, frequently feature three supporting female figures. Indeed, when the English firm of Wedgwood and Bentley reproduced Vien's design in black basalt in the 1770s, they added a socket to its lid

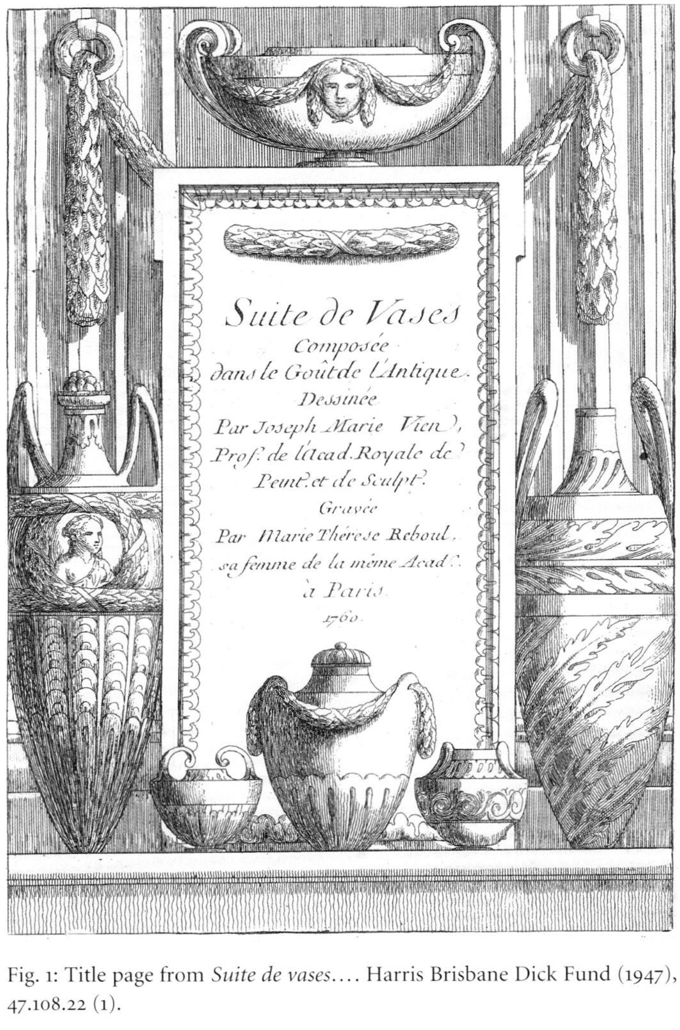

Fig. 1: Title page from *Suite de vases*.... Harris Brisbane Dick Fund (1947), 47.108.22 (1).

for holding a candle. The function of Vien's vessel was clearly open to variation among his contemporaries; the Chelsea factory, for example, reproduced it in soft-paste porcelain as a potpourri with a pierced lid for the emanation of cold scents.[3]

Like many of the earliest efforts in neoclassical vase design, Vien's has mannerist overtones, reflecting sixteenth-century prints that many late-eighteenth-century artists relied upon as sources of information, however dubious, about vases in antiquity (see cat. nos. 17, 31, 36). The strange fantasy of Vien's perfume burner—the chimerical women and the elastic, attenuating columns that sprout as handles from their heads—reflect mannerist preoccupations with mutability and the monstrous. Vien's figures, their backs following the contours of the decidedly neoclassical vessel, are related to similar, but more contorted figures in vase designs of sixteenth-century artists such as Polidoro da Caravaggio and Enea Vico.[4] As Vien and others reclaimed antiquity in the 1760s, they also rediscovered the work of pioneering late Renaissance artists whose imaginations had been equally impressed and inspired by the classical vase. H.J.M.

1 While the *Cupid Seller* is based on a Pompeian painting, Vien added details from his own imagination to the spare Roman composition.
2 The design, attributed to Raphael, survives in a sixteenth-century engraving by Marcantonio Raimondi, who reproduced many of the artist's works; see Miller 1999: no. 61, for this, and Marco Dente's engraving of the same design.
3 For Wedgwood and Bentley's and the Chelsea factory's vases after Vien's design, see Clifford 1978: 161, 171, pl. 83; and Young 1995: cats. D36–D37.
4 See Miller 1999: nos. 68–69.

12. VASE WITH ELEPHANT HANDLES

Designed by Ennemond-Alexandre Petitot (1727–1801); etched by Benigno Bossi (1727–1792)
1764
Etching, plate 16 from *Suite de vases*
Plate: 8 7/8 × 6 5/8 in. (224 × 168 mm)
Inscribed: [lower left] *Petitot in*; [lower right] *Bossi scul*
Rogers Fund, Transferred from the Library, 1985
1985.1024 (16)

Ennemond-Alexandre Petitot's *Suite de vases* can be understood as one in a long line of publications, from the early sixteenth century on, in which vessels serve as vehicles for the display of artistic inventiveness regarding form and ornament. This particular collection, comprising thirty-one plates, distinguishes itself by its origin as a royal commission. In 1753 Petitot, court architect for Duc Philippe de Bourbon-Parma, began designing large-scale statuary vases for the ducal gardens in Parma.[1] The first three designs in the *Suite* were realized in marble by Jean-Baptiste Boudard and placed in the gardens. Conservative in style, they incorporate a basic classical iconography of garlands, swags, ram's heads, and other motifs. To expand the series for publication, Petitot added new inventions that are more imaginative and whimsical. Several include naturalistic representations of animals, such as chickens, grasshoppers, turtles, lions (Fig. 1), and the elephants seen here.[2] Each creature interacts with the vase form in some manner, giving the viewer a sense that the animal is not just an applied ornament, but alive and participatory. In this plate, two female nudes sit on the upper curved ledge of the vase, each astride the head of an elephant. One straddles her animal's ear; each plays with an upturned trunk. The women's outstretched arms connect with the tips of the trunks to form the vase handles. This wit and vitality demonstrate an ingenuity that separates Petitot's designs from more pedestrian ones by his contemporaries.

The ornamentation of this vase exemplifies the diverse influences on neoclassicism that appear throughout the period, particularly that of Asia. Although elephants had been known to Europeans for centuries, they remained a quintessential symbol of eastern cultures and reflect the eighteenth century's fascination with the exotic.[3] Many other fanciful allusions to the Far East are incorporated into this design—the women's coiffures, twisted in large buns, resemble turbans; a fringed tapestry or carpet hangs from the lip of the vessel between the figures; they hold axe-shaped prods to suggest that they are "driving" the elephants.

As early as 1756, the Sèvres porcelain manufactory included an elephant motif in the design of handles on several of their

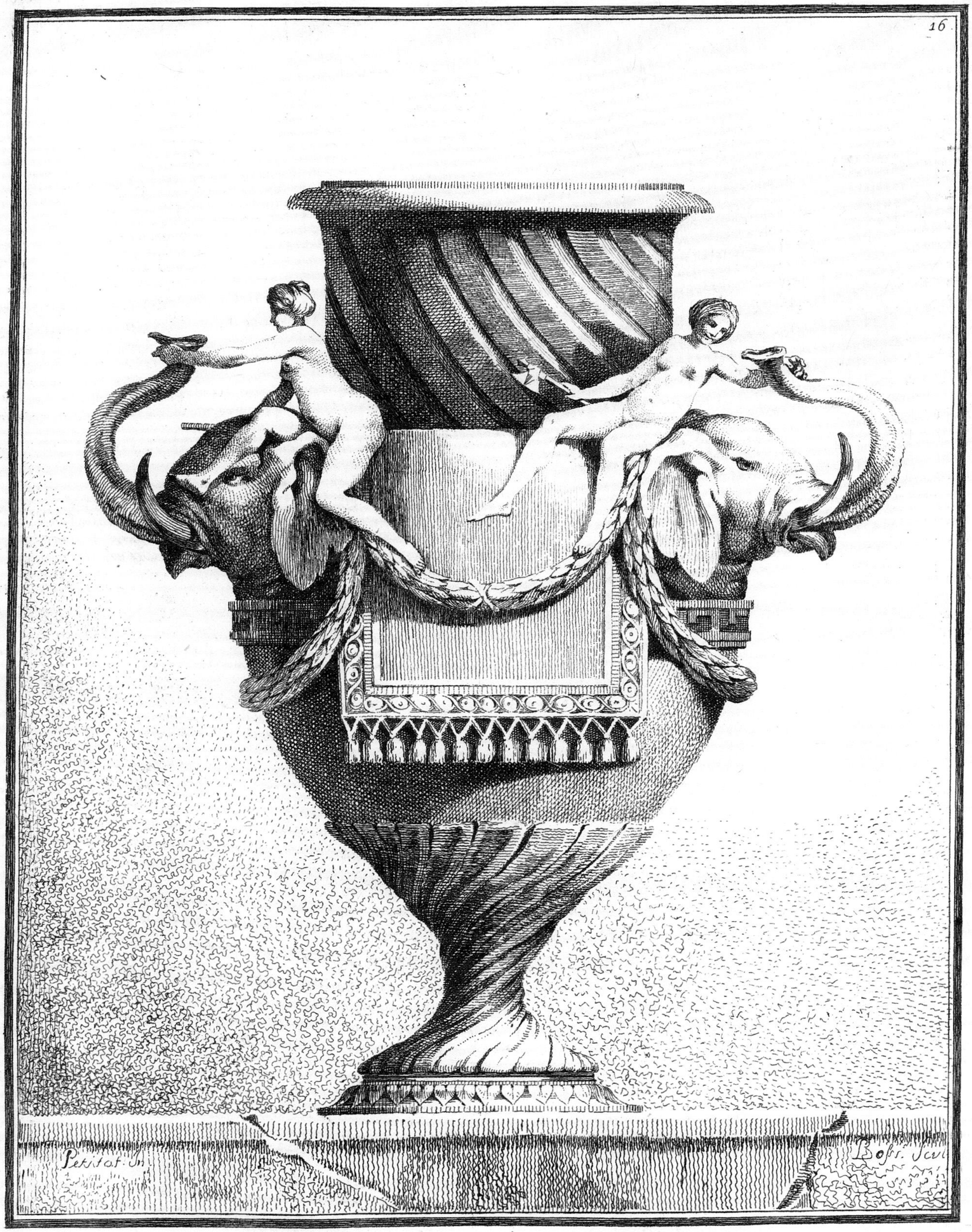

Cat. no. 12

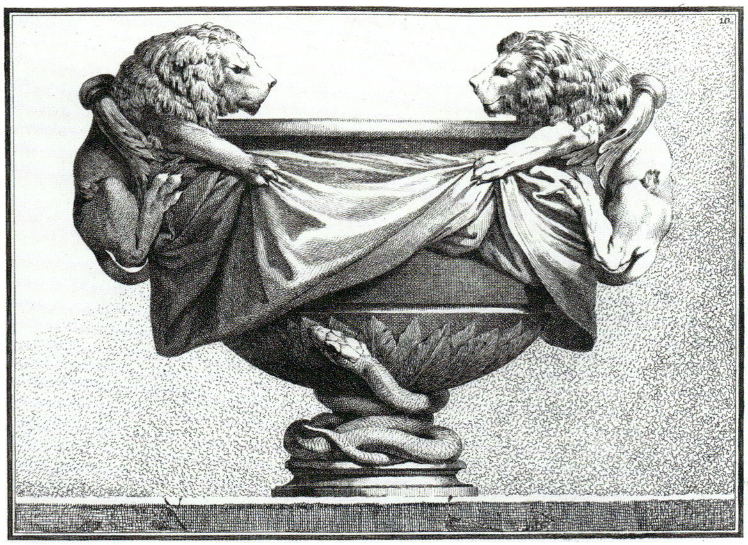

Fig. 1: *Vase with Lions*. Plate 10 from *Suite des vases*. Etching; plate: 6 3/4 × 8 7/8 in. (172 × 226 mm). Harris Brisbane Dick Fund (1940), 40.59.1 (24).

productions.[4] Because Petitot retained contacts in Paris, it is possible that he knew the Sèvres vase with elephant heads (*vase torchère à têtes d'éléphant*), or at least was aware of the design. He certainly frequented social circles where Sèvres porcelain was collected.[5]

Exotic references in Petitot's vase mingle with *goût grec* motifs; the frieze behind the rug is a geometric meander; a heavy garland hangs between the two elephant heads, continuing around the vase's body. Petitot successfully combined both sources in his design, which encapsulates the eclecticism of his imaginative style as well as the diversity of classical interpretations in the eighteenth century. E.W.M.

1. *Petitot, un artista del settecento europea a Parma* (Parma, 1997): 290.
2. A pair of eighteenth-century porphyry and red marble vases with gilt-bronze mounts fashioned after the lion-vase design in Petitot's *Suite* is in the J. Paul Getty Museum, Los Angeles; see *J. Paul Getty Museum Journal* 12 (1984): 265.
3. Louise E. Robbins. *Elephant Slaves and Pampered Parrots: Exotic Animals in Eighteenth-Century Paris* (Baltimore, 2002): 43.
4. A pair of *rose pompadour* elephant candelabra vases from about 1758 is in the Samuel H. Kress Collection at the Metropolitan Museum (58.75.90–.91).
5. Eriksen 1968: 70–71. The style and form of the Chinese example are very different than those of the Sèvres vases; the Asian piece features mannered animal trunks and heads (lacking definition) and has incised rather than molded decoration. Nonetheless, the association is in keeping with the period's interest in the East. If this chain of influence is true, then the Ming form might have served as ancestor to Petitot's design.

13. TWO DESIGNS FOR VASES

Ennemond-Alexandre Petitot (1727–1801)
About 1764
Pen and brown ink, brush and wash over black chalk
13 1/8 × 8 15/16 in. (333 × 227 mm)
Purchase, Robert and Phyllis Massar Gift, 1977
1977.644

The lively strokes, subtle tonalities, and sketchy contours of this drawing testify to the remarkable skill of the artist-designer Ennemond-Alexandre Petitot and provide us with an idea of how the preliminary drawings for his *Suite de vases* of 1764 may have looked (see cat. no. 12).[1] Comparing this sheet to some of the etched plates from the printed editions, one can see how Petitot's drawings may have been interpreted and translated into another medium by the printmaker Benigno Bossi. While these particular designs do not appear in *Suite de vases*, they must certainly be considered in the same vein, as they too emphasize human and animal interaction with the vase form.

In the larger design on the left, Petitot depicted an Atlas-like male struggling under the weight of an oval-shaped vessel, his great effort indicated by his dynamic pose, articulated musculature, and

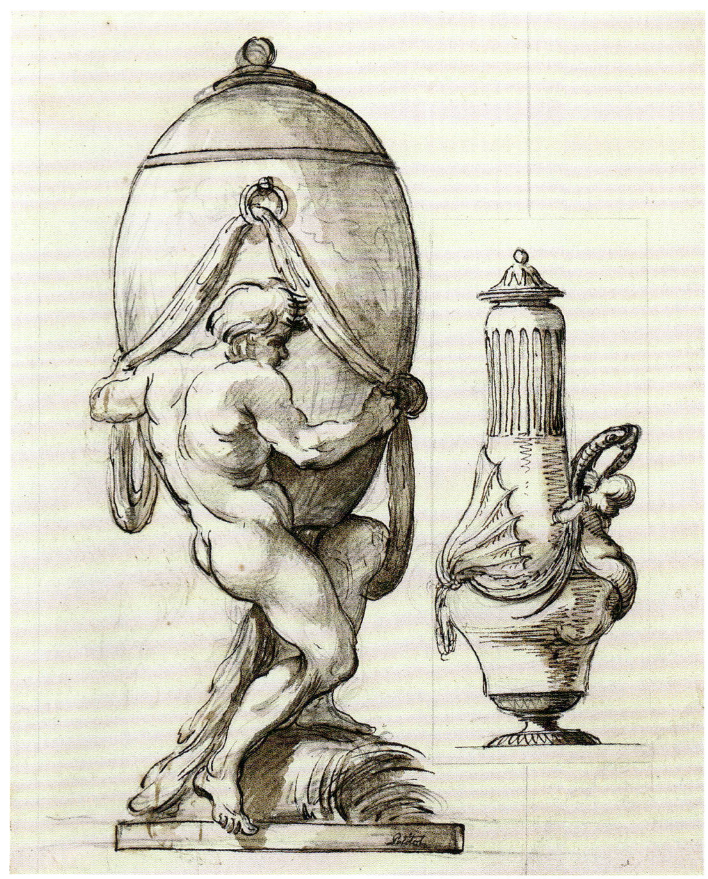

Cat. no. 13

concentrated facial expression. The figure attempts to balance the vase on one knee while bracing himself with his other leg. He reaches around the sides of the vessel to grasp a garland—more suggestive of drapery than a vegetal swag—strung through several rings on the vase,.

While the ornament and decoration of vases were the focus of the plates in *Suite de vases*, Petitot seemed more concerned here with studying movement and figural anatomy: the vase form seems secondary. The verso of this drawing essentially shows this design in reverse, following traces of ink that bled through the paper from the recto; however, Petitot altered the position of the rear leg, and gave the figure longer hair and a beard.[2]

The smaller image to the right is a completely different concept; its fanciful, elongated body is not rooted in classical forms. Petitot appears to have studied two different ornament possibilities: a fishtail applied to the left side of the vase, and to the right a childlike half-figure whose arms and lower body turn into leaf or vine forms. While drawings showing a single vessel with alternate proposals for ornament were often presented to clients for their selection, this seems to be a sketch in which the artist tried out several highly imaginative designs for himself. E.W.M.

1 Red chalk preliminary sketches for *Suite de vases* are known to have existed (see cat. no. 91).
2 Myers 1991: 164.

14. METOPE FRIEZE WITH CLASSICAL MOTIFS

Designed by Carlo Bianconi (1732–1802); etched by Angelo Ferri (eighteenth century)
About 1778
Etching, plate 7 from *Esemplare di alcuni ornati per la gioventù amante del disegno*
Image: 6 × 9 5⁄16 in. (152 × 237 mm)
Inscribed: [bottom left] *C. Bianconi del.*; [bottom right] *A. Ferri scul.*
The Elisha Whittelsey Collection, The Elisha Whittelsey Fund, 1976
1976.510.3 (7)

Carlo Bianconi began his career as a practicing artist, excelling at scenography (perspective painting for the stage), as well as creating architectural decoration for secular and ecclesastical settings. However, he developed a strong interest in the theory and teaching of design, becoming a member of the Accademia Clementina in Bologna in 1770.[1] From 1778 he served as the administrative head of, and essentially ran, the Accademia di Belle Arti di Brera in Milan. Its title translating roughly as "Some

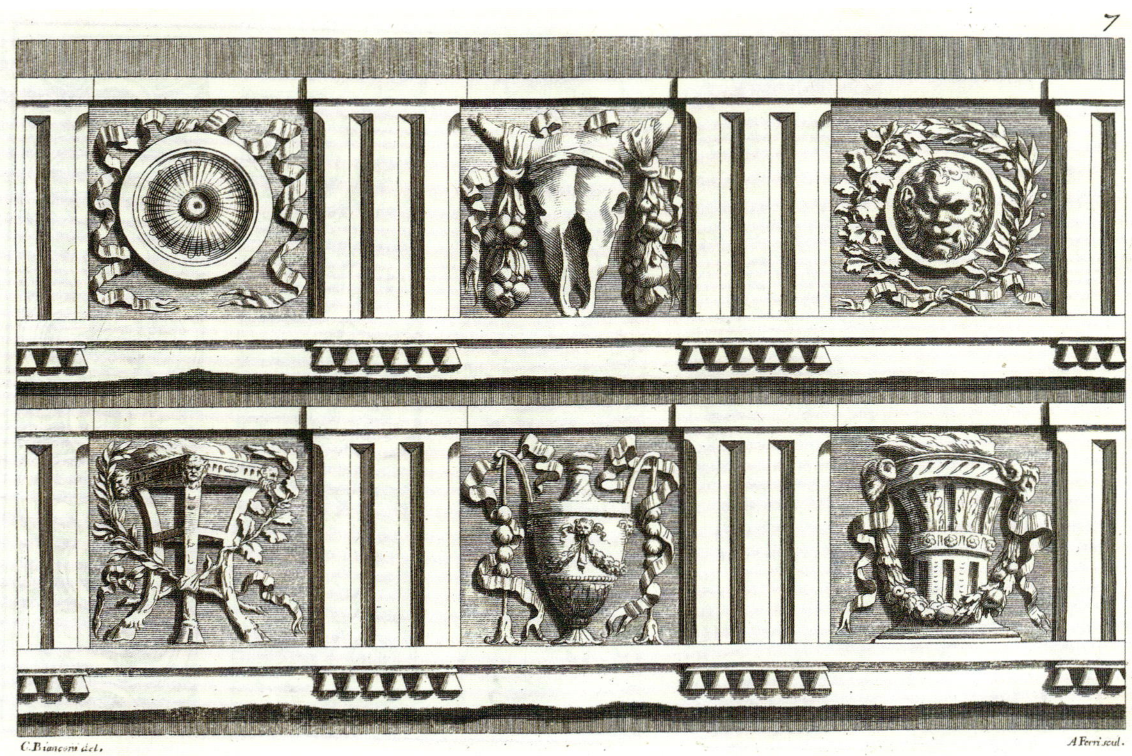

Cat. no. 14

Ornamental Models for Young Design Lovers," this volume probably was intended as a teaching tool. If so, its twenty-four plates, consisting mostly of classical architectural motifs in various forms and stages, vases, and a few vegetal and floral designs, functioned as an instruction manual conceived specifically for Brera students (see also illus. on p. 43).[2]

Here, Bianconi illustrated a section of a frieze, offering six options for metope designs. The motifs he selected—rosette within roundel, tripod, bucranium, vase, grotesque mask within a ring, and flaming altar—appear throughout the volume in various combinations.[3] This limited vocabulary signals Bianconi's firm emphasis on the antique, and may also indicate the somewhat limited extent of his creativity as a designer.

Bianconi's work draws on Renaissance interpretations of antiquity, and his designs, therefore, can be considered a filtered expression of classicism. All of the elements employed in this frieze are directly associated with ancient ritual sacrifice and were adapted first to a sixteenth- and then to an eighteenth-century sensibility.[4] The vase in the central lower metope, for example, is based on an ancient amphora, with its narrow neck and footed base. However, the addition of dynamic garlands and ribbons and dramatically elongated handles transforms the prototype into a fanciful representation quite removed from the original form. The incorporation of this vase into a classical architectural scheme, perhaps intended to decorate a fashionable building, validates its importance as an exemplary neoclassical motif.
E.W.M.

1 S. Samek Ludovici, "Bianconi," *Dizionario biografico degli italiani*, vol. 10 (Rome, 1968): 246–48.

2 The publication is not dated, but most probably it originated at the moment Bianconi assumed his teaching and administrative activities at the Brera in 1778. However, the fact that the book was published in Bologna, which he left for Milan never to return, gives pause to this dating. At the Bolognese Accademia Clementina, Bianconi was an elected member rather than an active teacher; thus, in this writer's opinion, he would not have created a pattern book for students at this time. All of Bianconi's other work was published in Milan after 1778.

3 The vase forms in plate 16 (illus. on p. 43) are ornamented with similar grotesque masks, rosettes, bucrania, and garlands. Plate 23 displays these motifs on a tripod, urn, and other objects.

4 An interior frieze of the Ara Pacis Augustae in Rome, dedicated in 9 B.C. after the Pax Augusta, includes bucrania with lush garlands suspended between them, curling ribbons, and rosette roundels; all are consistent with Bianconi's design lexicon. Certainly, he was aware of this ancient source for sacrificial iconography, even though he relied on later interpretations of it.

15. VASE WITH A SCENE BEFORE A DUEL

Sèvres Manufactory; shape possibly designed by Étienne-Maurice Falconet (1716–1791); figural painting attributed to Jean-Baptiste-Étienne Genest (1725–1789)
About 1765
Soft-paste porcelain
18 3⁄16 in. (46.2 cm)
Mark: crossed *L*'s, in blue
Gift of the Samuel H. Kress Foundation, 1958
58.75.74

Vessels of this form (*vases à bâtons rompus*), introduced by the Sèvres royal porcelain manufactory in 1763, are among the earliest made in the company's burgeoning neoclassical style.[1] Arresting even today for the boldness and angularity of its design, this vase exemplifies the originality of French designers working in the *goût grec* manner, the first stage of neoclassicism. Idiosyncratic, spirited elements like the sinuous, elongated S-curves of its whiplike handles, and the unusual gilded zigzag and diamond pattern of the pseudo-Greek meander, known as *bâtons rompus* or "snapped sticks," on its belly, stand in sharp contrast to the restraint and cool, "Augustan" tranquility of the neoclassicism that would develop in the next decade.

This vase's striking design does not have any conspicuous ancient prototypes or antecedents in prints. While Greek and Roman silver tableware may have inspired the angular elbows of the handles and ancient bronze and pottery vases feature similar abruptly reversing loop handles, the design of the body, like the *goût grec* vases of Jean-Charles Delafosse (see cat. no. 9), draws on classical architecture, not ancient vessels, for the thumb-grooved flutes and triglyph ornaments at its neck. Unusual as the form seems today, such vases were undoubtedly popular. Sèvres made two versions of this design in two sizes; the second, smaller type is missing the molded meander on the belly. A third, denuded variant, almost unrecognizable as a relation to the present vase, lacks the meander, handles, oak-leaf garlands, and triglyphs.[2] Many examples of the first two forms, which were made until at least 1773, are found in museums and at least one private collection.[3]

The name of all these variations, *vases à bâtons rompus*, is based on their similarity to a silhouette drawing labeled as such in an 1814 inventory of the factory's plaster models.[4] The expression *travailler à bâtons rompus* means "to work by fits and starts," which seems to describe the way the meander changes direction and reverses itself in short sections. This shape may have been designed by Étienne-Maurice Falconet, director of sculpture at Sèvres from 1757 until 1766, who was responsible for introducing many neoclassical designs to the manufactory's modelers.[5]

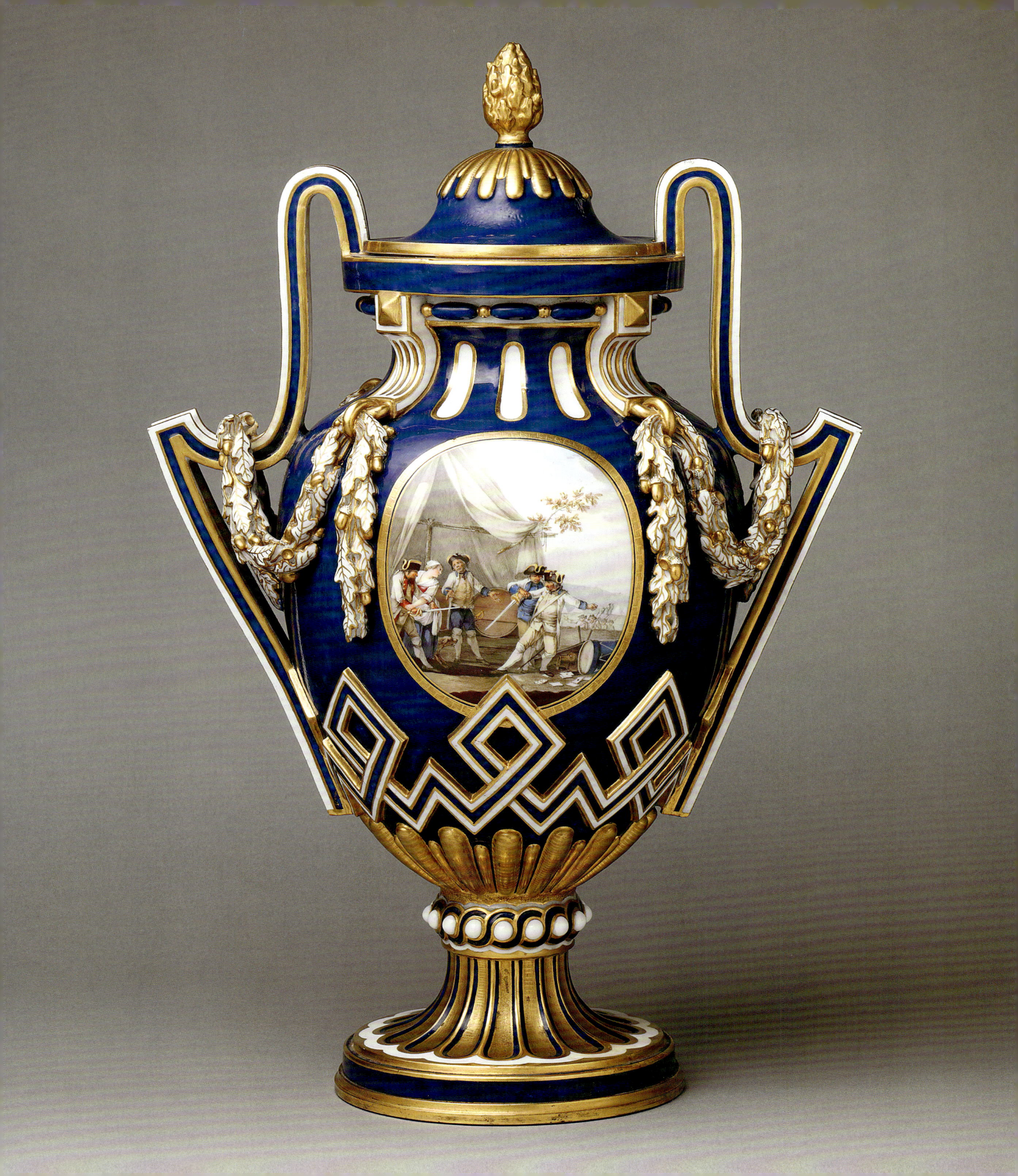

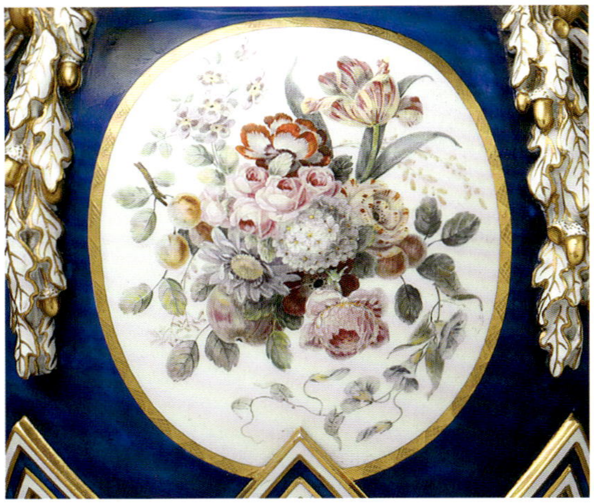

Fig. 1. Detail of the reverse of cat. no. 15.

Vases were among the first products at Sèvres chosen to reflect the new neoclassical style, since changes to single forms were easier and less costly than modifying the large table services of many different pieces that the manufactory also produced.[6] Vases—with their confirmed associations with antiquity—were an obvious place to begin shifting from the rococo vocabulary of sinuous, irregular curves and naturalistic ornament toward the symmetry, static quality, and restraint of the classical tradition. This early example combines innovations in its form with established decorative traditions that had long been successful for the factory. The disparity is not surprising. Decorative scene painting at Sèvres followed the developments of full-scale painting, a field in which the neoclassical style took hold less quickly than in other media. Architecture, sculpture, and the decorative arts, unlike painting, benefited from an arsenal of ancient forms and ornament on which to draw. Indeed, when this shape was designed in the early 1760s, the neoclassical style had only begun to emerge in painting.

Thus, while the form of this vase exemplifies the *goût grec* in its symmetry, hard-edged geometry, and ornament, its painted imagery is far less classical. The scene on the obverse, depicting an encampment and a sword fight between card-playing soldiers, is attributed to Jean-Baptiste-Étienne Genest, head of painting at Vincennes-Sèvres from 1753 to his death.[7] The reverse image illustrates a tangle of detailed flowers and leaves (Fig. 1), for which no author has been suggested to date. In both scenes, the picaresque subject matter, soft and colorful palette, and naturalism all reflect the rococo tradition still being practiced at Sèvres. An interesting hybrid, this vase provides a lesson in the readiness and pace with which different art forms embraced neoclassicism. "By fits and starts" is an appropriate name indeed. H.J.M.

1 Savill 1988: vol. 1, p. 245, noted that the form was introduced in 1763, was made in two sizes, and that the earliest surviving dated examples bear marks for 1764. She posited (246) a relationship between this design and Duplessis the Younger's Sèvres *vase grec à rosettes*. The *bâtons rompus* vase is discussed and illustrated in Dauterman 1960: 293, fig. 23; and catalogued in Dauterman et al. 1964: no. 46, figs. 168–69.

2 Savill 1988: vol. 1, p. 245, identified the three versions as shapes A, B, and C, respectively. An example of shape A, the museum's vase is the larger of the two sizes produced at Sèvres. The Metropolitan Museum also owns an example of shape B; see Dauterman et al. 1964: 226, no. 51, figs. 181–82.

3 Cognates to the Metropolitan Museum's version are in the following British collections: the British Royal Collection, Houghton Hall, Waddesdon Manor, and the Wallace Collection. Others are in the Musée du Louvre, Paris, and the State Hermitage Museum in St. Petersburg. Savill 1988: vol. 1, 245, nos. C270–71, noted that the last factory references to this shape date to 1773. For other surviving examples, see Dauterman et al. 1964: 217; and Savill 1988: vol. 1, nos. C245–46; and C250, n. 3. See also Ennès and Ducrot 1997: 44–46, 96, nos. 44–45.

4 Savill 1988: vol. 1, p. 245.

5 Eriksen 1974: 112, 371, pl. 275. See also Geoffrey de Bellaigue, *Sèvres: Porcelain from the Royal Collection*... (London, 1979): 109–10, no. 114, cats. 61–62.

6 Eriksen 1974: 110.

7 Originally given to Jean-Louis Morin, the painting on this vase recently has been reattributed by Linda H. Roth to Genest, the head of the painter's workshop at Sèvres who specialized in military encampment scenes; Linda H. Roth. "A New Attribution of Three Sèvres Vases," *MMA Journal* 37 (2002): 247–55. Roth indicated (n. 44) that no engraved source for this type of encampment scene has been located to date.

16. PAIR OF VASES WITH SATYR-HEAD MOUNTS

Sèvres Manufactory; mounts: French

About 1775 (vase), 1775–1800 (mounts)

Soft-paste porcelain and gilt bronze

11 3/8 in. (28.9 cm); diam. at handles 6 3/4 in. (17.1 cm)

Bequest of Mrs. Herbert N. Straus, 1978

1978.12.4–.5

Vases à monter, or vases for mounting, was the term used at Sèvres to describe the simple forms made to be set into elaborately sculpted gilt-bronze mounts.[1] In this pair, finely cast plinths, fluted rims, and festooned satyr-head handles embrace the smooth contours of the vessels, fired in a luminous green known as *pomme-verte* (apple green).

The fully modeled satyr-head handles are a variant of the mounts used on the earliest known Sèvres *vase à monter*, a mechanical vase once owned by Madame de Pompadour, now in the Wadsworth Atheneum, Hartford.[2] While factory records first note the production of vases for mounting in 1764 and the first sales in 1767, the Hartford vase is thought to date from 1763. The Metropolitan's pieces, which differ from the Hartford piece in featuring single swags of oak leaves instead of double swags of laurel leaves, are rare examples of this design. The Musée des Arts Décoratifs, Paris, owns a similar vase, and another is in private hands.[3]

This pair may have been ordered by Antoine or Jean Dulac; they were among France's most important eighteenth-century luxury merchants (*marchands-merciers*), who supported the production of *vases à monter*.[4] Acting as retailers, and sometimes as designers, these merchants were middle-men who brokered the manufacture of high-end goods, such as mounted vases and porcelain or lacquer-paneled furniture.[5] They were key to the development of goods that combined a variety of materials, since guild regulations limited the materials and techniques craftsmen used. Vases like these involved three distinct sources: porcelains from Sèvres, mounts that were modeled and cast by *fondeurs-ciseleurs*, and gilding and chasing by *ciseleurs-doreurs*.

With their heavy, leafy festoons and stern demeanor, the satyr heads (see Fig. 1) on this pair are related to those on a vase created by Jean-Charles Delafosse in the 1760s (see cat. no. 34).[6] The German designer Friedrich Kirschner also depicted a vase with satyr-head handles and swags similarly looped through scrolling horns in an undated etching of an egg-shaped vase with cover.[7] H.J.M.

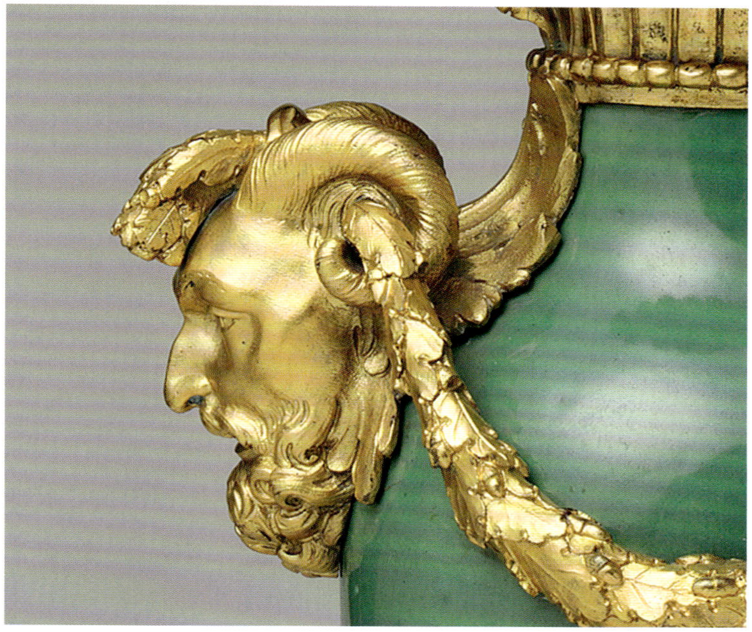

Fig. 1: Detail of cat. no. 16.

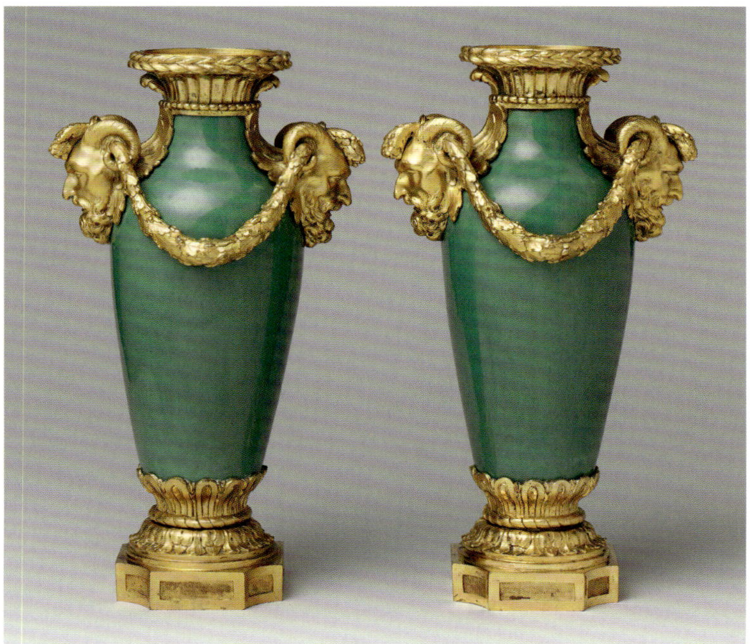

Cat. no. 16

1 The practice of setting Asian porcelain and other precious, rare, or curious objects in metal mounts was centuries old; but, the extensive production, marketing, and acquisition of mounted porcelains, as well as vessels in other materials, was a phenomenon of eighteenth-century European culture; see Francis J. B. Watson and Gillian Wilson, *Mounted Oriental Porcelain at the J. Paul Getty Museum* (Malibu, Calif., 1982); Francis J. B. Watson, *Mounted Oriental Porcelain* (Washington, D.C., 1986); and Pierre Kjellberg, *Objets montés: Du Moyen Âge à nos jours* (Paris, 2000).

2 The Wadsworth Atheneum vase features a miniature equestrian statue of Louis XV (modeled after a monument by Edmé Bouchardon), which emerges from the vase upon the removal of its lid; see Eriksen and Bellaigue 1987: 323–24; and Roth and Le Corbeiller 2000: 121–27, cat. 65.

3 See the example sold at Sotheby's Monaco, sale cat., June 17, 1988: lot 682; and Sotheby's Monaco, sale cat., Mar. 3, 1990: lot 179; cited in Roth and Le Corbeiller 2000: 121, 126, n. 18. Other examples of double-swag bronze mounts are found on vases at the Vyne, Hampshire; on a pair of potpourris at Waddesdon Manor; and on a pair of vases last known to be in a private Polish collection (illustrated in Sotheby's New York, sale cat., May 7, 1983: lot 87a); see Eriksen and Bellaigue 1987: 323–24. Eriksen 1968: p. 236, cat. 81, illustrates a potpourri *vase à monter* of circa 1767–70, with similar satyr-head and double-swag mounts.

4 "Dulac Md rue St. Honor, Invenit" is inscribed under the lid of the Wadsworth Atheneum vase; three dealers by that name worked in the rue Saint-Honoré, Paris, at the time the vase was made. Jean Dulac appears in the factory's sales registers from 1758 to 1776; see Roth and Le Corbeiller 2000: 124.

5 See Sargentson 1996.

6 Two similarly molded satyr-head handles rendered in porcelain rather than as gilt-bronze mounts are found on a Sèvres vase from 1770–75 in Palazzo Pitti, Florence; see Sven Eriksen, Sheila Tabakoff, and Kirsten Aschengreen Picenti, *French Porcelains in Palazzo Pitti* (Florence, 1973): 89.

7 Illustrated in Clifford 1978: 172, pl. 85; and Young 1995: 111, cat. E27.

Tradition and Innovation at the Royal Manufactories

Jennie L. McCahey

The adoption of the neoclassical style in France marked a radical departure from the curvilinear forms, asymmetrical motifs, and delicate refinement of the ebullient rococo that preceded it. However, not all French decorative artists immediately accepted the new taste for antiquity (*goût grec*) that emerged during the 1760s. The royal porcelain manufactory at Sèvres, for example, continued to respond to the taste for the rococo that prevailed among its royal patrons. The manufactory's struggle to maintain its successful tradition of vase production, while incorporating the fresh wave of classicism, becomes evident in a number of objects presented here that are transitional in form and decoration.

From the moment of its establishment at Vincennes in 1740, the company had been closely tied to King Louis XV and relied on the royal court for financial and artistic support. The king ordered the factory's transfer to Sèvres in 1756 and purchased all remaining shares to establish it as a royal manufactory. In the 1750s, both Jean-Claude Chambellan, known as Duplessis the Elder, director of forms, and Jean-Jacques Bachelier, chief of decoration, provided popular new vase models and designs that reflect the rococo taste, with elaborate piercing, sinuous curves, brightly colored grounds, and lighthearted figural scenes of children, putti, or flowers.

Despite the availability and influence of neoclassical engravings by Jean-Charles Delafosse and Ennemond-Alexandre Petitot, the emergence of the new style at Sèvres was subtle, appearing first in the 1760s in restrained, oval forms, and a preference for simple color combinations, such as deep blue, gold, and white. The new shapes incorporated classical architectural motifs such as gilded "Greek keys," crenellations, laurel or oak garlands, and fluting. The formerly asymmetrical reserve cartouches became simple ovals, but the painted scenes continued the pastoral tradition so popular with the court. These softly colored landscapes inhabited by cherubs, milkmaids, and other genre figures derive from paintings and engravings by seventeenth-century Dutch and Flemish painters and more contemporary French court artists, such as François Boucher.

Although the Sèvres factory's products continued to evoke rococo traditions, the ornamental vocabulary of neoclassicism appears more liberally on vases produced during the 1770s. Elaborate gilt-bronze mounts fashioned in the new style were attached to unpainted, simply shaped Sèvres vases at the behest of Parisian luxury merchants, who often introduced fashionable new trends in luxury objects. The thick, gilded tassels and cords, husk chains, and elaborate festoons that embellish the porcelain ware refer directly to classical antecedents, but the diffuse palette and bucolic mood of mythological scenes and landscapes (as well as the floral trophies on the reverses) indicate a certain reluctance to abandon the previous fashion entirely.

Only after the arrival of sculptor Louis-Simon Boizot in 1774 did the neoclassical mode at the manufactory come fully into its own. Vases now incorporated amphora-like handles and mythological figural supports or biscuit medallions reminiscent of marble reliefs and ancient cameos. With the 1785 appointment of Jean-Jacques Lagrenée as artistic codirector, the shapes and painted decorations of factory products exhibit a decidedly greater reliance upon more archeologically accurate classical models—as exemplified in the famed Rambouillet service (see cat. no. 24)—resulting in a look that was entirely in keeping with the delicate motifs and spirals of mature neoclassical (*goût étrusque*) designs appearing in other media.

Younger royal manufactories, such as Fürstenberg in Germany and Buen Retiro in Spain, were less burdened by a longstanding rococo tradition. Less exclusive as well, these companies enjoyed sales and income from both royalty and the public. Like the English Chelsea-Derby factory, they looked to Sèvres to keep abreast of the latest shapes, colors, or motifs, but they also consciously set out to become viable competitors, creating inventive and affordable objects that at times were more innovative than French examples in form and ornament.

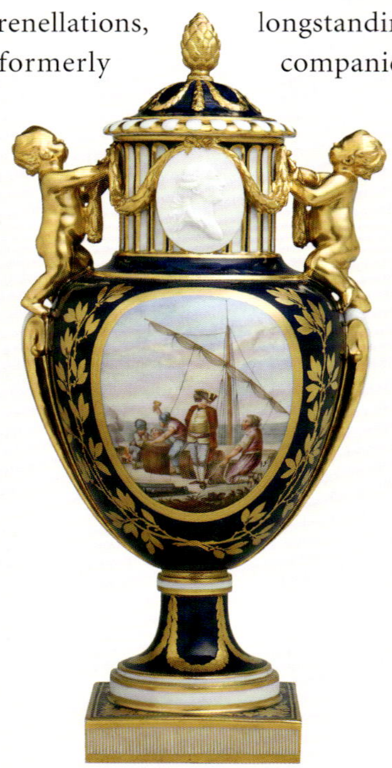

Cat. no. 22 with medallion of Louis XVI.

17. VASE WITH DOLPHIN HANDLES

Sèvres Manufactory; figural painting attributed to Charles-Nicolas Dodin (1734–1803); trophy painting attributed to Charles Buteux the Elder (1719–1782) or Louis-Gabriel Chulot (1736–1824)
About 1765
Soft-paste porcelain
13 ¼ in. (33.7 cm)
Mark: [incised inside foot, in glaze] *PT*
Gift of the Samuel H. Kress Foundation, 1958
58.75.67

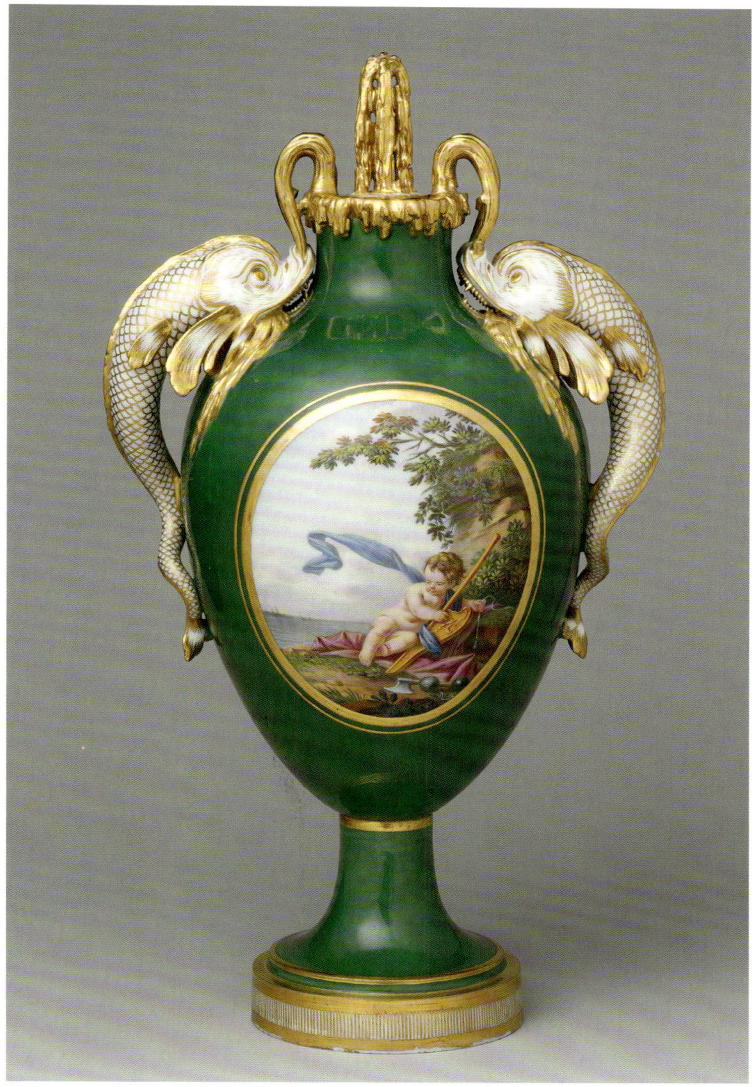

Cat. no. 17

As with many neoclassical vases, the distinctive design of this myrtle green, dolphin-handled vessel (*vase à dauphins*) from the royal porcelain manufactory at Sèvres was almost certainly suggested by one of the sixteenth-century Italian ornament prints to which French and English artists and manufacturers often referred for their classically inspired designs.[1] A similar vase, with conspicuous, scaly dolphin handles curving from its lip to its shoulders, was one of ten vessels in a well-known series of designs engraved in 1582 after frescoes by Polidoro da Caravaggio. Published and pirated repeatedly in the sixteenth century, these engravings were also known in eighteenth-century France. In fact, copies were reissued sometime before 1775 by the Paris printseller Jean-François Daumont, reflecting the demand for such images by neoclassical designers.[2]

Despite his dependence on Polidoro's unusual concept, the designer or modeler of this vase interpreted it with considerable wit. Now the dolphins are not just handles but participants in a sculptural drama, spurting arcs of water into a burbling fountain that forms the vase's lid. A mannerist spirit pervades this imitation of natural forms and play with materiality and substance, as the fountain "overflows" its edges, spilling gilded droplets onto the cool, myrtle green of the vase's neck. The initials "PT," which are incised in the glaze on the foot of this piece, may refer to a Sèvres factory employee known as Pierre-Aymand Tristant, who often finished modeling and chasing forms before their first firing.[3]

The technical complexity of the vases produced at Sèvres—a painstaking process that moved goods from the molding and curing rooms through repeated controlled firings, as they passed through the hands of glazers; specialized flower, trophy, and genre painters; and finally gilders—can distract one from the nature of their decoration. The carefully considered, sustained themes and iconography of some of these objects resulted in refined visual content. In this vessel, for instance, form, ornament, and imagery all reinforce a consistent theme, one of surprising subtlety. Here, the painted images in two oval reserves—depicting aspects of marine warfare—build on the vessel's water imagery and dolphin handles. In the main scene, a small child reclines at water's edge, his blue cloak blowing in the breeze, with ships visible in the distance. He cradles a rudder in his right arm and holds a plumb line in his left, while weapons lie at his feet.[4] The painting on the reverse displays more nautical symbolism: a trophy that includes a trident, bullrushes, a fishing basket, and an anchor hanging with pipes, horns, weapons, and armor (Fig. 1).[5]

The subject matter and form of the vase—its symmetry and decoration, with a plain ground surrounding simple oval cartouches—are neoclassical in spirit. Although the shape appears in factory records as early as 1757, the color and decoration of this version suggest that this piece might date to about 1765, representing one of the first efforts at Sèvres in the neoclassical idiom.[6] Still retaining some of the spirit of the rococo, the vase's

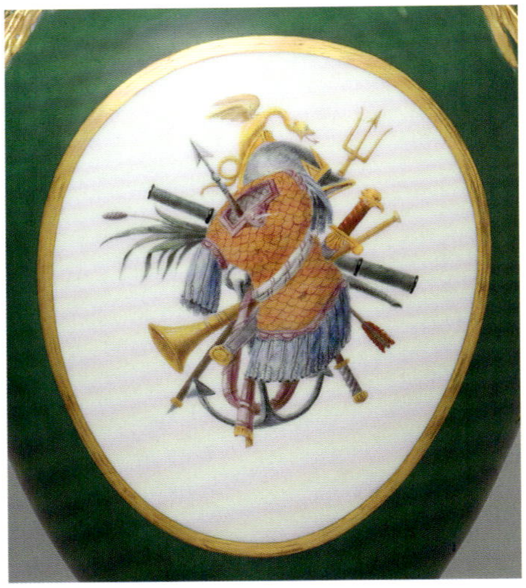

Fig 1: Detail of cat. no. 17.

playful pastoral and nautical trophy, the naturalism of the water spilling on the lid, and the vigor and movement of the dolphins make it an object of considerable charm and verve.[7] H.J.M./J.L.M.

1 Compare cat. nos. 11, 31, 36.

2 For Polidoro's vase designs, see Miller 1999: cat. 69. For other instances of the use in the eighteenth century of Polidoro's ideas, see Timothy Clifford, "Polidoro and English Design," *Connoisseur* 192 (Aug. 1976): 282–91. For more on Daumont, see cat. no. 87.

3 Carl Christian Dauterman, *Sèvres Porcelain: Makers and Marks of the Eighteenth Century* (New York, 1986): 229. While some Sèvres scholars consider the incised "PT" to denote the soft-paste formula, or *pâte tendre*, the initials "PT" or "Pt" appear on pieces before Sèvres discovered the formula for hard-paste porcelain and needed to distinguish between the two processes. This mark could also indicate two factory craftsmen, Pierre-Aymand Tristant and Philibert Choulet, known as Philibert. Both worked as *répareurs* and *tourneurs* of soft-and hard-paste porcelain; see Savill 1988: vol. 3, 1120.

4 The same scene by a different painter, possibly Étienne-Jean Chabry, appears on one of a pair of vases, about 1770, sold at Christie's New York, sale cat., Nov. 1, 1990: lot 22. The design of the semi-reclining putto may have been adapted from a design by François Boucher, perhaps from his suites of engravings of putti and children, which were purchased by Sèvres's artistic directors beginning in 1757. For an illustration of such an image, see Maureen Cassidy-Geiger, "Engravings and Etchings in the Manufacture Nationale de Sèvres: An Introduction to the Collection Formed in the 18th Century," *French Porcelain Society Journal* 14 (1999): 21, fig. 14a. Carl Christian Dauterman, "Sèvres Decorative Porcelain," *MMA Bulletin* 18 (May 1960): 263, cited a cup and saucer of 1772 with the same type of image, painted by Chabry, in the Victoria and Albert Museum, London (nos. 466, A1921). See Savill 1988: vol. 1, 357 (fig. C309), 384–87 (figs. C314–15).

5 Carl Dauterman attributed the trophy painting to Buteux, who joined the Sèvres Manufactory in 1756 as a porcelain painter, specializing in such groupings of attributes and objects; see Dauterman et al. 1964: 214–15, cat. 44, figs. 162–63. A recent note by David Peters in the curatorial files of the Metropolitan Museum's Department of European Sculpture and Decorative Arts suggests that it is by Louis-Gabriel Chulot.

6 In the Sèvres sales records, this shape is listed in at least eight entries made between 1757 and 1777, and may be a variation on a form known as *vase étrusque à cartels*; see Dauterman et al. 1964: 215. Dauterman (note 4): 293, discussed the dating of this piece. The dial case of a clock, or *cartel*, refers to versions of this form that have a clock in place of a painted reserve. A vase with clock, circa 1780–86, of the same shape with fountain decoration and flanking dolphins is in the Sèvres museum; see Brunet and Préaud 1978: 202, fig. 234. These authors stated that this form was published in the nineteenth-century Sèvres archives as a *vase oeuf anses dauphins*.

7 Other examples of the form survive in the collections of the Victoria and Albert Museum, London (Jones Collection, no. 145); Russborough, Ireland; and the J. Paul Getty Museum, Los Angeles. A related Sèvres vase type, in the form of a truncated, fluted column, features inverted dolphin handles and a wide, cascading fountain as its lid; examples can be found in the Wallace Collection, London, and in the Walters Art Gallery, Baltimore. See Carl Dauterman, *Sèvres* (New York, 1969): fig. 23; and Savill 1988: vol. 1, 281–86, cats. C284–85.

18. PAIR OF BEADED VASES WITH PASTORAL SCENES

Sèvres Manufactory; figural painting attributed to Charles-Nicolas Dodin (1734–1803); floral wreath painting attributed to Jean-Baptiste Tandart (1729–1816)
About 1766
Soft-paste porcelain
13 ⅛ in. (33.3 cm)
Gift of the Samuel H. Kress Foundation, 1958
58.75.83–.84

This pair of vases is very similar to a larger version of this model also in the collection of the Metropolitan Museum of Art (Fig. 1). The bulbous shape of all three is embellished with gilded studs on the folded strap handles and lids, modeled laurel garlands suspended from ribbons attached to the corners of the central panels or reserves, burnished gadroons at the base, as well as the blue ground color known as *bleu nouveau*. The factory issued this particular style of vase in several graduating sizes so they could be bought individually or assembled into garnitures (a set of three to five vases, in two differing forms, set on a mantel).[1]

The manufactory records describe this type of vase as "*à panneaux*" or "*à perles*," referring to the reserve panels and to the gilded pearl-like decoration on the projecting lids and handles at the side.[2] The basic model, with its lid, stepped shoulder, oval body, short stem, and applied and gilded gadrooning, exemplifies the early neoclassical decoration used at Sèvres from 1762 and is typical of the vases produced there during the second half of the decade.[3] However, the painted pastoral landscape decoration is most likely inspired by rococo prints reflecting seventeenth-century Dutch landscape paintings, thus indicating the transitional nature of the pair.[4]

Although the two vases are unmarked, the painted decoration is attributed to two artists—the landscapes to Charles-Nicolas Dodin and the floral wreaths on the reverse to Jean-Baptiste Tandart.[5] Dodin often worked from engravings to create detailed pastoral scenes on porcelain models. For this pair, he

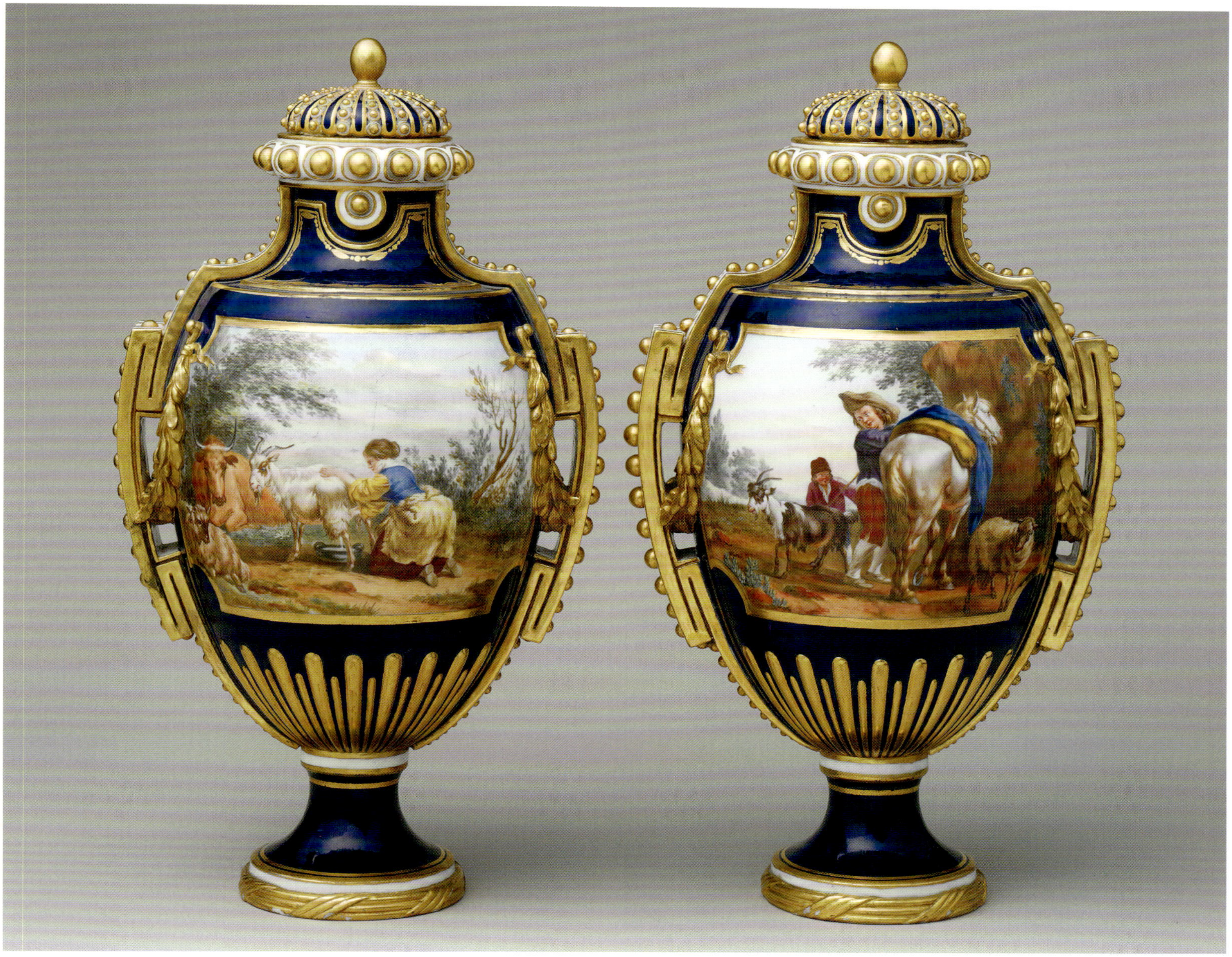

Cat. no. 18

depicted on one vase a young peasant standing next to his horse accompanied by a shepherd, a goat, and a sheep. The other shows a maiden milking a goat, as two cows look on. The figural compositions on the both vases are based on engravings by Johannes or Jan de Visscher after the work of Nicholaes Berchem, a seventeenth-century Dutch painter, printmaker, and draftsman. Berchem is known for depictions of idyllic peasant life in bucolic settings, which were popularized in eighteenth-century France through copies and printed reproductions.[6] The attribution to Dodin is substantiated by the similarity in the style and pose of the goats featured here to an image of sheep found on an example signed by the artist that is also in the Metropolitan (cat. no. 23). However, other Sèvres vases with scenes derived from Berchem and painted in a similar manner have been attributed to Antoine Caton.[7]

Depictions of idealized country life were popular among Sèvres's upper-class and royal clients, who entertained themselves by dressing in pastoral costumes and participating in or watching plays and entertainments on romanticized rural themes. The decision to feature a rococo scene on a vase displaying neoclassical elements reflects the reluctance of both the factory and patrons to relinquish the previous style entirely. Similarly, the painted interlaced floral wreaths of roses and cornflowers, a specialty of Jean-Baptiste Tandart, on the reverse of both vases (Fig. 2) combine the classical wreath tradition with the delicate technique

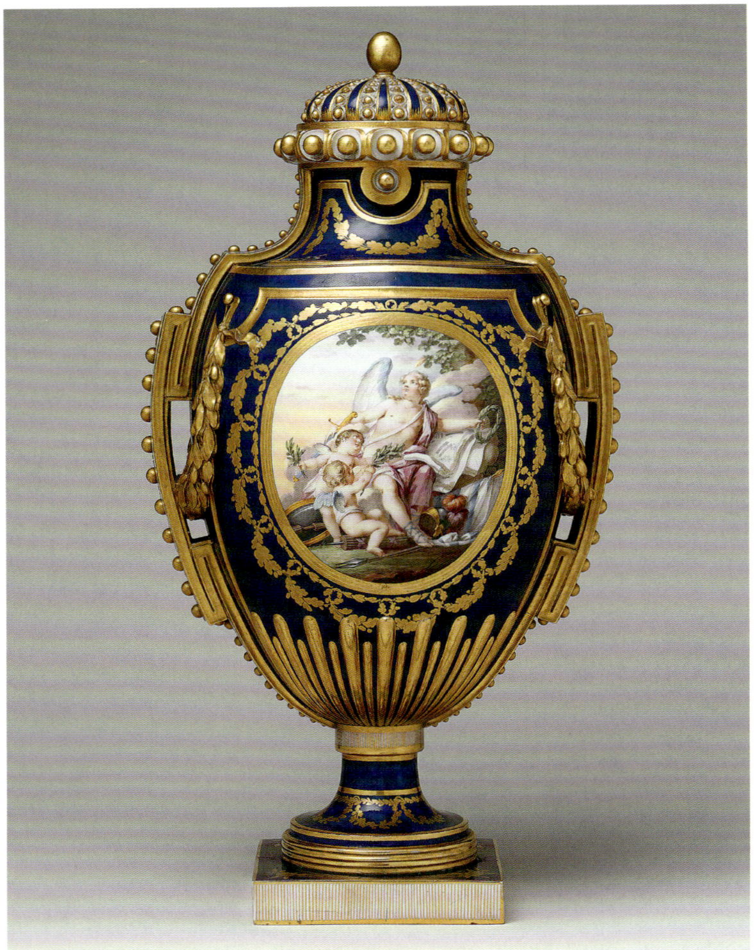

Fig. 1: *Beaded Vase with a Winged Allegory of Victory*, about 1789. Sèvres Manufactory. Soft-paste porcelain; 29 in. (48.3 cm). Gift of the Samuel H. Kress Foundation (1958) 58.75.80.

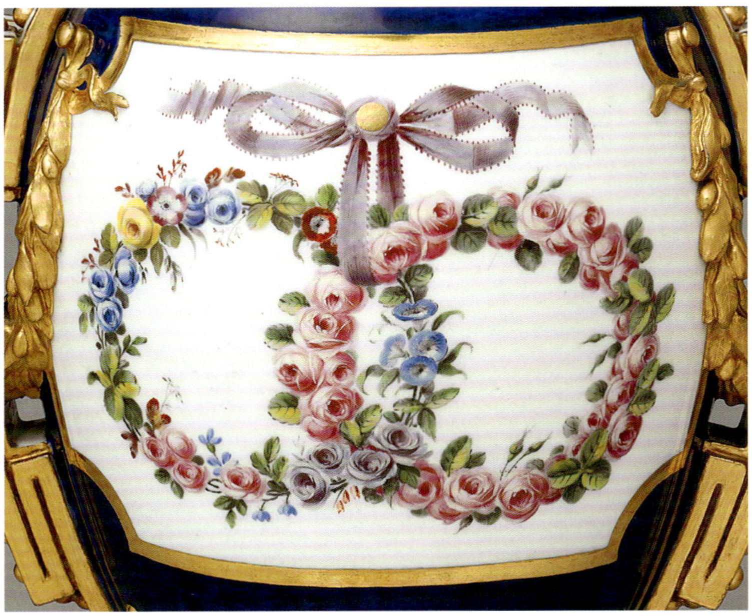

Fig. 2: Detail of the reverse of cat. no. 18.

and soft palette of rococo-style floral decoration still employed at the factory. Such wreaths, in use at Sèvres by 1765, proved to be very popular and appear on the reverses of many vases, along with either marine or pastoral vignettes.[8] J.L.M.

1 The earliest known examples of this form, with splayed foot, beaded decoration, and shaped panels, are dated 1766 and appear in several shapes and three sizes; see Savill 1988: vol. 1, 316, cats. 297–302. A similarly shaped pearl-studded vase with an oval reserve scene, circa 1766–70, is in the J. Paul Getty Museum, Los Angeles; see Adrian Sassoon, *Vincennes and Sèvres Porcelain: The Catalogue of the Collection* (Malibu, Calif., 1991): 106, cat. 21. Vases in this form can also be found in the Wallace Collection, London; Victoria and Albert Museum, London; Rijksmuseum, Amsterdam; Musée National de Céramique, Sèvres; Israel Museum, Jerusalem; and others. *Vases à panneaux* or *à perles*, were also copied by nineteenth-century porcelain factories, including Coalport and Limoges; see idem: 109. The Metropolitan Museum of Art also has a larger pair of vases in a version of this shape depicted in Dauterman et al. 1964: 215, no. 54a–b, figs. 164–67. It features military encampment scenes by Morin and on the reverse a trio of cornflower, rose, and laurel wreaths attributed to Tandart.

2 A plaster model from the 1814 inventory in the Sèvres archives describes this shape as "Panneau 1re 2me 3me gr." However, it was called *vase à perles* in other documents and likely both terms were used. References to this shape also appear in the glaze kiln records in 1768 and 1770 and in the *tourneurs* records of the 1770s; see Tamara Préaud in Sassoon (note 1): 106, cat. 21. The biscuit records of 1768 mention a *vase à perles*, probably in the smaller size; Savill 1988: vol. 1, 316. For vases of this type in sales, see Sassoon (note 1): 109; and Carolyn Gay Nieda, "A Sèvres Vase à panneaux" *Getty Museum Journal* 14 (1986): 129–30.

3 Sassoon (note 1): 106. The series's embrace of neoclassicism was likely influenced by the sculptor Étienne-Maurice Falconet, chief modeler at the manufactory from 1757; Eriksen and Bellaigue 1987: 116.

4 This pair and one *vase à panneaux* at the Getty Museum, dated to the 1760s, are the only known examples of this shape decorated with pastoral scenes; Nieda (note 2): 129.

5 Dauterman et al. 1964: 221, cat. 47a–b, figs. 170–73.

6 See Daniëlle O. Kisluk-Grosheide, "Cutting Up Berchems, Watteaus, and Audrans: A lacca povera secretary at the Metropolitan Museum of Art," *MMA Journal* 31 (1996): 87. The scene on the first vase copies part of a known pastoral engraving stamped "J. Visscher" after "N. Berchem"; Bertrand Jestaz, "Les Modèles de la majolique historiée: XVII et XVIII siècles," *Gazette des beaux-arts* 81 (1973): 114, figs. 9–10; and Bernard Rackham, *Victoria and Albert Museum: Catalogue of Italian Majolica*, vol. 1 (London, 1977): 388. The scene of the milkmaid on the other vase can not be linked to a specific print after Berchem, although Visscher engravings with similar scenes were used by the Sèvres painting studio; Roth and Le Corbeiller 2000: 137, fig. 69–1. Visscher's engravings of Berchem designs are also found on several other Sèvres vases from the mid-1760s; for examples see Eriksen 1968: 280 no. 101; Savill 1988: vol. 1, 236–37, cat. C267; Nieda (note 2): 133 (figs. 10–11), 244 (n. 30).

7 For vases with paintings attributed to Caton, see Roth and Le Corbeiller 2000: 137; and Ennès and Ducrot 1997: 91, no. 35.

8 Eriksen and Bellaigue 1987: 132; Roth and Le Corbeiller 2000: 136–37, fig. 69. A similar painted decoration by Tandart is featured on a saucer and covered cup from 1766 in the Metropolitan Museum (50.211.142). For a discussion of the cup, see Savill 1988: vol. 2, cat. C438. Similar reserve decoration with three interlinked wreaths is found on a pair of Sèvres Etruscan-style vases of about 1768 in the Musée du Louvre, Paris; see Ennès and Ducrot 1997: 96, cat. 45. See also Savill 1988: vol. 1, cats. C286–87 (ewers dated 1766).

19. MOUNTED TWO-HANDLED VASE WITH COVER

Sèvres Manufactory; mounts: French
About 1769–80
Soft-paste porcelain and gilt-bronze mounts
9 in. (22.9 cm)
Gift of Ann Payne Blumenthal, 1943
43.163.9

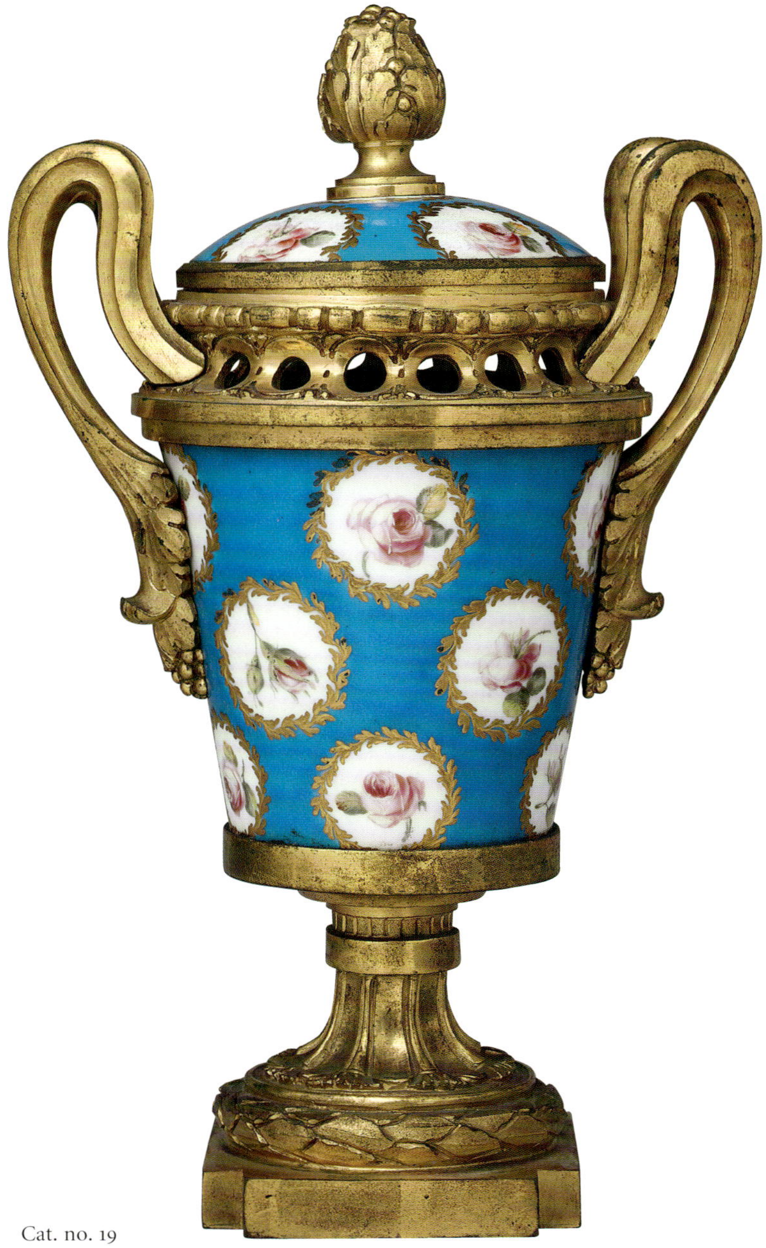

Cat. no. 19

The Sèvres manufactory began producing vases like this example, specifically for mounting in gilt bronze, by 1764, probably in response to orders from Parisian luxury merchants, or *marchands-merciers*.[1] This small potpourri, one of a pair of *vases à monter* in the Metropolitan Museum, was likely part of a garniture. Numbering three or five pieces, such a set could include vessels with either tapered cylindrical or egg-shaped bodies, in ascending size, and with pierced covers.[2] Surviving marked examples and other documentary evidence suggest a date for this vase of between 1769 and 1780.[3]

While the earliest mounted Sèvres vases were glazed in solid ground colors (see cat. no. 16), by 1770 decorators often painted scattered roses in asymmetrically shaped reserves on green, dark blue, and turquoise grounds bordered by gold-colored foliate wreaths. This rococo-style floral decoration was adjusted according to a more ordered, neoclassical approach: the disposition of the flowers was regularized, and they were framed in gilt laurel wreaths.[4] The type first appeared in the Sèvres product line in about 1767 and continued to be made well into the 1780s.[5] The decoration may have been the one referred to by the factory as *roses encadrées d'or* (roses encircled with gold) or *roses détachés* (free-floating roses).[6] The gilt-bronze mounts, however, fully embody a neoclassical vocabulary: a guilloche-pierced neck, scrolled handles terminating in acanthus leaves and grapes, fluted foot with a laurel and acanthus-leaf band, and square plinth with re-entrant corners. By limiting the neoclassical elements to the mounts, the design of this vase reflects a somewhat cautious approach to the new style.

Despite neoclassicism's growing influence and the popularity of marine and military reserve scenes on vases during the late 1760s, flowers remained a constant source of inspiration for Sèvres designers and reflected the taste of both the artistic director of factory, Jean-Jacques Bachelier, and his clients. In part because Sèvres operated within the machinery of state patronage and in the hands of older, conservative artists, the factory's adoption of neoclassical painted decoration was gradual.[7]

Vases were usually, but not always, mounted by *marchands-merciers*, who ordered them from the royal manufactory. Not only were these luxury-goods dealers, like Simon-Philippe Poirier and his partner, Dominique Daguerre, among the principal with models for vases.[8] They conducted business from premises in the rue Saint-Honoré, Paris, which had become famous as a district where high-quality furniture, clocks, porcelains, bronzes, and silks were sold to the court and a fashionable international clientele.[9] The existence of five different types of mounts on surviving examples of this potpourri vase form suggests that each merchant used his own designs for his mounts.[10]

Madame du Barry, the mistress of King Louis XV of France, was known as an early supporter of the neoclassical style, and among the furnishings Poirier delivered on September 4, 1770, to her residence, the Pavillon de Louveciennes, was a chimney garniture of five Sèvres vases of *bleu céleste* with small roses, mounted in gilt bronze. Its description reads, "2 vases made of urns, sky-blue grounds with medallions of flowers…2 ditto smaller, 1 ditto in the form of an egg"; this closely matches the present example and related surviving ensembles.[11] J.L.

1 Possibly, the idea of using mounts on porcelain originated with *marchand-merciers* like Simon-Philippe Poirier and Jean Dulac, who had successfully promoted Chinese porcelain mounted in gilt bronze; see cat. no. 16. For a green inverted bell-shaped vase linked to Dulac and similar to the present example in scale and mounts, see Eriksen and Bellaigue 1987: 122, 324.

2 For examples, see Hermann Ball and Paul Graupe Berlin, *Erich von Goldschmidt-Rothschild*, sale cat., Mar. 23. 1931.: lot 206 (a 1769 garniture with a green ground); Eriksen 1974: 363, fig. 242 (three-piece garniture including a pair of vases with a shape similar to the Metropolitan's potpourri); Roth and Le Corbeiller 2000: 156, cat. 74 (*bleu céleste*, five-piece floral-patterned garnitures in the J. Pierpont Morgan Collection, Wadsworth Atheneum, Hartford, and Musée Cognac Jay, Paris); Christie's New York, sale cat., May 22, 2002: lots 311–12 (two pairs of potpourris, one with a similar shape, the other egg-shaped with the same decoration from 1770). Other examples with dark blue (*gros bleu* or *bleu nouveau*) and green (*vert*) grounds, without floral decoration but with identical mounts and forms, were sold on the market; see Christie's London, *Cholmondeley Family Collection*, sale cat., Dec. 8, 1994: lots 36–37; and Christie's, *Earl of Sefton*, house sale cat., Sept. 17–20, 1973: lot 908.

3 Although no molds exist at the Sèvres factory for this shape, the *gobelets à monter* listed as ready for glazing in the 1764 factory inventory may refer to this popular model. By 1780 more than one hundred *gobelets à monter* had been sold; see Roth and Le Corbeiller 2000: 156.

4 Brunet and Préaud 1978: 141, fig. 53.

5 A number of artists painted this motif. Jean-Baptiste Tandart depicted similar roundels of pink roses on a cup and saucer of about 1769; see Roth and Le Corbeiller 2000: 214, cat. 101. The marks of François Binet and Jacques-François-Louis de Laroche appear on Wadsworth Atheneum vases; see Eriksen and Bellaigue 1987: 132; and Roth and Le Corbeiller 2000: 156–57. Binet's marks also appear on a larger similarly decorated potpourri at the Huntington with folded gilt-bronze handles and a gadrooned mount above the foot; see Robert Wark, *French Decorative Art in the Huntington Collection* (San Marino, Calif., 1979): 99, fig. 116.

6 These terms are listed in factory records referring to a cup and saucer by Tandart (see note 5); Roth and Le Corbeiller 2000: 213. Tandart received overtime payments in 1769 for painting similar roses encircled with gold and free-floating; Roth and Le Corbeiller 2000: 213.

7 Henry Hawley, *Neo-Classicism: Style and Motif* (New York, 1964): 12.

8 In 1782–83 the factory produced eleven urn-shaped *vases à monter Daguerre*, copied from a wooden model supplied by Daguerre and intended specifically for use with gilt-bronze mounts. Designer and bronze maker Duplessis the Younger published two books of vases (see cat. no. 28), supplied models and designs for mounts to the Sèvres factory, and purchased vases for mounting. Duplessis's father, who was responsible for the *tourneurs* and *répareurs* at Sèvres until 1774, also provided vase models; see Savill 1988: vol. 1, 468; Eriksen and Bellaigue 1987: 113–14, 122; and Roth and Le Corbeiller 2000: 156.

9 Poirier and Daguerre are associated particularly with furniture that utilized porcelain plaques commissioned from Sèvres; see Sargentson 1996: 1, 10.

10 Roth and Le Corbeiller 2000: 157.

11 "*2 vases en urne fond bleu de ciel a médaillons de fleurs…2 idem plus petit, 1 idem en forme d'oeuf…*"; see Eriksen and Bellaigue 1987: 132–33; Savill 1988: vol. 1, 28, 31, n. 67; and Roth and Le Corbeiller 2000: 156.

20. TASSELED VASE WITH COVER

Sèvres Manufactory; figural painting attributed to Jean-Louis Morin (1732–1787), trophy painting attributed to Charles Buteux the Elder (1719–1782)
About 1775
Soft-paste porcelain
14 1/16 in. (35.7 cm)
Gift of the Samuel H. Kress Foundation, 1958
58.75.75

This soft-paste porcelain vase is one of a pair whose shape corresponds to a Sèvres form registered in the nineteenth century as *vase à gland* (fringed tassel vase). The manufactory's inventory of 1769 also refers to a model for a vase of this shape called "vase à glands 2e gr" and refers to the creation of new molds, implying that a larger size, perhaps vase "vase à glands 1re gr" existed.[1]

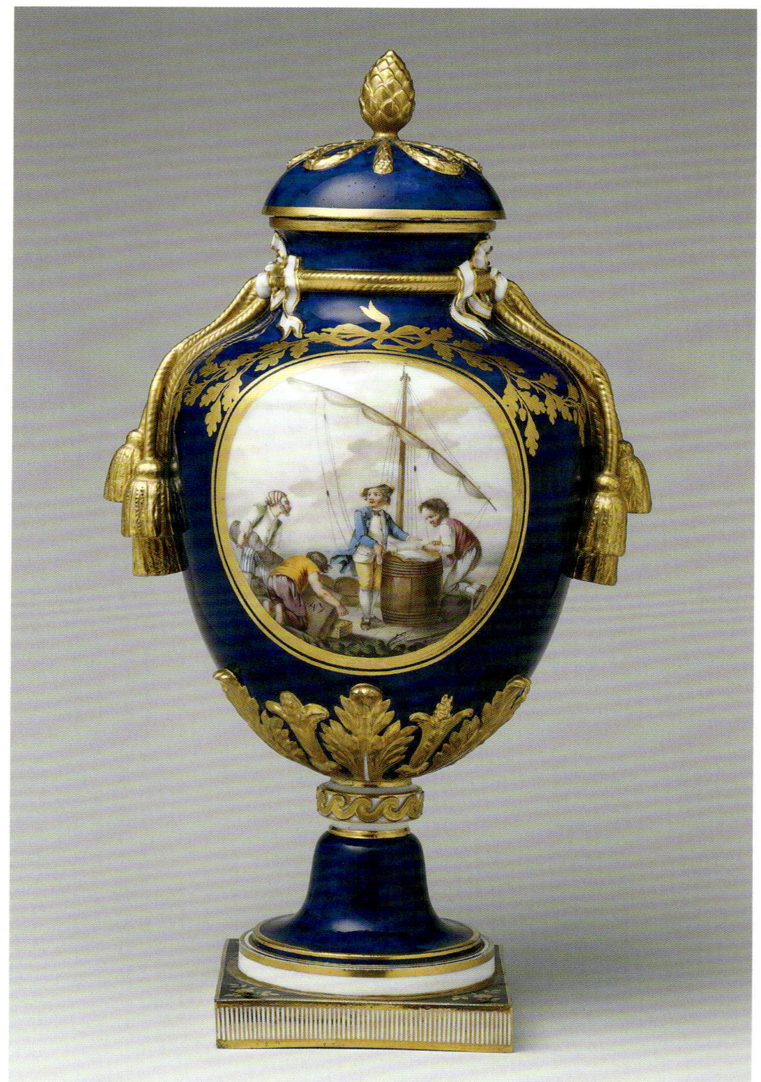

Cat. no. 20

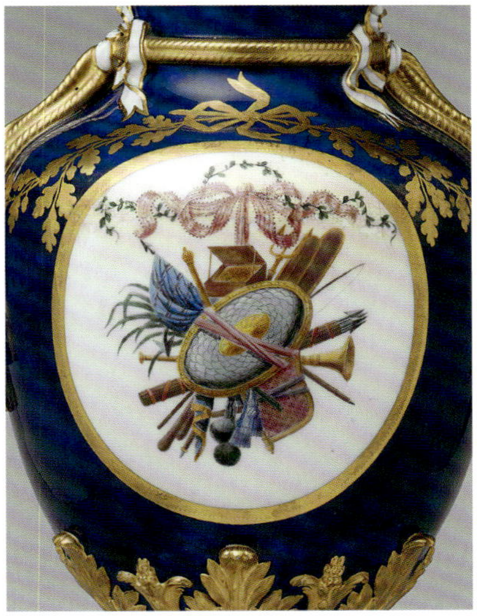

Fig. 1: Detail of the reverse of cat. no. 20.

The vase's exuberant molded decoration, with elaborate gilding and circular reserve scenes painted in soft colors, reflects the mixture of rococo and neoclassical elements characteristic of the transitional Sèvres style that existed until the factory adopted a more severe neoclassical decorative manner in the 1780s.[2] The domed cover's silhouette is softened by gilded ribbonlike swags in low relief that fall from the pinecone knop. Molded gilt cords and pendant tassels drape over the shoulders and temper the ovoid form of the body. Below, applied gilt acanthus leaves reiterate the undulating patterns of the lid's swags.

The painting of the reserve medallions has been attributed to Jean-Louis Morin, a major figure at Sèvres whose hand can also be seen in cat. nos. 21 and 22. Together with Charles-Nicolas Dodin, Morin was considered the most important painter at the factory by 1781 and recognized for the marine and quayside views—his specialty—found on many Sèvres vases from the third quarter of the century.[3] The finely detailed, softly colored dockside reserve scene on this tasseled vase exemplifies Morin's compositions that depict several figures engaged in various maritime activities and incorporate strong compositional lines created by leaning masts and sails. Many drawings and etchings based on paintings of quayside views were common sources of design inspiration for eighteenth-century artisan-decorators. These marine images were repeatedly used on vases or tableware of various sizes and shapes by the same artist; a version of the scene depicted on the obverse here, also painted by Morin, is featured on a *jardinière* from 1770 in the Metropolitan Museum's collection.[4] Morin may well have relied on engravings or paintings from the manufactory's collection that he received from the resident head of the factory studio; however, no specific models for this dock scene has yet come to light.[5] Those artists whose work may have inspired Morin include seventeenth-century Dutch painters Rudolf Backhuisen, Johannes Lingelbach, David Teniers the Younger, or his contemporary, the French painter Claude-Joseph Vernet.[6]

The reverse features a medallion containing a nautical trophy that is attributed to the decorator Charles Buteux the Elder, who became the factory's most prolific painter of such scenes (Fig. 1).[7] Buteux's charming reserve contains oars, a shield, trumpet, anchor, trident, fishing rod, antique nautical compass, and sprig of vegetation, all tied together by a magenta ribbon that dangles from a bow. In the composition, the artist chose to highlight or contrast color rhythms rather than illustrate objects in true proportion to one another. Like Morin's nautical view on the obverse, Buteux's delicate palette remains a typical rococo element that demonstrates the Sèvres factory's slow development toward a fuller neoclassical decoration.[8] J.L.M.

1 Brunet and Préaud 1978: 185, cat. 182. The present vase is considered an example of the second, smaller size B. The model of this *vase à gland* in the Wallace Collection, London, does not possess tassels or the pinecone finial, but has the same form and acanthus leaves at the base, as does the Metropolitan's example. Another tasseled vase, shape A with round knop, from the Wallace collection is discussed in Savill 1988: vol. 1, 365–67. This 1781 vase features a green ground and is 43.2 cm in height. A vase in the same shape with a turquoise ground, c. 1770 is in the Huntington Library, San Marino, Calif.; see Robert Wark. *French Decorative Art in the Huntington Collection* (San Marino, Calif.): 1979: 89, fig. 108. For another *vase à gland*, shape A, with tassels from 1768, see Sotheby's New York, sale cat. of the collection of Dr. Anabella Brown, Apr. 23, 1977, lot 54. A correspondingly shaped *vase à gland*, painted by Jacques-François Micaud around 1768, with tassels, guilloched collar at the foot, gilt acanthus leaves, a *bleu fallot* ground, and rococo painted flowers, was exhibited at the International Ceramic Fair and Seminar, London, June 15–18, 2001, by Adrian Sassoon.

2 See Dauterman et al. 1964: 234–35, cat. 56, figs. 199–202.

3 See Dauterman et al. 1964: 187. Morin studied surgery before 1754, when he began to work at the Vincennes ceramics factory.

4 The *vase à glands* and *jardinière* (58.75.102) marine scene of four sailors on a wharf preparing cargo appears on a Sèvres vase, c. 1780, in the Victoria and Albert Museum, London, and a vase year-marked 1757 at the Walters Art Gallery, Baltimore; see Dauterman et al. 1964: 225, cat. 50, fig. 180.

5 Geoffrey de Bellaigue, "Sèvres Artists and Their Sources II: Engravings," *Burlington Magazine* 122, 932 (Nov. 1980): 751.

6 Savill, 1988: vol. 2, 1052. Vernet was well known in the eighteenth century for his painting series, commissioned by King Louis XV, depicting the ports of France; see Carl Christian Dauterman, "Sèvres Figure Painting in the Anna Thompson Dodge Collection," *Burlington Magazine* 118 (Nov. 1976): 761.

7 Buteux's characteristic trophy style included using a shield or other object as the central focus surrounded by related attributes. The juxtaposition of the periwinkle blue and rose in the central shield depicted in this trophy, as well as the use of berried vines are also distinguishing hallmarks of Buteux's style. See Dauterman et al. 1964: 181–82.

8 The decoration on another *vase à gland* from the same period also demonstrates this curious mixture of styles, combining gilded tassels and acanthus leaves with a neoclassical mythological scene of Mars and Venus painted with rococo delicacy and palette; see Brunet and Préaud 1978: 182, cat. 152, inv. 12; and Savill 1988: vol. 1, 366.

21. VASE WITH EGG-AND-DART RIM

Sèvres Manufactory; shape modeled by Jacques-François Deparis or Paris (1735–1797), figural painting by Jean-Louis Morin (1732–1787), trophy painting by Nicholas Petit the Elder (1723/24–1814), gilding by Étienne-Henry Le Guay (1719/20–1799)
About 1775
Soft-paste porcelain
12 7/8 in. (32.68 cm)
Marks: crossed *L*'s and *LG* (for Le Guay), in purplish gray
Gift of the Samuel H. Kress Foundation, 1958
58.75.77

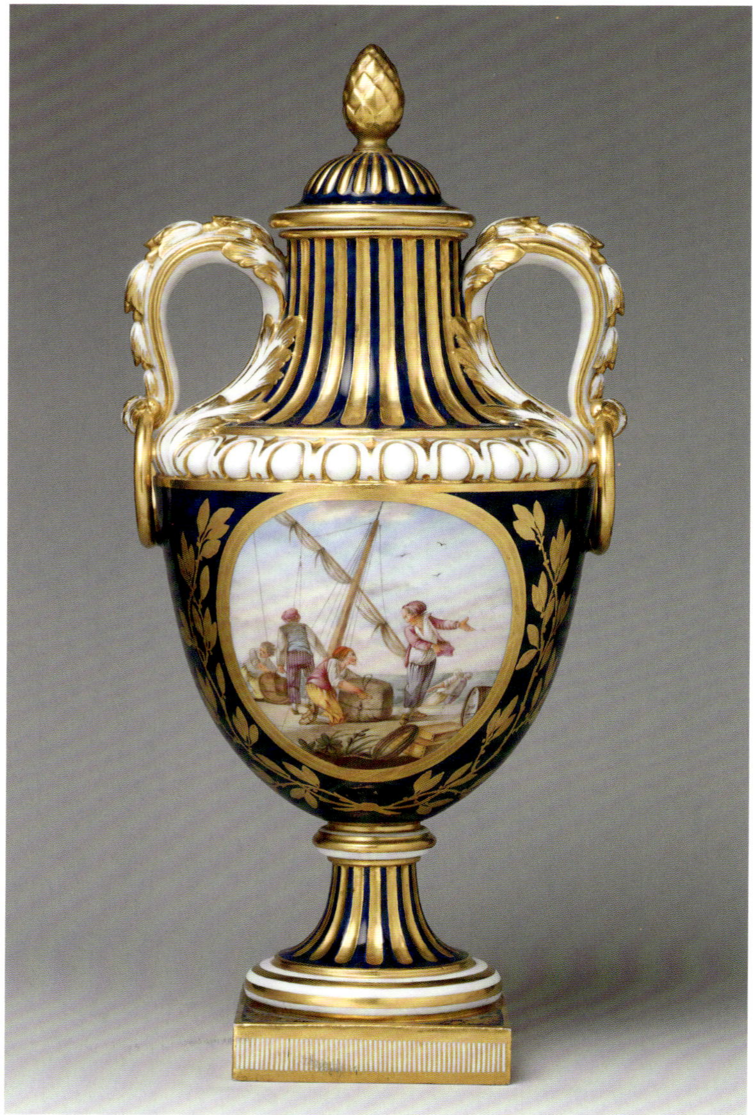

Cat. no. 21

The shape of this vase, derived from the classical amphora and given a tapering foot, is exceptional among Sèvres vases because of the prominent egg-and-dart band that encircles the rim of its shoulder.[1] The emphasis on this neoclassical motif, the fluted neck, the foot set on a square plinth, and the modeling of the handles as acanthus fronds lined with husk-leaf chains all exemplify the neoclassical treatment of vase forms by the Sèvres factory during the third quarter of the eighteenth century. The vase, one of a pair, is among the smallest featured here.

Based on an inscription found on a Sèvres archives drawing, the form of this model was called *vase de côté de Paris, No. 2* in the nineteenth century, but in the eighteenth, this shape also may have been known simply as: "ornamental vase number 4."[2] The form was probably modeled by Jacques-François Deparis or Paris, head of the *répareurs* (modelers) at Sèvres in 1774.[3] The dock scene of the large reserve, in which a worker gestures to his comrades, is attributed to Jean-Louis Morin and recurs on other examples signed by this painter.[4] The nautical trophy on the reverse—a playful combination of a basket of eels, red coral, and a corked bottle in the center—is attributed to Nicholas Petit. The gilder is thought have been Étienne-Henry Le Guay, who, like his father, began his career as a *tourneur* at the porcelain manufactory at Saint-Cloud.[5] Rivaling later methods of division of labor, the forming of the ceramic body of this vase and its surface decoration represent the combined efforts of numerous specialists at Sèvres; *mouleurs* and *tourneurs*, who molded or threw vessels; *répareurs*, who finalized modeling and chased the forms before they were initially fired in a kiln; and a figure painter, flower painter, and gilder who finished the colorful surface. Incised, stamped, scratched, or painted marks on the vessels indicate artists who worked on the object. J.L.M.

1 The egg-and-dart molding is similar to the lid rims on a pair of Sèvres Louis XV ormolu-mounted Taillandier vases with covers (*vases canneles à bandeau*) of about 1771, sold at Christie's London, sale cat., Dec. 12, 2000: lot 35. This pair has a severe neoclassical form and a reserve painting after a work by François Boucher, in a soft palette, thus exhibiting the same combination of rococo and neoclassical elements seen in the present example. A pair of vases (*ovale mercures*), circa 1767, at the Walters Art Gallery, Baltimore, also feature a gilded egg-and-dart band just below the lid; see Brunet and Préaud 1978: 85, pl. 33.

2 Brunet and Préaud 1978: 200, cat. 239; and Dauterman et al. 1964: 235.

3 Brunet and Préaud 1978: 200; and Savill 1988: vol. 3, 985. Deparis arrived at the Vincennes porcelain manufactory in 1746, when he was eleven, and worked as an apprentice, making handles and spouts. In 1752 he became the assistant to Duplessis the Elder and replaced him after his death.

4 See cat. no. 20 for more information about Morin. See also Dauterman et al. 1964: 188, 235–37, fig. 203; and Roth and Le Corbeiller 2000: 322, fig. 160.

5 Savill 1988: vol. 2, 1045.

22. VASE WITH GILT PUTTI AND ROYAL MEDALLIONS

Sèvres Manufactory; shape designed by Jacques-François Deparis or Paris (1735–1797), figural painting by Jean-Louis Morin (1732–1787), trophy painting by Nicholas Petit the Elder (1723/24–1814), gilding by Étienne-Henry Le Guay (1719/20–1799)
About 1778
Soft-paste porcelain and biscuit
15 ¼ in. (38.7 cm)
Marks: crossed *L*'s and *LG* (for Le Guay), in lavender; [incised within ball of foot] *39* and *.10*
Gift of the Samuel H. Kress Foundation, 1958
58.75.79

The characteristic form of this vase, which belongs to a group named *vases de Paris* after its designer, Jacques-François Deparis or Paris, has a cylindrical neck similar to a fluted column that rises from its ovoid body. This form first appeared in 1778 as *vase Paris enfants* or *colone de Paris.*[1] Deparis, a modeler (*repareur*) and head of the soft-paste studio at Sèvres after 1774, is often credited with combining materials in a novel fashion. Here, unglazed biscuit porcelain was used for the classical-style medallions of Louis XVI (front) and Marie-Antoinette (back) at the neck (Fig. 1); gilt putti appear in lieu of handles, emphasizing the vase's ornamental function.[2] They, together with the column-shaped neck, medallions, swags, pinecone finial, and golden laurel branches, signal the design's classicizing intention, a goal underscored by its color scheme, which is restricted to blue, white, and gold.

Biscuit (unglazed white porcelain), which was introduced at Vincennes in 1751, was first used for royal portrait medallions on vases made for King Louis XV in 1762.[3] Such uncolored profile heads recall ancient cameos, as well as classical decoration on cameo glass, which the English potter Josiah Wedgwood copied in his jasper ware ceramics after 1775. Certainly, Sèvres designers were aware of Wedgwood's jasper ware medallions, since leading French cabinetmakers often incorporated them into neoclassical-style furniture, and combined them with inset soft-paste floral plaques made at Sèvres.

Several other features reflect the more emphatic adoption of the neoclassical style at Sèvres at this time. Between 1780 and 1790, the manufactory began to produce vases that were closer to the amphora shape and had wider openings with constricted necks and handles, either scrolls or, as in this case, figural forms. Furthermore, the *bleu lapis* color that had been a factory staple for over a decade was replaced in about 1763 by a richer, more purplish (and therefore regal) *bleu nouveau*, which is also referred to as *bleu du roi*, or king's blue.

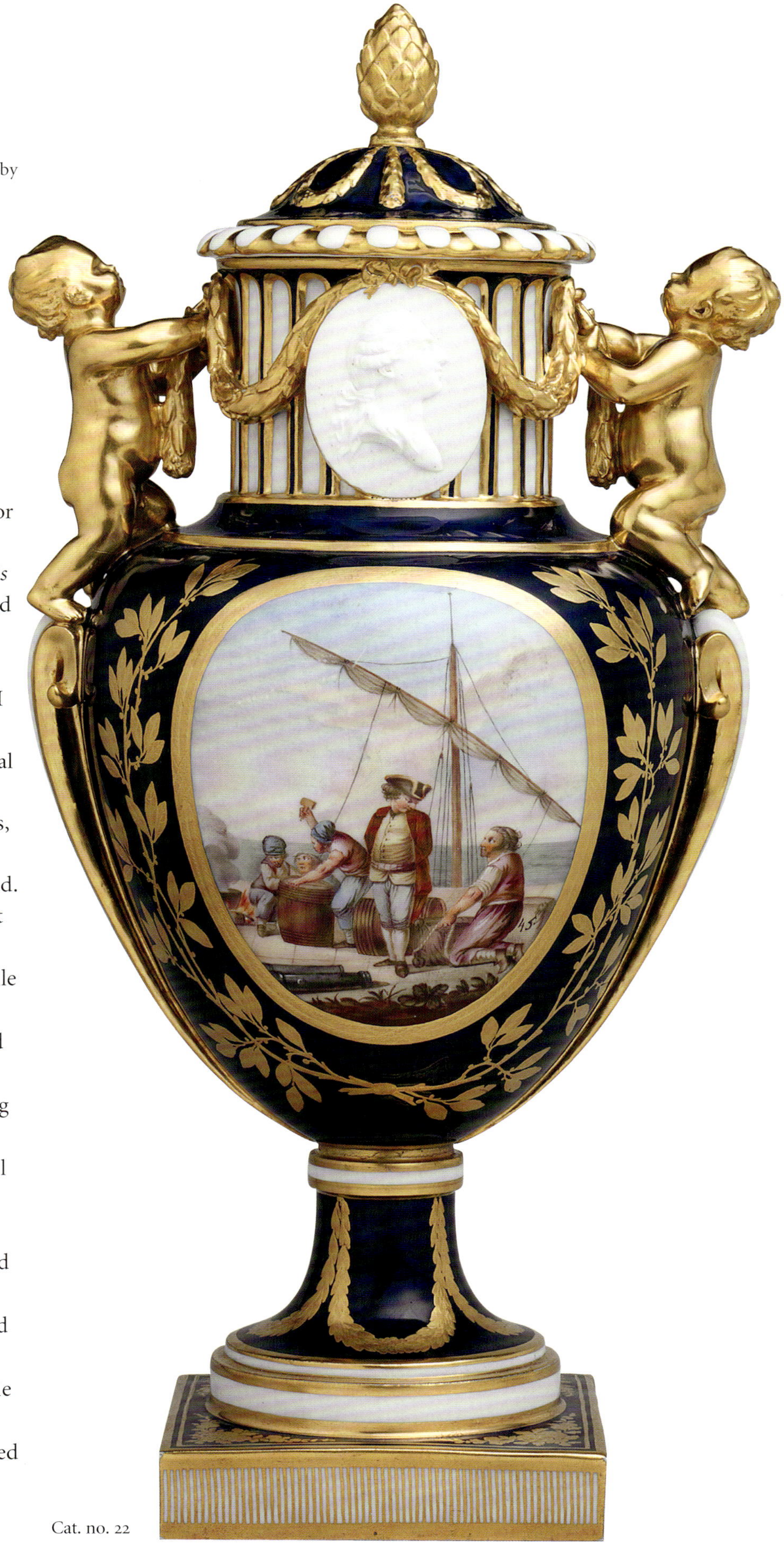

Cat. no. 22

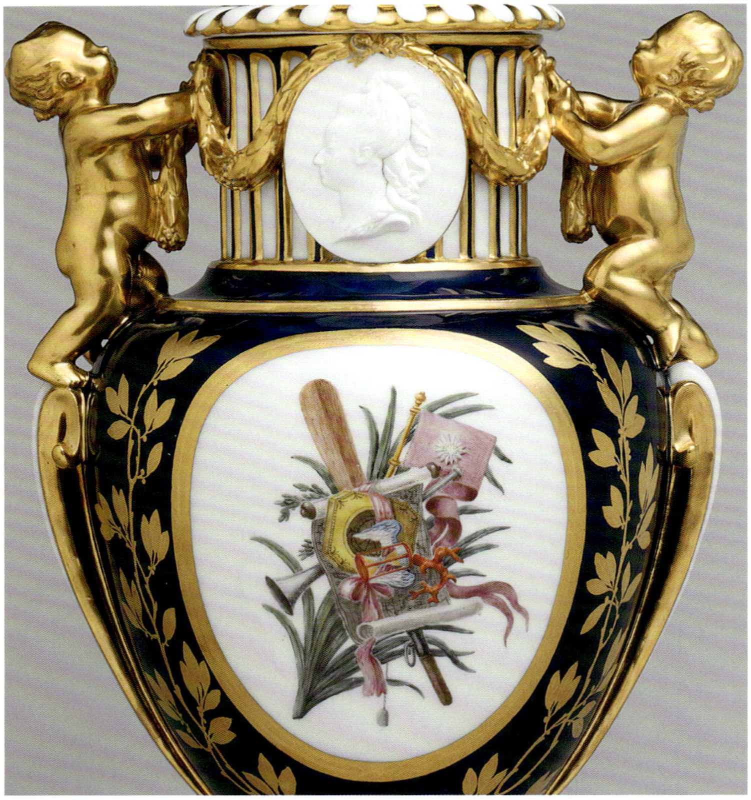

Fig. 1: Detail of the reverse of cat. no. 22.

The dock scene is attributed to Jean-Louis Morin,[4] while the marine trophy (Fig. 1), with rushes, map, compass, oar, spyglass, and other attributes is attributed to Nicholas Petit, based on the presence of the artist's "signature" branch of red coral.[5] A cursive lavender mark, "LG," indicates that the gilder was Étienne-Henry Le Guay.[6] Although gilding was growing in importance at Sèvres, it was only during the later, Napoloeonic era that gold would cover a surface entirely, as if almost to disguise the ceramic material; the development of gilders signing their work after 1770 was one step in this gradual evolution. J.L.M.

1 Savill 1988: vol. 1, 436.
2 The medallions resemble those on *vases oeuf Louis XVI* from 1777. They were probably prepared by the sculptor Louis-Simon Boizot, head of decoration at Sèvres between 1785 and 1800, after other models; see Brunet and Préaud 1978: 197, fig. 230.
3 Dauterman et al. 1964: 238.
4 These same principal figures appear on a signed *vase flacon à cordes* of 1771 in the Tuck Collection at the Petit Palais, Paris; see Brunet and Préaud 1978: 186, fig. 185. The quay motif is repeated with an additional figure on a *vase à glands* that is the mate to cat. no. 20.
5 Carl Christian Dauterman, "Sèvres Decorative Porcelain," *MMA Bulletin* 18 (May 1960): 295.
6 Savill 1988: vol. 2, 1045.

23. VASE WITH STRAPS

Sèvres Manufactory; figural painting by Charles-Nicolas Dodin (1734–1803), floral painting attributed to Henry-Martin Prévost the Younger (act. 1757–97)
About 1770–75
Soft-paste porcelain
12 $^{15}/_{16}$ in. (32.9 cm)
Marks: [on base, above, in rose] crossed *L*'s, *K*; [on base, below, in rose] *HP*
Gift of the Samuel H. Kress Foundation, 1958
58.75.69

At the Sèvres manufactory, basing the subject or composition of painted decoration on an engraved reproduction of a famous work was an established practice. Charles-Nicolas Dodin, whose mark, "K," appears on the underside of this vase, one of a pair, was a skilled painter active at the factory for nearly fifty years. Dodin often referred to engravings as sources for his painted decorations on porcelain; for this vase, he turned to Antoine Watteau's *Happy Shepherd*, which was engraved by Louis Crépy in 1729.[1] Prints such as Crépy's helped disseminate the rococo style in France; copies of Watteau's paintings, especially his *fêtes galantes*, often appear in eighteenth-century decorative objects. Dodin painted this same scene on a ceramic plaque that was incorporated into in a circular tea-table top produced for Madame du Barry in 1774, possibly about the same time as this vase was made.[2]

The mark "HP" in rose on the base likely refers to Henri-Martin Prévost as the maker of the flower painting on the reverse (Fig. 1). Although predominantly a gilder, Prévost, who started at Sèvres in 1757, was known to execute floral images in the later years of his employment at the factory, where he worked until 1797.[3]

While both the figural and floral paintings are firmly rooted in rococo tradition, the verticality of the vase's swelling form, the moldings on the foot, and much of the applied decoration directly refer to neoclassical designs.[4] The fluted neck and volutes derive from ancient architecture; the laurel swags at the vase's shoulders and pinecone finial also are classicizing motifs. Furthermore, although not derived from antique sources, the vase's studded strap handles, similar to the beading on cat. no. 18, possess the heavy quality of early neoclassicism, or the *goût grèc*.

The coexistence of rococo and neoclassical features is typical of the hybrid forms created at Sèvres at this time and demonstrates the factory's entrepreneurial interests in catering to its noble patrons, who required the most fashionable goods, by reflecting their taste and carefully introducing new approaches. Thus, this particular example capitalizes on the popularity both of Watteau's painted scenes and the iconic neoclassical form, the ornamented vase. E.W.M./J.L.M.

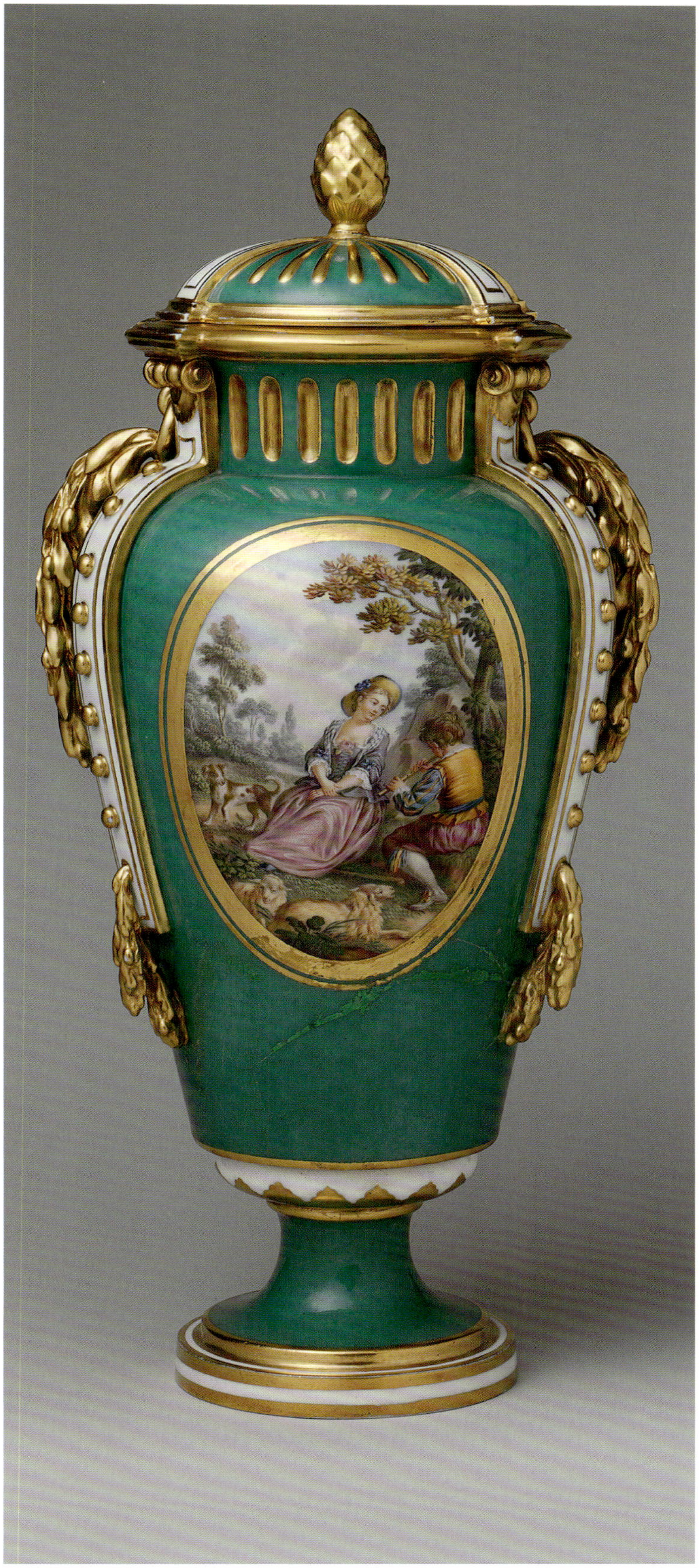

Cat. no. 23

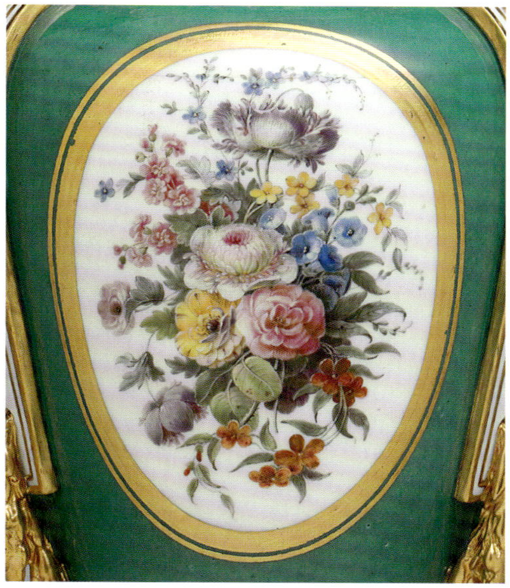

Fig. 1: Detail of the reverse of cat. no. 23.

[1] Guillaume Glorieux. "Les Arts décoratifs sous l'influence de Watteau," *L'Estampille/L'Objet d'art* 341 (Nov. 1999): 45. The print that served as a model for Dodin was probably from the edition published by Edme-François Gersaint, a dealer and patron of Watteau's in Paris; see idem: 47–48.

[2] Ibid.: 46.

[3] The attribution of the mark "HP" is problematic, because several artists at the manufactory had the same surname (Prévost) and employed the same mark. The flower painting on this vase was traditionally assigned to Prévost the Elder, but, because his involvement at the factory is only documented from 1754 until 1758, it seems improbable that he painted the flowers as late as 1770.

[4] This form *à bandes*, executed in two sizes, first appears in the inventories at Sèvres in 1769; see Eriksen 1968: 178, cat. 86; and Savill 1988: vol. 1, 392. Two vases of this type are included in a Versailles inventory of 1773 from M. le Comte de Provence valued at 360 and 720 *livres*. The form is a simplified version of the *vase Falconet de côte* with an added fluted collar; see Eriksen 1968: cat. 86b.

24. BOWL FROM THE DAIRY AT RAMBOUILLET

Sèvres Manufactory; shape probably designed by Louis-Simon Boizot (1743–1809), decoration designed by Jean-Jacques Lagrenée the Younger (1739–1821)
About 1787
Hard-paste porcelain
Marks: [on underside] crossed *L*'s (factory mark); *kk*, in puce; incised *Rn*
3 ⅜ × 9 ½ in. (8.6 × 24.1 cm)
Purchase, Mrs. Sid R. Bass Gift, in honor of Mrs. Charles Wrightsman, and Anonymous Gift, 1997
1997.518

Queen Marie-Antoinette's dairy at the Château of Rambouillet, one of the last royal commissions prior to the Revolution of 1789, initiated a new archeological form of neoclassicism in France, one more reliant on actual classical prototypes than before. The furnishings and porcelain commissioned for the dairy were inspired by "Etruscan" vases and the excavations at Pompeii and Herculaneum. The Comte d'Angiviller, Director General of the King's Works and a keen proponent of classical art, intended to strengthen the commercial viability of the Sèvres factory by exploiting the current taste for the antique and copying ancient examples, for example, from Pierre d'Hancarville's illustrated catalogue of Sir William Hamilton's vase collection (see cat. no. 1).[1] In order to help reform the factory's designs, d'Angiviller appointed Jean-Jacques Lagrenée the Younger as Sèvres artistic codirector in 1785 to work with Jean-Jacques Bachelier, who had held this position since 1749. Lagrenée had studied in Rome at the French Academy from 1763 to 1768, and was well known for his neoclassical paintings and prints (see cat. no. 29).[2] The precise date of the order for the Rambouillet porcelain service, which comprised sixty-five pieces, is unknown. A portion of it was delivered on May 25, 1787, possibly in anticipation of a visit by the queen; the remainder was sent May 15, 1788.[3] While 108 pieces were planned originally, the number had been reduced by the time the second delivery took place, one year later. Of the present shape, forty-eight blank examples were made, of which possibly two were decorated and ultimately delivered to the dairy.[4]

The decoration of the service, for which Lagrenée was principally responsible, followed the precepts laid out for the project by Jean-Jacques Hettlinger, inspector and codirector of the factory. In a letter of May 1785 Hettlinger recommended to use "Etruscan" decoration on antique shapes without any gilding.[5] Opaque ground colors (*fonds*) appear sparingly throughout the service, allowing the milky white of the pieces' bodies to stand out instead; replacing gilding are black or puce lines, known as *ornements étrusques*.[6] However, the colored bands and other decoration were not based on any ancient prototypes; rather, the service's palette—pale lilac, gray, blue, green, and yellow—is much closer to colors popularized by designers like Robert Adam for contemporary interiors.[7] As the low walls of this bowl did not permit figural decoration, it bears only ornamental

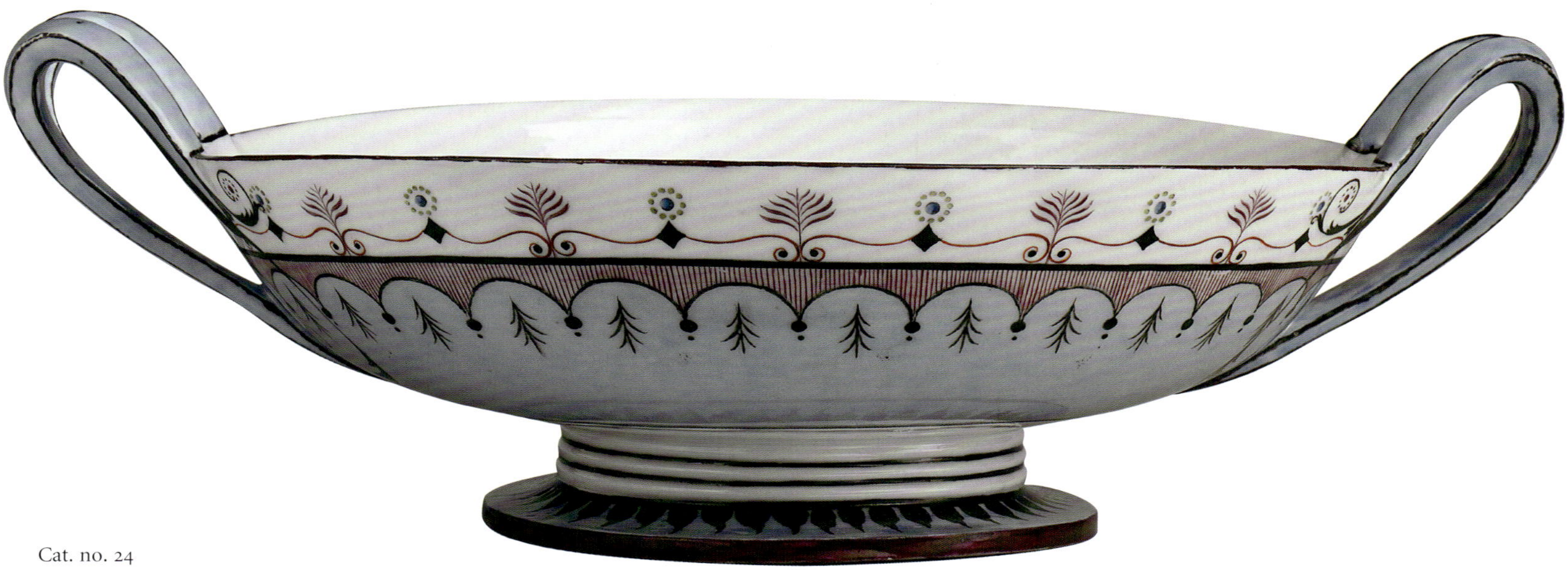

Cat. no. 24

motifs. Many of these—the circle of dots enclosing a larger dot, palmettes linked by arches, delicate scrollwork combined with palmettes, lampets around the foot, and use of parallel hatching—are found in the Hamilton/d'Harcanville catalogue. Dotted borders and stylized leaf sprigs, as well as upturned "Etruscan" handles, are also seen on examples from the Baron Dominique Vivant Denon collection of 525 "Etruscan" vases that the factory had received on June 6, 1786, to serve as a vast design resource for its employees.[8] The delicacy and disposition of the motifs are also related to the so-called "arabesques," part of the decorative vocabulary that Lagrenée developed from studying wall-paintings in Roman baths and that he published in his *Recueil de différentes compositions frises et ornaments* (1784).[9] The leaf sprig is a reduced version of a motif he used as a border around the figure of a standing woman holding an urn, made in imitation of red-figure vase painting (see cat. no. 29, Fig. 1).

The present vessel takes the form of a Greek kylix and was one of three types of small bowls designed for the dairy, all intended as stands for large drinking cups known as *gobelets cornets* or *grands gobelets.*[10] Like the majority of the service, it is made of hard-paste porcelain. Both a working drawing and a plaster model of the *jatte à anses relevées* (bowl with upturned handles) survive in the Sèvres archive. The shape was also called a *jatte anses de cuire refendu* (bowl with split-leather handles). The hard-paste records in June 1791 list an *écuelle Boizot anses fendu* (Boizot's bowl with split handles), suggesting that Louis-Simon Boizot, a neoclassical sculptor in charge of modeling, was responsible for this shape.[11] Boizot, along with Lagrenée, Hubert Robert (the supervising designer of the dairy) and Charles-François Bolvry (head of the hard-paste workshop), all contributed shapes for the service, most based directly on ancient prototypes.[12] The three coils on the foot of the Metropolitan's bowl are derived from classical pottery and common to many of the forms in the Rambouillet service.

The concept of the dairy, often a component of an ornamental farm (*ferme ornée*) in a large picturesque garden, served as a destination in the increasingly popular aristocratic practice of garden tours. It also allowed a patroness or patron to offer rural products to visitors and to enjoy a simple, natural diet, as recommended by the philosopher Jean-Jacques Rousseau in *Émile* (1762). One aristocrat, the Duc de Croÿ, described such a light repast: "ices, fruits and all kinds of dairy products."[13]

Two "bowls with upturned handles on a sandstone gray ground [with] only Etruscan ornament" appear on the May 25, 1787, delivery list for Rambouillet and two "bowls with upturned handles and light blue ground [with] Etruscan ornament" are found in a list of finished pieces designated for the "*Seconde table.*"[14] However, only two examples of this shape were ultimately delivered and probably correspond to the sandstone coloration. The Metropolitan's example may have been intended for the second console at the side of the dairy, but was never used or delivered.[15] J.L./J.L.M.

1 In 1786 the factory received copies of d'Hancarville's volumes (Hamilton/d'Hancarville 1767–76), as well as the second volume of *D'Antiquités* by the Abbé de la Chaux. See Eriksen and Bellaigue 1987: 125; Selma Schwartz, "The Sèvres Porcelain Service for Marie-Antoinette's Dairy at Rambouillet: An Exercise in Archeological Neo-Classicism," *French Porcelain Society* 9 (1992): 3; and David Irwich, *Neoclassicism* (London, 1997): 234.

2 Lagrenée's brother, at the time director of the French Academy in Rome, claimed that his sibling's appointment would give the factory "more of a taste for the Italian than it has had up to now"; see Schwartz (note 1): 5. Lagrenée's appointment was in keeping with the policy of the *travaux d'encouragement* (commissions of encouragement) that featured in the art reforms initiated by d'Angiviller; see Barthélémy Jobert, "The Travaux d'encouragement: An Aspect of Official Arts Policy in France under Louis XVI," *Oxford Art Journal* 10, 1 (1987): 3–14; and Schwartz (note 1): 5, 13.

3 Selma Schwartz, "The 'Etruscan' Style at Sevres: A Bowl from Marie-Antoinette's Dairy at Rambouilllet," *MMA Journal* 37 (2002): 262.

4 More blanks of the bowl and cup shapes were created than were needed, indicating that the factory may have produced the remainder in hopes of commercial sales; ibid.: 264.

5 Much of the service is decorated with human and animal forms, consistent with the dairy's overall iconographic program of mythological themes and pastoral idylls. These are bordered by colored bands embellished with ornamental motifs and arranged in a manner characteristic of Greek pottery. See Schwartz (note 1): 6, 8.

6 Schwartz (note 3): 9.

7 Ibid.: 263.

8 Carolyn Young, "Marie Antoinette's Dairy at Rambouillet," *Magazine Antiques* 158 (Oct. 2000): 551, fig. 2.

9 In 1787 Lagrenée purchased three sets of engravings by Giovanni Ottaviani of Raphael's decorations in the Vatican, a primary source for grotesque decoration; see Schwartz (note 1): 6, 9; and Young (note 8): 550.

10 These were slightly flared, tall cups with two curved handles; Boizot designed the shape and Lagrenée the decoration. Only a few of the ewers and vases from the service were based directly on copies of classical prototypes, but most pieces incorporated classically inspired handles and other elements; see Schwartz (note 3): 263.

11 Schwartz (note 1): 14.

12 Shapes created for the dairy service appeared again on an "arabesque" service in 1797 and a number of "Etruscan" services in the 1790s; see Brunet and Préaud 1978: figs. 286–87, 289; Eriksen and Bellaigue 1987: 350; and Young (note 8): 553.

13 "Une collation de glaces, de fruits and de toutes sortes de laiteries"; Schwartz (note 1): 2.

14 Schwartz (note 1): 23–25, app. 2, delivery list, May 25, 1787: "jatte à anses relevées fond grès ornements étrusques seulement…jatte à anses relevées fond petit bleu ornements étrusques."

15 Schwartz (note 3): 265.

25. GARNITURE OF FIVE VASES WITH ROMAN EMPEROR MEDALLIONS

Fürstenberg Porcelain Manufactory
About 1768
Hard-paste porcelain
94.4.351: 7 7/8 in. (20 cm); 94.4.352: 9 1/2 in. (24.1 cm); 94.4.353: 11 1/2 in. (29.2 cm); 94.4.354: 9 1/2 in. (24.1 cm); 94.4.355: 7 1/2 in. (19.1 cm)
Marks: [inside the foot] *F* in blue script under the glaze on all pieces; 94.4.351: illegible impressed mark; 94.4.352: impressed *FB2* over *1*; 94.4.353: impressed *B31*; 94.4.354: incised *FB* [illegible]; 94.4.355: no further marks
Gift of Henry G. Marquand, 1894
94.4.351–.55

Comprising five covered potpourri vases, this garniture may be the earliest surviving set in the classical style made by the Fürstenberg manufactory. The longest continuously operating porcelain works in Germany after the Meissen manufactory in Saxony, Fürstenberg began production in a facility on the Weser River in 1747, and within a decade was making coffee, tea, and chocolate services, as well as incense burners, tureens, and potpourri vases.[1] In the late 1760s, Fürstenberg introduced garnitures or sets of vases meant to be displayed together. By this time, the factory had caught on to the new neoclassical style, which had already captivated the Wedgwood and Sèvres factories, and began incorporating Roman and Greek motifs into a variety of porcelain ware, including busts, medallions, figures, services, and, most significantly, vases.[2]

Indeed, the three distinct models for this garniture display a tour de force of neoclassical ornament. Fürstenberg, as runner-up and fierce competitor to Meissen, must have been eager to prove its prowess and expertise in the classical style and appears to have spared no expense or subtlety on this set. While each model shares the same basic formula—glazed white porcelain with pierced covers or necks; profiles of Roman emperors painted in gray on a pink ground; heavily gilded and molded neoclassical garlands, satyr- (or grotesque-) head handles and decorative bands—the shape and treatment of each piece differ significantly.

Like many neoclassical vases, the largest form of the five presented here adapts an ancient prototype, in this case a bell krater with lug handles. As the central piece, it is the most ornate. The high-domed cover is richly decorated with gilded and raised leaves radiating from a vegetal finial, and above, with a pair of trophies painted with brown, yellow, gray, purple, and blue over-glaze enamels. Below the rim of the cover are two rows of round holes on the vase's molded, gilded neck. Double swags of garlands in high relief loop through the horns of satyr heads, which serve as grips. The round, fluted foot sits on a square plinth with gilded sides showing a slightly raised meander pattern, a common motif on Greek and Greek-style vases. A close look at the gilded borders and bands reveal miniscule patterns painted or etched on the surface.

Unlike earlier Fürstenberg vases, this potpourri has two *Schauseiten*, or display sides.[3] The central motif of each is a profile medallion of a Roman emperor, painted over the clear glaze: in this case, on one side, Trajan wearing a crown and, on the other, Julius Caesar sporting a laurel wreath.[4] Both heads are depicted in grisaille, on a vibrant purplish pink ground framed by green holly leaves dotted with red berries and tied with a yellow ribbon. Adjacent to the medallions, under the satyr heads, are painted trophies, larger than those on the vessel's cover, featuring various bows, arrows, axes, swords, spears, horns, batons, and shields.

The decorative schemes of the smallest pair of vases are clearly related to the tallest vase in their use of medallions, garlands, and a round pedestal foot on a square plinth. The closet model for these narrow-necked, round-bottomed vessels is the Greek *oinochoe*, to which the designer added prominent pierced covers. The garlands, less gracefully modeled than their counterparts on the others, are from a different mold. The outsized grips—not satyrs but turbaned heads, with flamelike licks of foliage forming handles—leave no room for painted trophies. The scaled-down medallions, compressed into the area below the garland, depict four different emperors in the same manner as the pair on the center vase.[5]

The mid-size pair shows still other variations of now-familiar neoclassical motifs, but they also include new elements.[6] Most distinctive are the vertical U-shaped handles and the fully gilded domed lid, which incorporates elongated teardrop cutouts in its design. The vases each sit on an eight-sided plinth, slightly taller than the square bases of the other vases but similarly gilded. Adjacent to the painted medallions are painted trophies, similar to those on the main vase. Curiously, the vessel does not have a classical antecedent but rather an eastern one: the straight-sided neck and rounded, elongated form relate to Song Dynasty porcelains and demonstrate the Asian touches that marked Fürstenberg's brand of neoclassicism. In general, the garniture's gilded handles, covers, and bases mimic the gilt-bronze mounts added to Asian porcelains earlier in the eighteenth century.[7] This relationship to Chinese imports is most readily seen in the middle pair of this group.

One might think that the diverse motifs and treatments between the forms suggest that these vases were not intended to be displayed together. Several factors however support their origin as a garniture. First, the quality of the porcelain, gilding, and particularly the painting is consistent throughout the series. Secondly, at least one example of each shape bears a series and a model mark, along with the painted blue "F," the insignia of the manufactory.[8] The "B" year mark on this series indicates that it was one of the earliest inaugurated by the factory; the design can

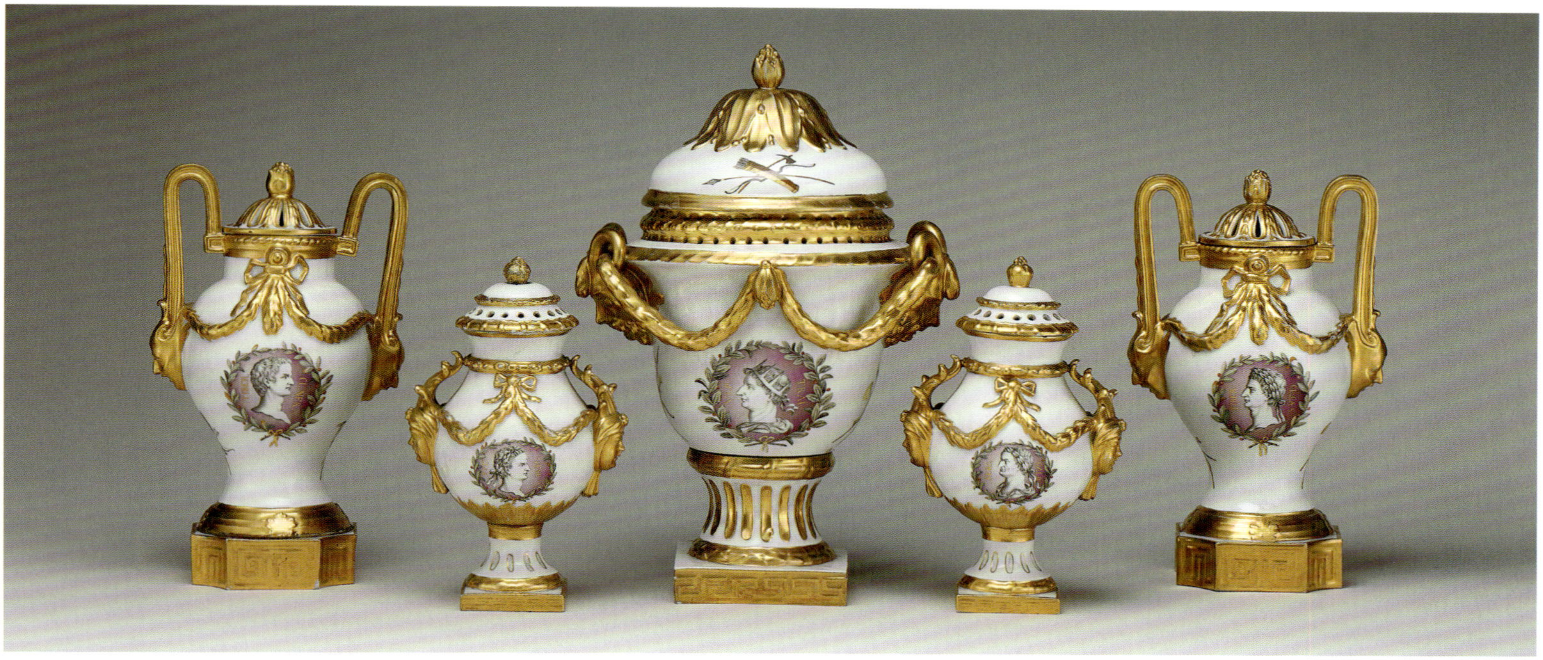

Cat. no. 25

be assigned to about 1768.[9] Noteworthy too is the inclusion of the potpourri, a form familiar to German nobility from the rococo period.[10] When such vessels, often strategically located on a mantelpiece, were filled with a mixture of petals and salts, a sweet aroma emitted through their perforated lids would fill the room.[11]

A 1785 pricelist records this and other Fürstenberg garnitures, including three with seven pieces and five others with five pieces, under the heading "Aufsätze von Urnen nach Antiken auf Kamine oder Plateaux" (display pieces in the shape of urns after the antique for chimney-pieces or shelves).[12] These later sets are also in white and gold, but tend to be more restrained in the use and depth of the relief of the neoclassical ornament.[13] While popular at the time of their creation, many garnitures no doubt were separated over the years as fashions changed and fortunes were dispersed. This may well have been the fate of a single unmarked "B 3" vase from the present series, which survives in the collection of the Porzellanmuseum, Fürstenberg.[14] C.H.

1 Beatrix Frfr. Wolff Metternich, *Porzellan Museum Fürstenberg* (Braunschweig, 1992): 62.

2 See "Klassizistische Motive und ornamentaler Dekor," in *Fürstenberger Porzellan: von Rokoko bis zum Historismus* (Bremen, 1986): 205.

3 Metternich (note 1): 62.

4 The inscriptions in gold within the pink medallions on 94.4.353 read: *VLP/TRAI.* and *CIVL./CAES.* The handles relate to those of a central vase in a garniture from Sèvres, dating circa 1765–70, in the Musée du Louvre, Paris, and illustrated in Ennès and Ducrot 1997: 46, cats. 44–45.

5 The inscriptions in the medallions on 94.4.351 read: *SERG./GALB* and *CAI./CALIG.* The inscriptions in the medallions on 94.4.355 read: *AVI/VITEL.* and *TIT./VESP.*

6 The inscriptions in the medallions on 94.4.352 read: *OCTAV./AUGUST.* and *TIBER./CL.CAES.* The inscriptions in the medallions on 94.4.354 read: *TIBER/CL.NERO* and *NERO/CLAUD.C.*

7 See Francis J. B. Watson, *Mounted Oriental Porcelain* (Washington, D.C., 1986); and Pierre Kjellberg, *Objets montés: Du Moyen Âge à nos jours* (Paris, 2000).

8 See Christian Scherer, *Das Fürstenberger Porzellan* (Berlin, 1909): 157–58.

9 The letter represents the series and the number of its position within the series. Fürstenberg, seeking to specialize in "Aufsätze von Urnen nach Antiken auf Kamine oder Plateaux" (display pieces in the shape of urns after the antique for chimney-pieces or shelves), had implemented a system of coding new vase designs using letters of the alphabet by 1768; see *Journal von und für Deutschland. Zwenter Jahrgang. Erstes bis letztes Stück* (1785): 91–92, repr. in Sigfried Ducret, *Fürstenberger Porzellan* (Braunschweig, 1965): vol. 2, 290–91. Angelika Lorenz, *Weissen Gold aus Fürstenberg: Kulturgeschichte im Spiegel des Porzellans 1747–1830* (Braunschweig/Münster, 1989): 337, mentioned a 1770 pricelist with five garnitures, which may have included the "B" series. See Bodo Hedergott, "Fürstenberger Vasen und Servicen vom Empire bis zum Biedermeyer," *Weltkunst* 52 (1982): 2301. Fürstenberg ran through the alphabet within a decade and started marking new vase designs with double letters. Howard Couts, "The Fürstenberg Porcelain Factory," *Apollo* 129 (1989): 416, posited 1785 as the date for reaching the end of the alphabet. He is the only scholar consulted to suggest a date later than 1780.

10 Lorenz (note 9): 337.

11 Scherer (note 8): 158.

12 *Journal von und für Deutschland* (note 9).

13 Susanne Schroeder linked a seven-piece garniture bearing the series mark "G" to Wedgwood antecedents modeled by William Hackwood in 1773 in the collection of Anna Amalia, Duchess of Saxony-Weimar-Eisenach. She further suggested that the duchess' brother, Crown Prince Carl Wilhelm Ferdinand, who was one of the first Germans to make the Grand Tour, took an interest in promoting the neoclassical style at Fürstenberg when he became head of government in Braunschweig ten years after this garniture was first made. See Susanne Schroeder and Petra Damaschke, *Tafelrunden: Fürstenberger Porzellan der Herzogin Anna Amalia in Weimar* (Munich, 1966): 82.

14 This vase is illustrated in Ducret (note 9): vol. 2, 239, fig. 339, attached to a tall square plinth, but it was also photographed without one in Metternich (note 1): 63.

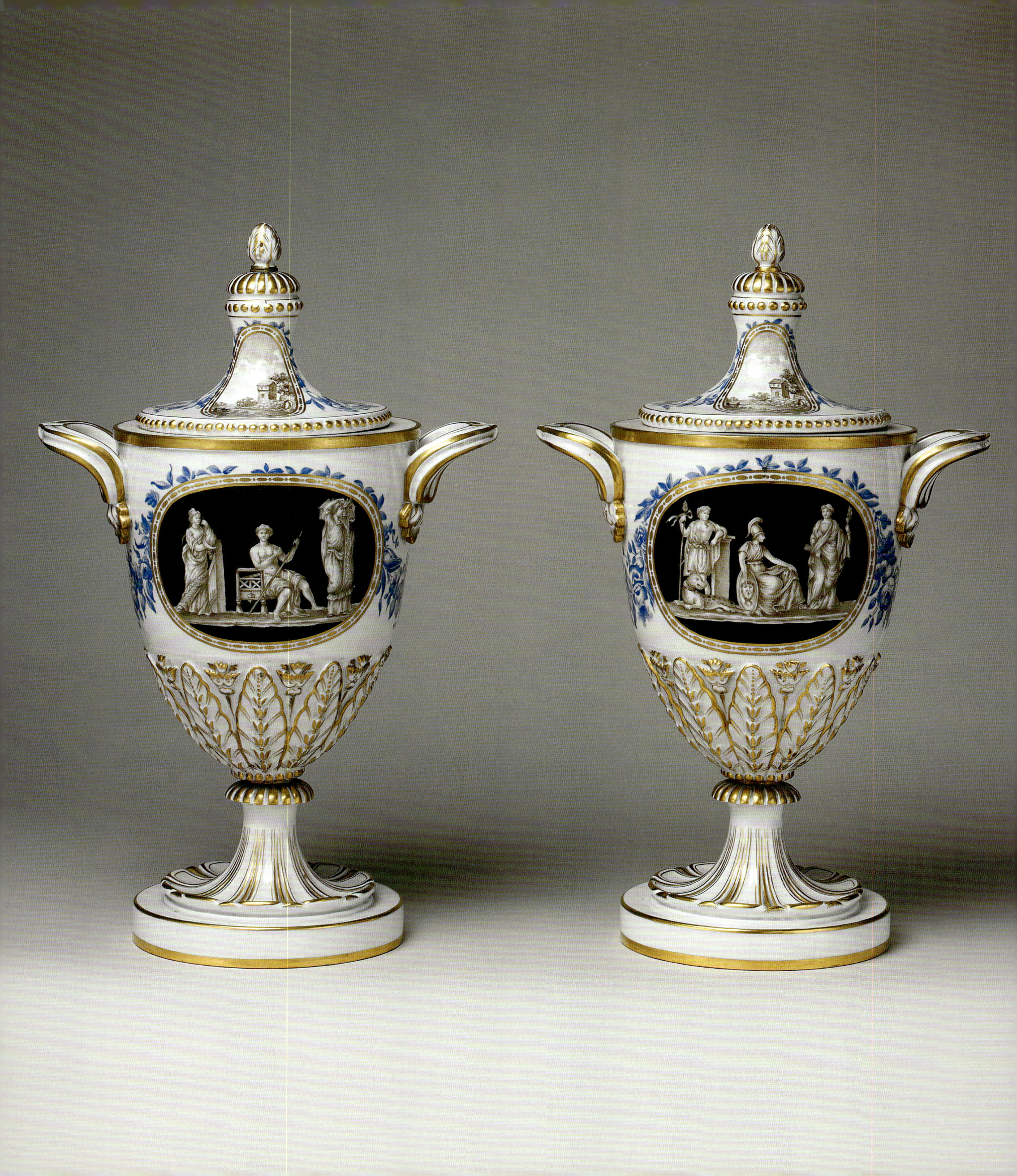

26. PAIR OF VASES WITH COVERS

Real Fabrico de Buen Retiro
About 1784–95
Soft-paste porcelain
18 ¾ in. (47.6 cm)
Gift of Mr. and Mrs. Luke Vincent Lockwood, 1942
42.30.17–.18

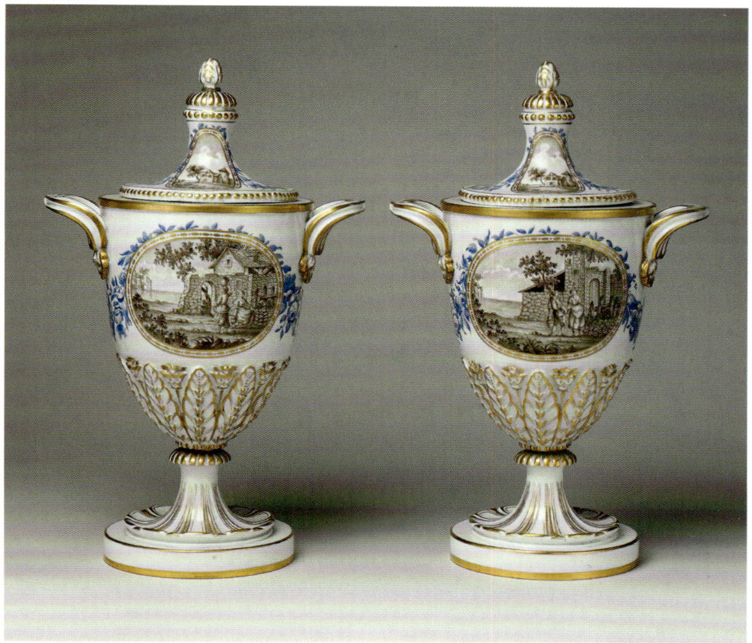

Fig. 1: Reverse of cat. no. 26.

The basic urn-shape of this pair of neoclassical vases from the Spanish Buen Retiro factory is derived from the Greek bell-krater.[1] While exhibiting a limited range of colors, the vessels are elaborately painted, with gilding employed to highlight edges and molded elements. The use of molded lancet leaf borders interspersed with foliate elements at the bottom of the bodies is a typical feature of Buen Retiro vases, and replaces the gadrooning often seen in comparable English examples, although this ornamental device was also used at the Spanish factory. Later, under the increasing influence of international neoclassicism, Buen Retiro's designers gradually abandoned the use of molded decoration in favor of smooth surfaces; vases were produced in a much more severe neoclassical style, strongly influenced by Wedgwood jasper wares.[2]

The artists at Buen Retiro, like those at Sèvres, typically illustrated their vases with mythological or pastoral themes, often based on rococo designs. Here, oval reserves on the fronts contain mythological figures rendered in a neoclassical manner, while the backs and lids of each vase are adorned with rural scenes in sepia tones that are more closely related to earlier styles. The frontal views are not an exact match, but the basic compositions are similarly conceived so as to appear like ancient cameos or vases. The black background appears to have been added last and fixed with a very low-temperature firing. It is possible that the factory was hoping to achieve a low-gloss surface resembling that of ancient Greek vases.[3] The scenes and figures are difficult to decipher: one group shows a seated Athena, identifiable by her helmet and shield, flanked by Dionysus and perhaps Ariadne; the other includes Artemis standing to the left of a seated man who seems to be gesturing toward a woman carrying a small house.

When the Bourbon king of Naples and the Two Sicilies ascended to the throne of Spain in 1757 as Charles III, he literally took the Neapolitan royal porcelain manufactory at Capodimonte with him. He traveled to Spain accompanied by fifty-three of its artists, along with eighty-eight tons of equipment and porcelain paste, with which he founded a porcelain works at Buen Retiro in 1760. From the factory's inception in 1759, Charles III's enthusiasm for classical themes insured their popularity alongside rococo-inspired subjects.[4] During its early years, Buen Retiro produced numerous mythological figure groups inspired by seventeenth-century baroque classicism, which was never entirely abandoned by Bourbon court artists. The pastoral scenes on this pair of vases (Fig. 1) reflect the lingering influence of the French rococo artist François Boucher, whose themes and paintings frequently inspired Buen Retiro's products.

Unlike other European royal manufactories, including Sèvres and Meissen, Buen Retiro originally had only one customer, the king. It concentrated on porcelain rooms and figural groups. The first major commission the factory undertook, between 1760 and 1765, was the creation of a porcelain room in the rococo style for the Palacio Real of Aranjuez. From 1766 to 1771, a second porcelain room, this time in the neoclassical style, was installed in the Palacio Real in Madrid. The panels of this second room incorporate vases similar in form and decoration to the ones shown here, as well as to objects with Boucher-inspired pastorals.[5] After Charles III's death in 1789, the factory's warehouse opened to public, with the intent to offset through sales the great expense of making porcelain. Although Buen Retiro continued to focus on ornamental porcelain pieces, as opposed to table services, freestanding vases now became the factory's most important product.[6]

From 1784 to 1803, Carlos Gricci served as director of Buen Retiro. He developed an excellent glassy white paste for the production of vases that became increasingly neoclassical in form and decoration. The technical challenges faced by the factory in this period—evident in its products' limited palette, unsophisticated painting, and simple gilding—are highlighted by comparing the wares of Buen Retiro with those of Sèvres. The good-quality paste used for the present pair places their origin in the second period, but their resemblance to the vases of the porcelain room in Madrid suggest they were produced at a fairly early date.[7] J.L.

1 The present objects were first published in Parke-Bernet Galleries New York, *Collection of Mrs. Paul E. Gardner*, sale cat., Oct. 31, 1941: no. 268; and Louise C. Avery, "A Gift of Porcelain," *MMA Bulletin* 30, 7 (1942): 164.

2 Madrid, Museo Arqueológico Nacional, *Manufactura del Buen Retiro, 1760–1808* (Madrid, 1999): 98, 117, 295, 299, 307.

3 My thanks to Jeffrey Munger of the Metropolitan's Department of European Sculpture and Decorative Arts for this suggestion.

4 Maria Jesús Sánchez Beltrán, *La porcelana de la Real Fábrica del Buen Retiro* (Madrid, 1998): 50–51. For more on Buen Retiro, see Arthur Lane, *Italian Porcelain* (London, 1954): 52.

5 Maria Jesús Sánchez Beltrán, "La Porcelana del Buen Reito en el Palacio Real de Madrid," *Reale sitios* 24, 94 (1987): 50–53; and idem (note 4): 50.

6 Sánchez Beltrán (note 4): 30–34, 59; and Lane (note 4): 55.

7 The dating of Buen Retiro products is complicated by several factors, the destruction of the factory building and records in 1812 being chief among them. In addition, prior to 1803, difficulties in developing a satisfactory paste resulted in objects of inconsistent quality, obscuring any clear technical developments. Scholars recently have attempted to establish a chronology for objects during the manufactory's first epoch (1760–84) by using the documented porcelain rooms as a basis; see Sánchez Beltrán (note 4): 45–46.

27. AMPHORA-SHAPED VASE WITH PARIS AND OENONE

Chelsea-Derby Manufactory; figural painting attributed to Richard Askew (about 1730–1798); landscape painting by Zachariah Boreman (1738–1810)
About 1775
Soft-paste porcelain
10 3/8 in. (26.4 cm)
Mark: *No. 71*, incised
Bequest of John L. Cadwalader, 1914
14.58.67

While more than three-quarters of Chelsea-Derby's sales came from the production of functional breakfast, tea, and tableware sets, the factory's vases are notable because they emulated the neoclassical style of the Sèvres manufactory, which was highly regarded in England. Direct connections between Chelsea-Derby and Sèvres certainly existed through transmission of styles, forms, and even factory employees: Andrew Planché, a Chelsea-Derby figure modeler from 1748 to 1766, had worked under Duplessis the Elder, chief modeler at Sèvres from 1748 to 1774.[1] This covered, amphora-shaped vase was a product of the ceramic factories owned by William Duesbury during the Chelsea-Derby period, the years between 1770—when he purchased the Chelsea factory in London and maintained it in addition to another in Derbyshire—and 1784, when the Chelsea porcelain factory was closed and its craftsmen transferred to Derby.[2]

Neoclassical forms and designs of Sèvres vases, including shapes from the mid-1760s such as the beaded vase featured here (cat. no. 18) and the style of figural painting contained in oval reserves, were also imitated by Chelsea-Derby.[3] Like Josiah Wedgwood, Duesbury adapted ancient classical forms to create his vase shapes, including the modified *oenochoe*, known at Wedgwood as number 60 (see cat. no. 31) and various krater and kylix forms.[4]

Here, a softly colored painted image, the royal-blue areas on the lid and foot, and the prominent striated gilding were all intended to evoke the Sèvres style for English consumers. The oval reserve on the front depicts a mythological scene, Paris carving the names of Oenone and himself on beech tree, derived from a painting by Angelica Kauffman, a leading neoclassical artist active in England and Rome.[5] A noted porcelain figural painter, Richard Askew, most likely copied Kauffman's composition from an engraving after it by Francesco Bartolozzi, while the landscape on the reverse is by Zachariah Boreman. Their collaboration exemplifies the tendency of eighteenth-century porcelain painters to specialize in figure or landscape decoration.[6]

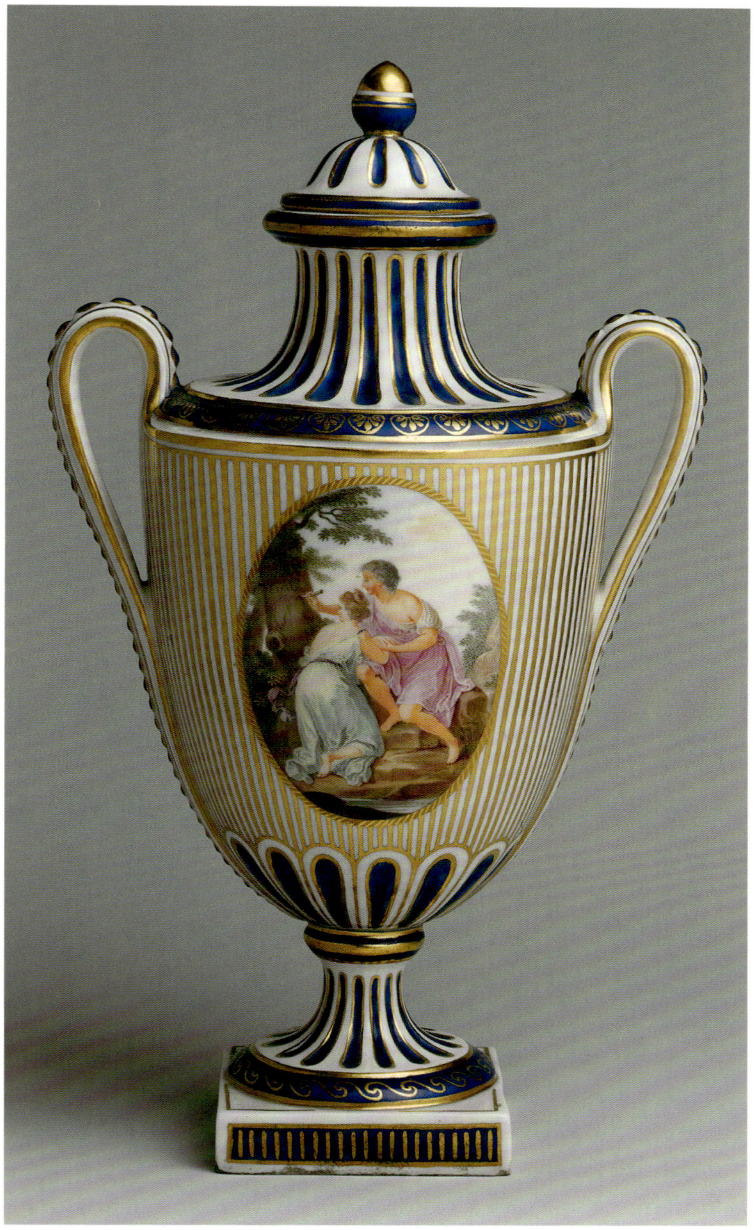

Cat. no. 27

Vases accounted for a mere two percent of the sales in the Chelsea-Derby period, but they illustrate the transitional shift in eighteenth-century taste, from the asymmetrical outlines of the rococo to a more linear, restrained manner. Vases of this type tended to be sold in sets, or garnitures, usually numbering three to five pieces. The Metropolitan Museum owns a vase matching this example, but the two originally might have been made as part of a larger group. The catalogue, by a Mr. Christie, of Derby's annual six-day auction on May 23, 1785, on the Pall Mall describes a garniture of five "elegantly finished" vases, "beautifully enamel[ed] in compartments on one side with pastoral groups and on the other landscapes, in a pure self colour, elegantly ornamented with fine blue gold."[7] Another set of five in this sale seems to have had similar figural decorations: "the centre one full 12 inches high...the two side pieces to correspond, on one is represented a group of Damon and Delia, on the other a group of Paris and Oenone, and two ewer shape Vases to complete the set."[8]

Duesbury was granted a royal appointment in 1775, and his company retained its popularity until the late 1790s. His products caught the attention of other competitors: at an auction of "Derby and Chelsea Porcelane" held by Christie and Ansell in 1782, Josiah Wedgwood bought "an elegant Etruscan shape vase and pedestal."[9] E.S.

1 John Twitchett, *Derby Porcelain 1748–1848: An Illustrated Guide* (Woodbridge, Eng., 2002): 19. Planché may have come into contact with Duplessis between 1740 and 1748, but may have not completed his internship.
2 Twitchett (note 1): 24.
3 Ibid.: 146, pl. 100. A Chelsea-Derby three-vase garniture, formerly on the London art market, from about 1775, includes two side vases adapted from the Sèvres *vase à perles* form and a figural scene painted by Richard Askew.
4 John Twitchett, *Derby Porcelain*, foreword by Anthony Hoyte (London, 1980): pl. 29, fig. 168; and a pair of ewers with satyr masks from the William Bemrose garniture, in Anthony Hoyte, *The Charles Norman Collection of 18th Century Derby Porcelain* (Abingdon, Eng., 1996): 26.
5 Twitchett (note 4): 153. Duesbury's catalogues advertised vases "admirably finished from those sublime compositions of A. Kauffman." Kauffman's works were in great demand; see Gilbert Bradley, Judith Anderson, and Robin Barkla, *Derby Porcelain: 1750–1798* (London, 1990): 96.
6 For more information on the working conditions of the painters, see Hilary Young, *English Porcelain 1745–95* (London, 1999): 110–12, 137–41. Boreman was a watercolor painter before joining the Chelsea operation; Askew, moving freely between the trades of enameling copper and ceramics, was the more itinerant, transferring from Derby to Birmingham with the promise of more work and improved living conditions.
7 Twitchett (note 4): 142. The catalogue describes such a set of vases as selling for £9 19s. 6d. in 1785.
8 Twitchett (note 1): 26.
9 Gilbert Bradley, *Derby Porcelain 1750–98* (London, 1990): 95–96.

31. EWER WITH SATYR MASK

Josiah Wedgwood (1730–1795), Etruria, Staffordshire
About 1769–80
Cream-colored earthenware with slip decoration, lead-glazed slip decoration, and gilding
11 in. (27.9 cm)
Mark: [bottom of plinth] stamped within circle around screw, *WEDGWOOD & BENTLEY: ETRURIA*
Rogers Fund, 1940
40.165

Josiah Wedgwood maximized the production and range of his vases by varying the materials, decorative treatments, and ornament of basic designs. Among the best demonstrations of the adaptability of a single shape is Wedgwood and Bentley's leafy-lipped ewer, known as number 60 in the factory Shape Book.[1] The many surviving examples of this form illustrate the great variety made possible by the technical sophistication and flexibility that characterized Wedgwood's industrialized production.[2]

This cream ware ewer features agate-like surface decoration with gilded ornament in imitation of hardstone vessels with gilt-bronze mounts. Wedgwood's decorators produced this and other variegated finishes by trailing and blending colored slips on the surface of unfired clay forms. They achieved other hardstone effects by layering and wedging various colored clays together to produce marble-like veining in the clay body itself.

Wedgwood and Bentley offered this ewer in several sizes and finishes, including a mottled surface resembling granite and marbling in several colors with and without gilding. Optional ornament included palmette collars, applied medallions, and engine-turned stripes or fluting. The firm also used a variety of plain and molded plinths in cream ware and black basalt, including an unusual round example with garlands.[3]

Like Wedgwood and Bentley's *Vase with Lion-Head Handles* (cat. no. 36), this ewer derives from Renaissance designs inspired by antiquities. Many designs for ewers of this form were published from the second quarter of the sixteenth century by Italian engravers like Agostino Veneziano and Enea Vico, and later by a number of northern European artists.[4]

With its shell-like spout, pedestal foot, and high looped handle terminating in a satyr mask (Fig. 1), the ewer featured here relates in particular to a fanciful 1543 design by Vico, one of a celebrated set of twelve engravings of vases.[5] Although we do not know whether Wedgwood or Thomas Bentley actually saw this print, they did have access to the design: in 1724 Bernard de Montfaucon reproduced it in his monumental encyclopedia,

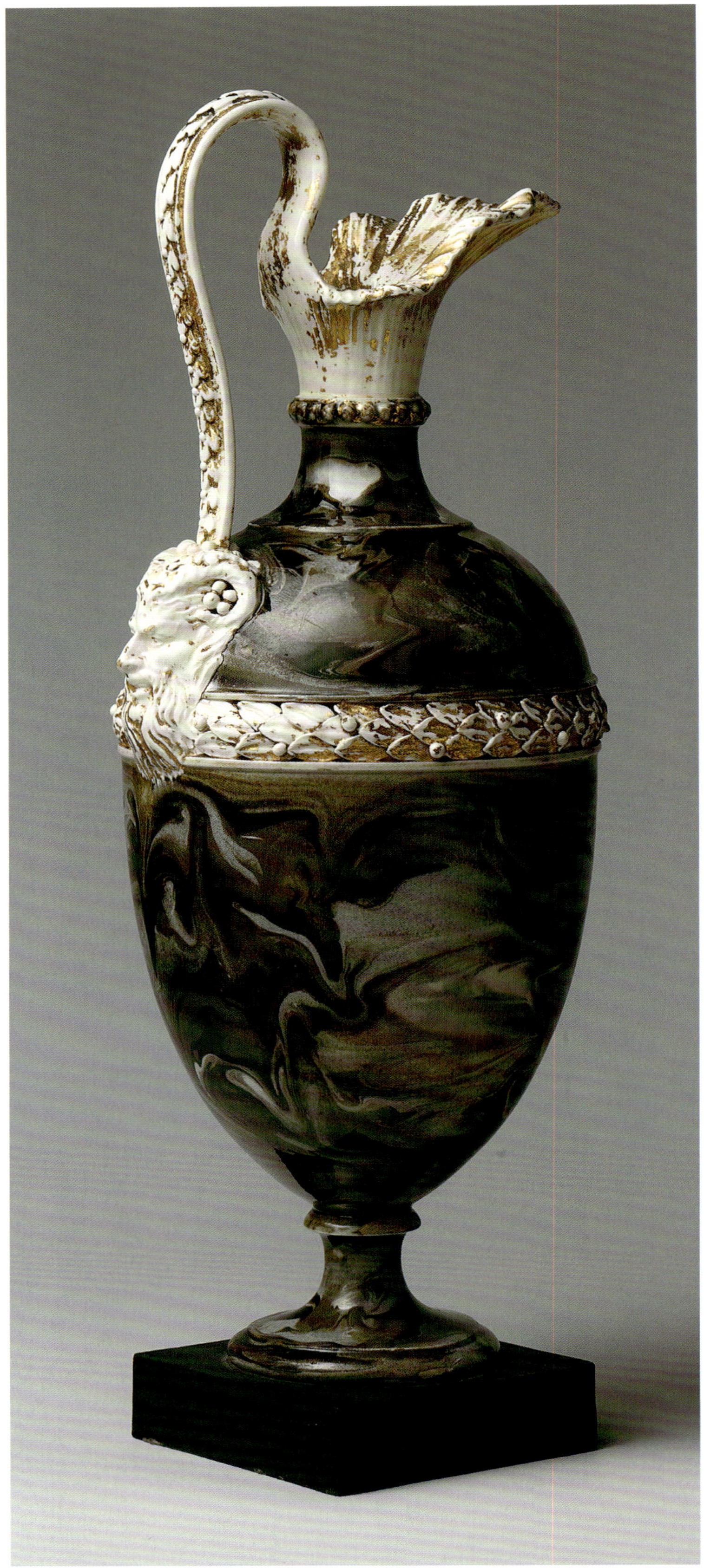

Cat. no. 30

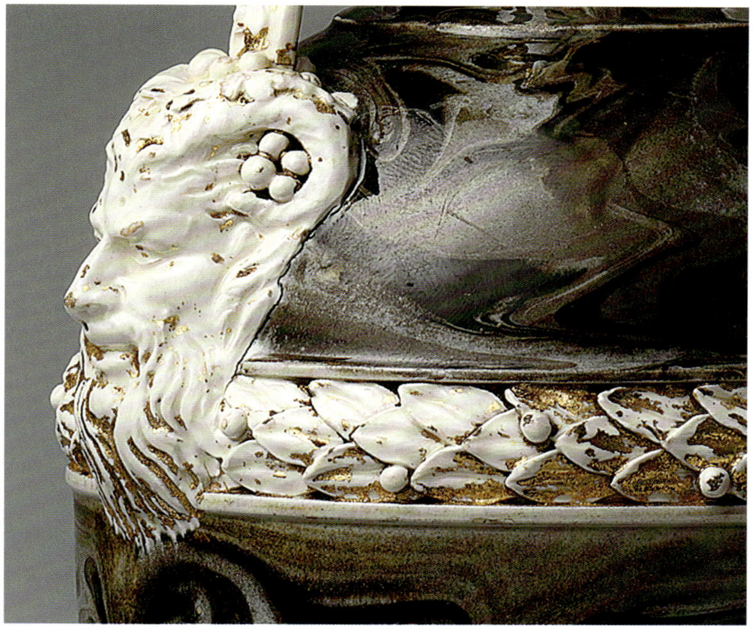

Fig. 1: Detail of cat. no. 31.

Österreich. Museums für Kunst und Industrie: Erwebungen seit dem Jahre 1871* (Vienna, 1889–1917): vol. 1, 195; vol. 2, 179, 181.

[5] Vico's engraving is illustrated in Miller 1999: cat. 68a, pl. 4.

[6] Bernard de Montfaucon, *L'Antiquité expliquée et représentée en figures...*, 5 vols. (1719; rev. ed., 1722); and idem, *Supplément au livre de l'antiquité expliquée...*, 5 vols. (1724). Vico's design appears as plate 13 in the second volume of the *Supplément*, "Prefericules, vases pour les sacrifices." Montfaucon did not identify Vico as author of the design, noting only that it was engraved in Rome in 1543. With Vico's ewer, Montfaucon reproduced two of Veneziano's designs for vases. An August 1770 list of books owned by the Wedgwood and Bentley partnership includes all of Montfaucon's volumes; see Johnson 1990: 13–23; and Johnson 1991: 34.

[7] The shapes and imagery of many Wedgwood and Bentley designs can be traced to Montfaucon's plates; see Adams 1992.

[8] Young 1995: cats. D41–D42.

L'Antiquité expliquée.[6] This richly illustrated publication, which Wedgwood and Bentley owned by 1770, was an important source for many of the firm's neoclassical designs.[7]

Several other manufactories produced similar ewers in various materials, including a nearly identical soft-paste porcelain version manufactured by the Derby factory. Matthew Boulton made similar ormolu-mounted hardstone vessels that are very close to sixteenth-century protoypes.[8] A black basalt ewer attributed to the Church Works pottery (cat. no. 38) closely resembles both Wedgwood's and Boulton's designs, inspired by imaginative mannerist interpretations of classical antiquities. H.J.M.

[1] Edwards 1994: 53, 62; and Young 1995: 82, cat. D41.

[2] Wedgwood's understanding of the potential of interchangeable ornaments and varied decorative finishes dates to the early years of his collaboration with Bentley; see Wedgwood to Bentley, Sept. 26, 1769, E25-18261, Keele University Library, quoted in Finer and Savage 1965: 81.

[3] For examples of the many variations on this form, see Grant 1967: pls. 7:3, 9:3, 10:2; Reilly 1989: vol. 1, pl. C78, figs. 450, 544, 544a; Edwards 1994: figs. 25, 46; and Young 1995: cat. D40. Wedgwood and Bentley also manufactured a simpler version of the form, changing the shape of the body. Another unmarked variation of this ewer may be what Wedgwood described as his firm's "Dolphin Ewer"; see Reilly 1989: vol. 1, figs. 545–46.

[4] For related designs by Hans Vredeman de Vries, Virgil Solis, and Peter Flötner, see Franz Ritter, *Illustrierter Katalog der Ornamentstichsammlung des K. K.

32. "SUGAR DISH" WITH LAUREL FESTOONS

Josiah Wedgwood (1730–1795), Etruria, Staffordshire
1769–80
Cream-colored earthenware or white terracotta stoneware
7 ¼ in. (18.4 cm)
Mark: [on plinth] impressed within a circle around screw, *WEDGWOOD & BENTLEY:/ETRURIA.*
Rogers Fund, 1909
09.194.6

Sponged with metallic oxide colors and accented with richly gilded details, this low, lidded vessel is typical of the first phase of Wedgwood and Bentley's neoclassical vase trade in England.[1] Along with the cream-colored earthenware known as cream ware and black basalt vessels, vases that imitated various hardstones—granite, agate, and porphyry are some of the names used for this variegated ware (see also cat. no. 37)—were the earliest products of Josiah Wedgwood's famous partnership with Thomas Bentley, established in 1767 for the production of ornamental pottery.

Although Wedgwood and Bentley's variegated vases—also produced at rival Staffordshire firms (see cat. no. 35)—allude to the carved marbles of classical antiquity, their immediate prototypes were French hardstone vases with gilt-bronze mounts. Despite its diminutive size, this vessel has a robustness reflecting the substantial forms and lush gilding that characterize the *goût grec* manner originating in France.

The form may be associated with what Wedgwood referred to as a "sugar dish Vase." These vessels, which were made in several variations, were in production by 1769, when Wedgwood ordered a large quantity for the firm's London showroom.[2] While the

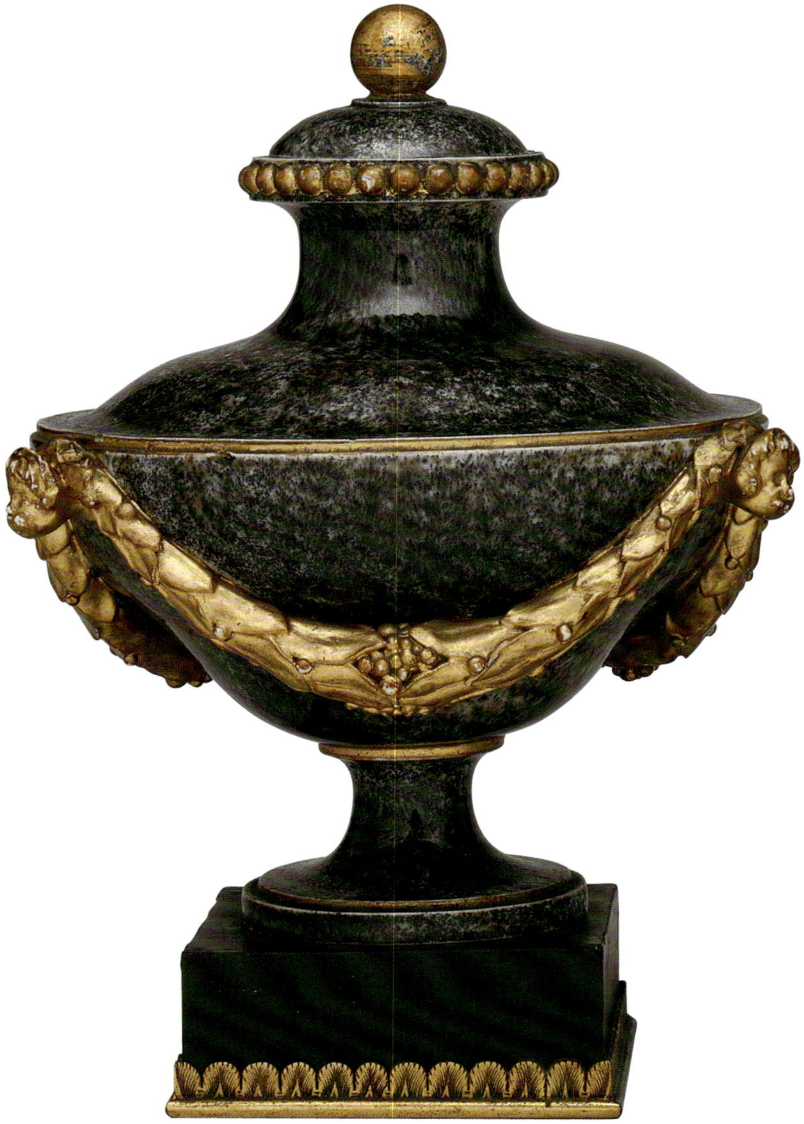

Cat. no. 32

order confirms his faith in the popularity of this model, he shared his concerns about the shape with his partner: "Some of the Ladys say the same thing that occurred to you and me about the Sugar dish Vases, that they are like the things on the tops of Clock cases, or Beds heads. They certainly are not Antique and that is fault enough to D—n them with most of our customers."[3]

Gilded ornament was also controversial, and the taste for variegated ware was not as long-lived as that for the black basalt that often accompanied it in the showroom display. Persistent sales, however, meant that gilded vases were produced long after initial criticisms, which seem to have emerged almost with the introduction of variegated ware. As early as February 1769, Wedgwood told Bentley: "Marbleing with Gold is hissed universally, so we must have no more done in that way.... Etruscan [black basalt] Vases are the run at present, though the others go off by degrees, and serve very well to make a variety, and prevent that *sameness* which must be rather disgusting if we confined the shew to one sort only."[4]

Wedgwood, aspiring to artistic excellence and technical superiority as well as profits, found himself torn between the criticisms of discriminating patrons and his awareness that gilding attracted attention that resulted in sales. After a visit from Sir William Hamilton in 1772, Wedgwood wrote that he hoped to "make a good use of his hints...by greatly reduceing, if not totally banishing this *offensive Gilding*," but he confessed to Bentley that he did "not find it an easy matter to make a Vase with the colouring so natural, vairied, pleasing & *un-pot-like* & the shape so delicate, as to make it seem worth a great deal of money, without the additional trappings of handles, ornaments, & Gilding."[5]

Wedgwood, keenly attuned to both elite and popular patronage, continued to balance his artistic skill with commercial aspirations; even as he labored to perfect the jasper vases that required no gilding for their dramatic decorative effect, the firm carried on the production of variegated ware into the 1780s. H.J.M.

[1] This vase, one of a pair in the Metropolitan Museum, was published with its mate in Byron A. Born, "Josiah Wedgwood's Queensware," *MMA Bulletin* 22, 9 (May 1964): fig. 7; and Reilly 1989: vol. 1, fig. 464. Reilly noted that the brass gallery moldings, mounted at the base of the plinths, are later in date than the vases. Although the museum catalogued the pair as cream ware, Reilly identified them as white "terracotta" stoneware, a term Wedgwood used alternately with "crystalline" to describe a light-colored, vitrified stoneware body that superceded the use of cream ware for variegated vases after 1772. See Reilly 1989: vol. 1, 70, 341, 357; and Adams 1992: 49.

[2] For this order, see Reilly 1989: vol. 1, 348. The term "sugar dish" appears in a letter from Wedgwood to Bentley, Nov. 26, 1777; see Reilly 1989: vol. 1, 349, pl. C82, figs. 456, 536, 555. Reilly identified this form as Shape 89 in the Wedgwood Factory Vase Drawing Book (fig. 464).

[3] Wedgwood to Bentley, Feb. 23, 1769, E25–18231, Keele University Library, Staffordshire; quoted in Finer and Savage 1965: 71.

[4] Wedgwood to Bentley, Feb. 6, 1769, E25–18227, Keele University Library, Staffordshire; quoted in Finer and Savage 1965: 70.

[5] Wedgwood to Bentley, Feb. 6, 1769, E25–18365, Keele University Library, Staffordshire; quoted in Reilly 1989: vol. 1, 347; and Young 1995: 102–03.

33. LIDDED VASE WITH SATYR MASKS

Josiah Wedgwood (1730–1795), Etruria, Staffordshire
About 1769–80
Cream-colored earthenware with lead-glazed, mottled metallic oxide colors and partial gilding, on a black basalt plinth
13 $\frac{7}{8}$ inches (35.2 cm)
Mark: [on a wafer on the base] *WEDGWOOD & BENTLEY: ETRURIA*
Rogers Fund, 1940
40.68

The bold, sculptural quality and robust gilded ornament of this granitelike cream ware vase differ in character from the light and delicate low-relief ornament with which the name of Wedgwood is normally associated. The source of its design and assertive character is the work of French architect Jean-Charles Delafosse, who was key to the development of the *goût grec* in the 1760s and first half of the 1770s.[1] His work is characterized by luxuriant, ropelike festoons such as those on this vase. It combines the heavy, baroque-inspired ornament typical of the manner with a sleek form, adapted from authentic Greek vase shapes, that reflects the character of later eighteenth-century neoclassicism.

For the design of this vase, Wedgwood and Bentley looked to Delafosse's influential work, *Nouvelle Iconologie historique.* According to a library list drawn up by Thomas Bentley, the firm acquired the *Nouvelle Iconologie* not long after its publication in 1768.[2] One of its plates, an allegorical representation of Germany, includes a bold design for a vase that served as the model for the vessel featured here. The unknown Wedgwood and Bentley designer of this vessel borrowed from Delafosse's example the high looped handles, horned satyr masks, and laurel festoons, altering their relative proportions and balance.

Wedgwood and Bentley were not the only entrepreneurs to recognize the potential of this and other Delafosse engravings as design sources in the vase-crazed culture of the 1770s. The same vase appears in the frontispiece of Francis Vivares's 1770 *New Book of Vases* (see cat. no. 34). But unlike Vivares's faithful rendition, Wedgwood and Bentley's vase has a character very different from that of its prototype. While its ornament clearly derives from Delafosse's image, the full contours of its body and its generously proportioned, round finial, lend this vase a lyricism that contrasts with the severity of Delafosse's architectonic design. In these departures, the present vase makes more explicit reference to classical forms. The body, lid, and curving loop handles reflect the influence of various Greek vase shapes: the Attic neck amphora, the hydria, and the kylix.

For all the liberties its anonymous designer took with its source, the vase is notably cogent and well-integrated. It is to his credit that he was able to absorb the highly distinctive work of Delafosse and create an object with a character of its own. H.J.M.

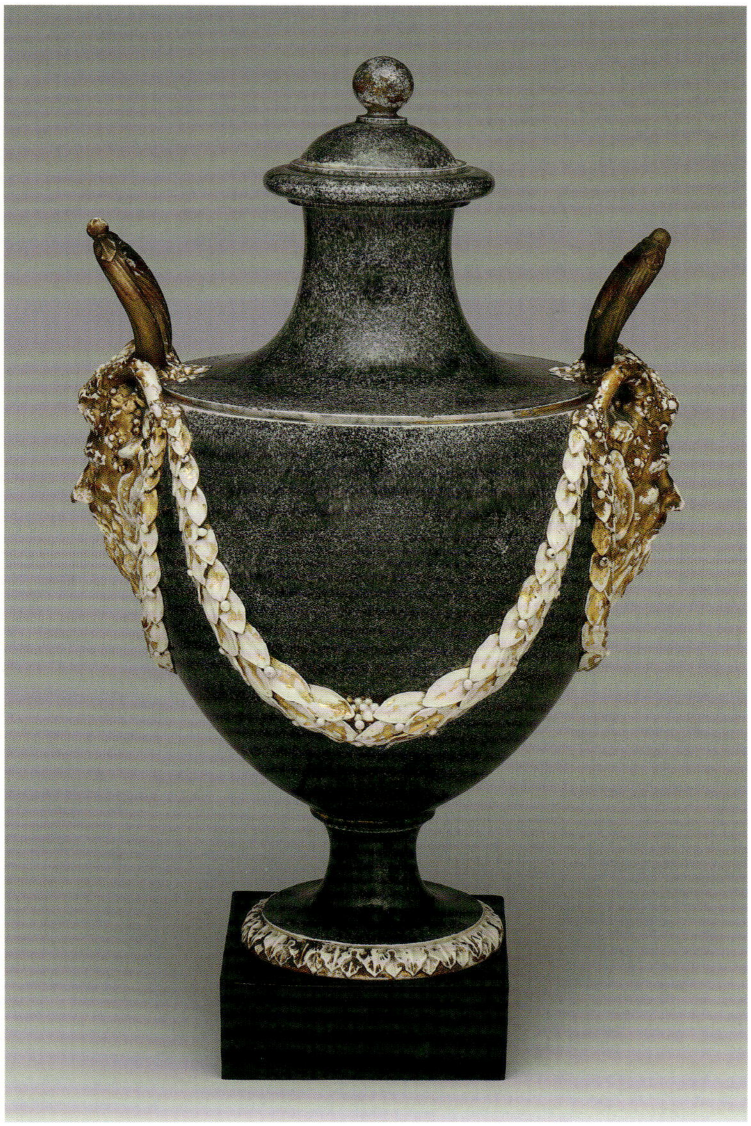

Cat. no. 33

1 The Victoria and Albert Museum, London, has a similar, slightly larger, dark gray, mottled cream ware vase, from about 1770; see Young 1995: cat. E2, pl. 17.
2 Drawn up on August 10, 1770, Wedgwood and Bentley's library list records: "Iconologie Historique par de la Foß[ss]e"; see Johnson 1990: 13–23; and Johnson 1991: 34.

34. DESIGN FOR A VASE

Engraved by Thomas Vivares (1735–1821) after a design by Jean-Charles Delafosse (1734–1789)
1770
Engraving, plate 1 (frontispiece) of *A New Book of Vases*
Plate: 7 ¾ × 5 ¹¹⁄₁₆ in. (196 × 145 mm)
Inscription: [on each plate, bottom left and right] *Delafosse delint./Thos. Vivares Sculpt.*; [frontispiece, bottom center] *A New Book of Vases / Published and Sold by Fran.s Vivares 1770*
Harris Brisbane Dick Fund, 1928
28.94.36 (1)

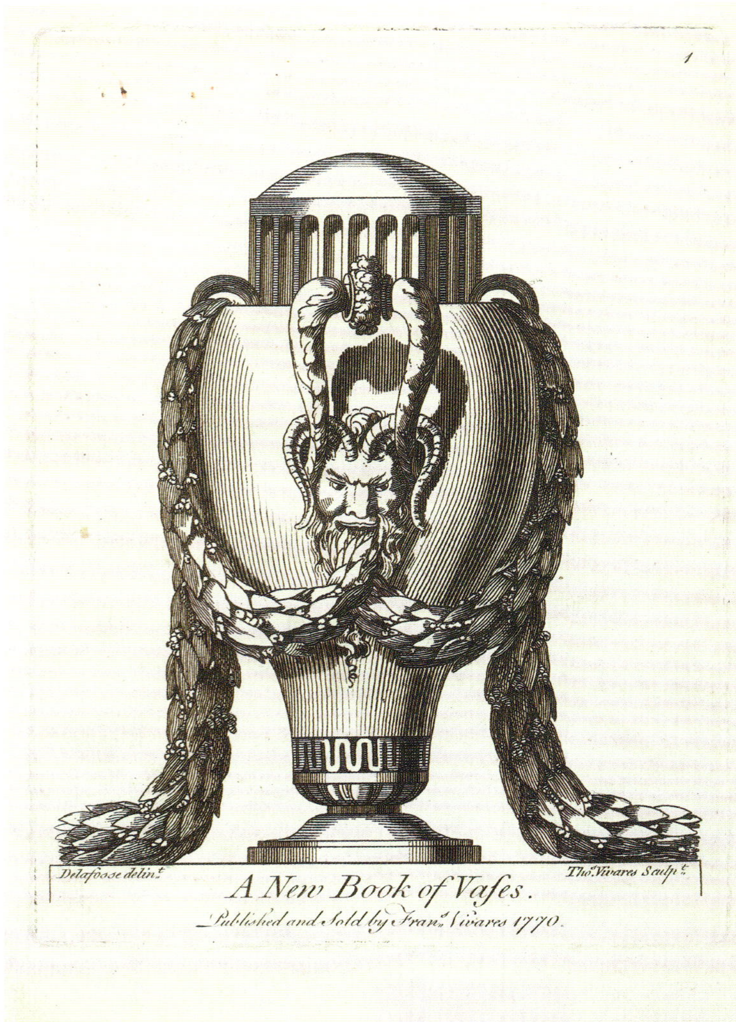

Cat. no. 34

Shortly after the 1768 publication of Jean-Charles Delafosse's *Nouvelle Iconologie historique*, the London-based engraver and publisher Francis Vivares adapted a selection of its plates to produce this collection of six of the French artist's designs (see also cat. no. 10, Fig. 1). Published in 1770, they capitalized on the extraordinary passion for vases that was sweeping across fashionable English society, where "Vases was all the cry."[1] At the same time, the faithful renderings of Delafosse's designs, engraved by Vivares's son Thomas, helped to spread the sober and monumental style of 1760s French classicism to an English audience.[2]

While Delafosse's publication combined bold concepts for architectural features and objects—including doorways, frames, tables, and vases—with emblematic imagery to illustrate broader themes, the younger Vivares isolated the vases from the plates, producing reversed copies of Delafosse's designs which were less than half the size of the originals.[3] A slim collection of practical prototypes, the now-rare *New Book of Vases* was probably inexpensive, intended for use by artists, craftsmen, and other designers.

Thomas Vivares produced accomplished renderings of Delafosse's designs. While he generally followed the originals closely, his discriminating changes contribute successfully to his versions' sober and monumental character.[4] This is amply demonstrated in the frontispiece featured here, based on Delafosse's emblematic representation of Germany (see cat. no. 33). By deepening contrasts, sharpening contours, and especially lowering the center of gravity through subtle shifts in proportion, Vivares imbued his plates with the impression of timelessness that was a key goal of neoclassical art. H.J.M.

1 Josiah Wedgwood to Thomas Bentley, May 1, 1769, E25–18240, Keele University Library, Staffordshire; quoted in Reilly 1989: vol. 1, 363.

2 Although Francis Vivares credited Delafosse for the designs, he probably did so to benefit from Delafosse's reputation as well as the enduring cachet of French design, since it is likely that the publication was not authorized by the artist. In 1771 Vivares also published reversed and slightly altered copies of vase designs by Edmé Bouchardon that had been engraved and produced by Gabriel Huquier in Paris sometime before 1731. Since Vivares did not identify their source, it is more than likely that he pirated them; see Clifford 1978: 168.

3 Five of Vivares's six plates are based on Delafosse's symbolic representations of nations, and the sixth on an allegory of abundance. In order, Vivares's plates correspond to plates 38, 37, 64, 33, 46, and 50 in the *Nouvelle Iconologie* (Delafosse 1768), illustrating respectively Germany, Switzerland, Abundance, Portugal, Sweden, and Thrace.

4 Only plate 2 of the *New Book of Vases* exhibits significant variation from its model: Vivares changed the ram's head mask illustrated in plate 37 in Delafosse 1768 to that of a satyr, and made the wings of the female figure somewhat more elaborate. The vase shown here also lacks the finial illustrated in Delafosse's version of the design.

35. VASE WITH RAM'S HEADS AND HUSK GARLANDS

Humphrey Palmer (act. 1760–78), Church Works, Hanley, Staffordshire
1770–78
Cream-colored earthenware (cream ware), with lead-glazed slip decoration and traces of gilding
9 ½ in. (24.1 cm)
Mark: [bottom of plinth] *MADE BY H. PALMER*
Bequest of Mrs. Mary Mandeville Johnston, 1914
14.102.362

Highly successful pottery entrepreneurs like Josiah Wedgwood and Matthew Boulton shared the lucrative market for ornamental vases with scores of lesser-known competitors in England and Europe. The well-conceived, Roman-inspired design and fine execution of this vase demonstrate why its manufacturer, Humphrey Palmer of the Church Works pottery in Hanley, Staffordshire, was one of Wedgwood's most significant competitors. Even Wedgwood grudgingly recognized the quality of Palmer's wares: "P-s black vases," he wrote in 1769, "I saw one at Mr. Bents. yesterday. The Body is very good—the shape & composition very well.... Upon the whole it was better than I expected—we must proceed or they will tread upon our heels."[1]

Palmer opened his pottery sometime before 1760 and became one of the early successful producers of black basalt ware. By the late 1760s, he was producing both ornamental vases and medallions and useful wares, such as teapots.[2] Based on surviving examples, Palmer's products also included transfer-printed and lead-glazed cream ware, in addition to variegated wares such as this vase. Although he manufactured high-quality pieces and undercut Wedgwood's prices, Palmer was ultimately unsuccessful and went bankrupt in 1778 (see cat. no. 38).[3]

Palmer's ram's-head vase, which he produced in at least two versions, is a thoughtful adaptation of Roman prototypes.[4] In Roman art, carved ram's heads draped with floral garlands typically ornamented marble altars and cadelabra. By the eighteenth century, the motif was frequently applied to vases, as is the case here.[5] The three evenly spaced ram's heads supporting festoons on Palmer's vase specifically reflect three-sided ancient antecedents, such as tripod forms. H.J.M.

1 Wedgwood to Thomas Bentley, Oct. 9, 1769, E25–18266, Keele University Library, Staffordshire; quoted in Reilly 1989: vol. 1, 441–42; and Edwards 1994: 52–53.
2 Although no utilitarian black basalt wares bearing Palmer's mark, "H. PALMER: HANLEY," survive, Wedgwood referred to black teapots by Palmer in letters to Bentley dated Sept. 13, 1772, and Feb. 25, 1778, E25–18403 and E25–18814, Keele University Library, Staffordshire; quoted in Edwards 1994: 220.
3 For the most recent studies of the Church Works, see cat. no. 38, n. 5.

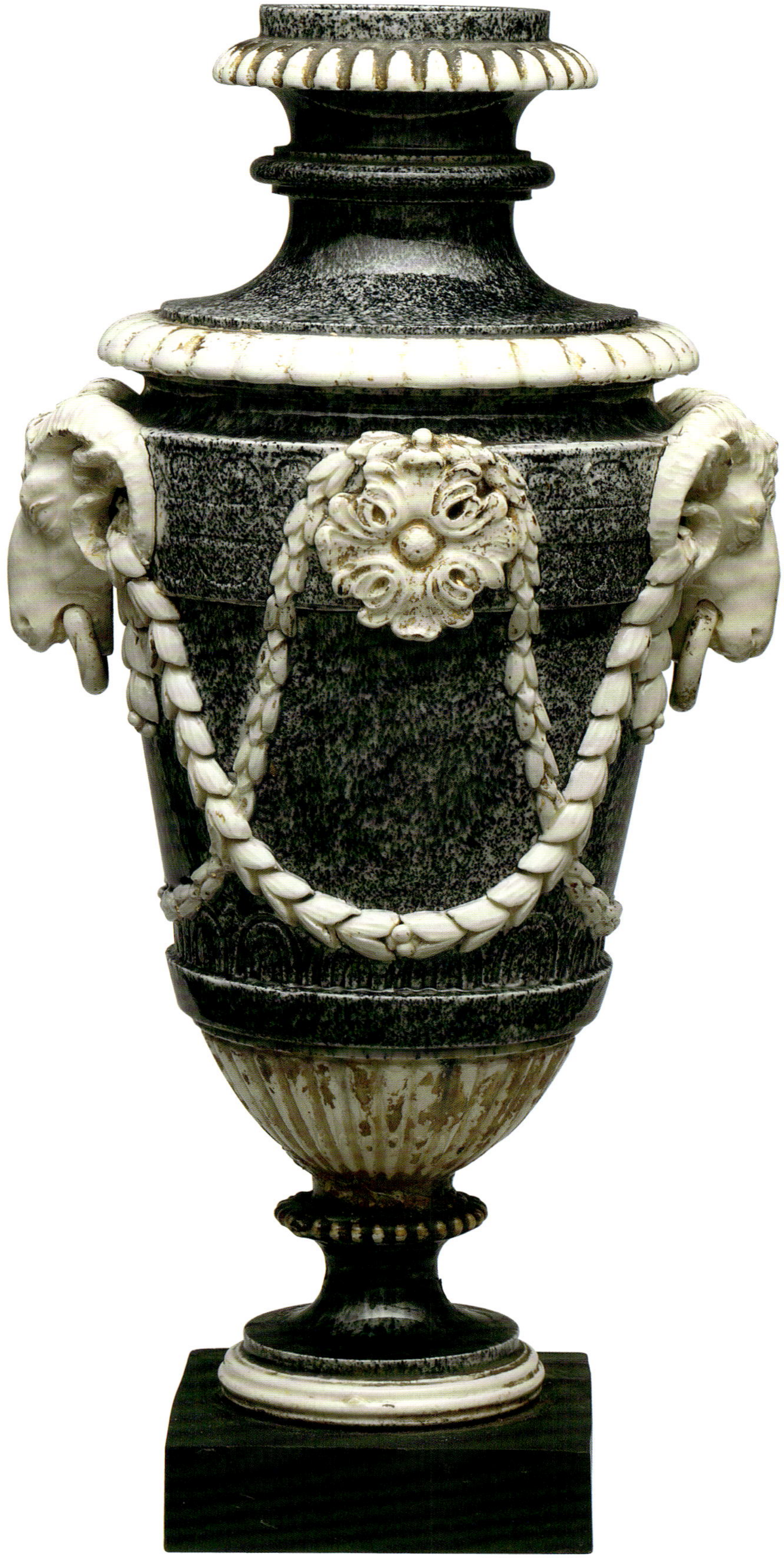

Cat. no. 35

4 In the collection of the City Museum and Art Gallery, Stoke-on-Trent, is a mottled, blue-green ram's-head vase by Palmer, somewhat different in design than that of the Metropolitan Museum's example; see Edwards 1987: pl. 2.

5 Palmer's vase and a corresponding one by the obscure Fenton firm of S. Greenwood (British Museum, London, Franks Coll., K30) relate closely to vases by Robert Adam for the niches of the Porters' Lodges at Syon House, Middlesex; see Grant 1967: pl. 85: 2; Reilly 1989: vol. 1, 414; Edwards 1994: 163. Adam published the Syon interiors in Adam and Adam 1773: pl. 2. About 1780 William Duesbury's Derby factory produced soft-paste porcelain vases with ram's heads that appear to be based on Adam's design; see John Twitchett, *Derby Porcelain*, foreword by Anthony Hoyte (London, 1980): fig. 108.

36. VASE WITH LION-HEAD HANDLES

Josiah Wedgwood (1730–1795), Etruria, Staffordshire
About 1775
Black basalt; stoneware, stained, with molded, engine-turned, and applied ornamentation
16 ¾ in. (42.5 cm)
Mark: [bottom of plinth] *WEDGWOOD & BENTLEY*
Gift of Ferdinand Hermann, 1910
10.85.1

In addition to emulating celebrated antiquities such as the Roman cameo glass Portland Vase (cat. no. 8), or Sir William Hamilton's Greek pottery (cat. no. 5), Josiah Wedgwood and his designers borrowed freely from the work of other artists inspired by the classical past. This restrained black basalt urn, for instance, is one of a group of Wedgwood and Bentley vases with lion handles, all of which relate to plate 2 of Jacques Stella's *Livre de vases*, a 1667 publication that Wedgwood owned and utilized for several of his firm's designs.[1] Wedgwood actively collected prints and publications for source material, "picking up every design, & improvement for a Vase work," with the advice of his partner, Thomas Bentley, who sent him Stella's book in August 1770.[2]

Wedgwood's reliance on print sources was not unusual. The Scottish architect Robert Adam, for instance, probably used the same Stella engraving for his design of a stucco panel at Syon House, near London.[3] Certain related features of the Wedgwood and Adam vases, which are not present in the print by Stella, suggest that the potter may have been influenced by the architect's design. Undoubtedly, Wedgwood would have known it from its 1773 publication in the first volume of Adam's *Works in Architecture.*

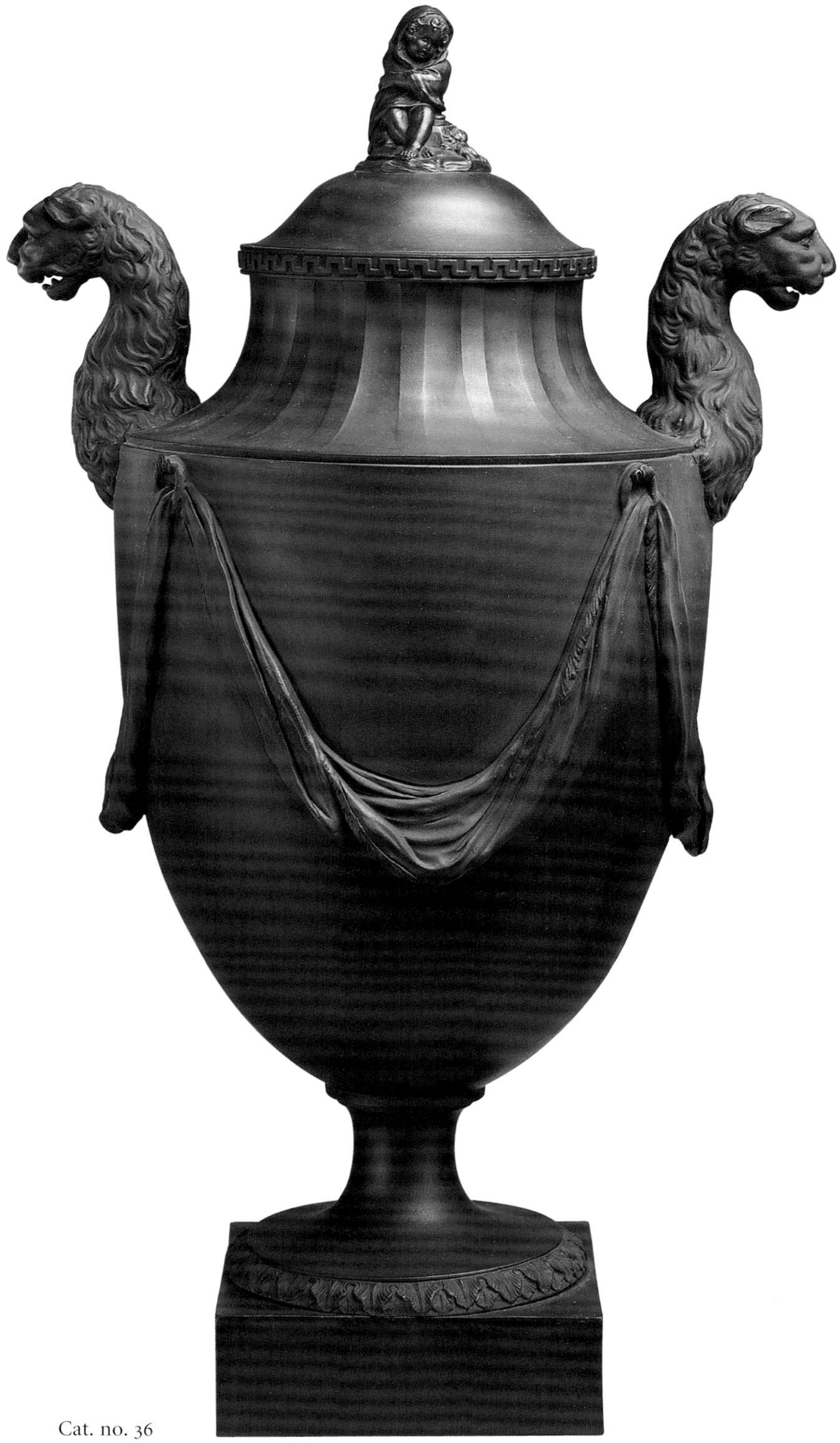

Cat. no. 36

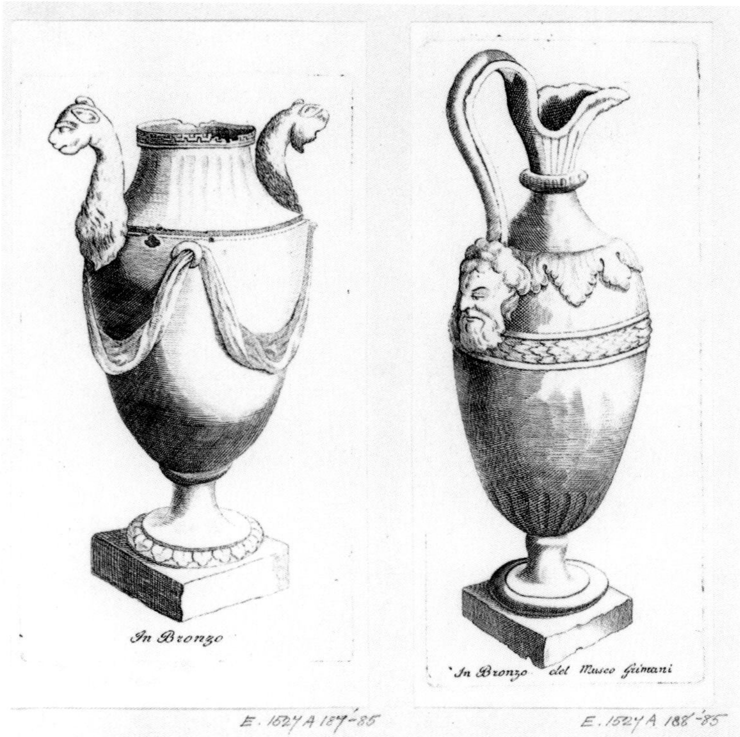

Fig. 1: *Two Vases from the Museo Grimani*, nineteenth century. Italian. Victoria and Albert Museum, London, E. 1527 A. 187–1885 and A. 188–1885. Photograph V&A Picture Library. Note the similarity of the ewer on the right to cat. no. 31.

With its distinctive and unusually long-necked lion handles, the present example also illustrates the extent to which eighteenth-century vases reflected the influence of designs published in the first half of the sixteenth century. Wedgwood's and Adam's designs, and Stella's as well, are dependent on a set of twelve engravings by Agostino Veneziano, which had been well known since its publication in 1531. Such sixteenth-century models had enduring appeal; several objects featured here illustrate their influence (cat. nos. 17, 31).[4]

With the exception of its lid, the present lion-handled vase is also virtually identical to one illustrated in an anonymous set of nineteenth-century Italian vase prints (see Fig. 1).[5] While the series appears to illustrate antiquities, scholars have suggested that it may in fact represent Wedgwood designs, pointing out that the plates depict recognizable Wedgwood forms and that, by the nineteenth century, Wedgwood vases were occasionally mistaken for antiquities themselves.[6] If this is the case, it is a testament to the persuasiveness of Wedgwood's neoclassicism that in some instances his pieces came to define antiquity. With the outstanding success of his firm, Wedgwood contributed as much to the classical idiom as he borrowed. H.J.M.

1 Jacques Stella, *Livre de vases, Inventé par Mr. Stella, Chevalier, et peintre du Roy*... (1667). This volume, etched by Stella's niece and published posthumously by his sister, reproduces fifty of his designs. See Reilly 1989: vol. 1, 397, which also illustrates the Metropolitan Museum's vase (fig. 534); and Young 1995: 104–06. Grant 1967: pl. 16:2, illustrates a similar, unmarked vase that differs primarily in the treatment of its drapery swags. Several variations of the Wedgwood and Bentley lion-handled vase have been published; see Grant 1967: pls. 19:3, 20:2; Geoffrey Wills, *Wedgwood* (Secaucus, N.J., 1980): 52; Reilly 1989: vol. 1, figs. 530, 572; Adams 1992: 64–65, 204; and Young 1995: cat. E11. The Church Works pottery, under Humphrey Palmer, also manufactured a related design; see Grant 1967: pl. 65:1. Timothy Clifford demonstrated that at least four of Stella's concepts inspired Wedgwood vases; see Clifford 1978: 164–65, pls. 71–73.

2 Wedgwood to Bentley, June 6, 1769, E25–18150, Keele University Library, Staffordshire, quoted in Reilly 1989: vol. 1, 347.

3 Both Timothy Clifford and Elizabeth Adams pointed out the similarity between Adam's design and Wedgwood and Bentley's lion-handled vases. See Clifford 1978: 165, pl. 72; and Adams 1992: 64–65. For Adam's publication of his interior designs for Syon House, see Adam and Adam 1773: pl. 7. Similar vases can be seen on the mantel in an 1784 interior by Michelangelo Pergolesi featured here (cat. no. 81).

4 Sixteenth-century vase series were reissued and pirated until well into the seventeenth century and beyond; see David Landau and Peter Parshall, *The Renaissance Print, 1470–1550* (New Haven/London, 1994): 299; for vase prints specifically, see Miller 1999: 244–54; and Timothy Clifford, "Polidoro and English Design," *Connoisseur* 192 (Aug. 1976): 282, 285, 287–89.

5 Illustrated in Young 1995: cat. E14.

6 Young 1995: 107. In the series, vases are depicted with wear and damage. Inscriptions identify some as bronze, and others as part of the collection of the Museo Grimani, Venice, which housed antiquities collected in the sixteenth century by Cardinal Domenico Grimani and his nephews Cardinal Marino Grimani and Giovanni Grimani. The Grimani antiquities now form part of the collection of the Museo Archeologico, in the Palazzo Grimani, Venice.

37. MEDALLION VASE WITH FEMALE MASKS

James Neale (1740–1814), Church Works, Hanley, Staffordshire
About 1778–80
Black basalt, with molded and applied ornament
9 5/8 in. (24.4 cm)
Mark: [on base] *I. Neale Hanly*
Rogers Fund, 1932
32.127.2

Many examples survive of this well-balanced form, with its applied central medallion and paired female masks draped with festoons of bellflowers. Bearing the marks of a succession of Staffordshire firms that grew out of the partnership between Humphrey Palmer and James Neale, numerous variations of this type attest to the enduring appeal of its design in the 1770s and 1780s.[1]

Inspired by the necked and footed shapes of ancient Greek vessels like the amphora and hydria, this black basalt vase represents the most characteristic and easily recognized product of the Church Works pottery at Hanley (see also cat. no. 38). Introduced during Palmer's ownership of the firm (about 1760–78), the medallion vase form continued to be produced by the Church Works under Neale and into his partnership with Robert Wilson, which ended in 1792. Featuring the mark "I. Neale Hanly," this vase was made sometime during Neale's ownership.

As in this example, Church Works medallion vases typically feature a coin or cameo-like portrait "hanging" from the frieze with loop and rivet, reflecting a contemporary passion for collecting such medals among the learned and fashionable alike. The scenes and portraits on Church Works medallions, such as the Roman emperor seen here, Apollo flaying Marsyas, or the Minotaur derive from ancient coins and later Renaissance medals. These would have been familiar to eighteenth-century artists and manufacturers through publications such as Bernard de Montfaucon's illustrated 1719 encyclopedia of ancient cultures.[2] In the space on the reverse of the vase, corresponding to the medallion, is a rosette.

Also characteristic of Church Works medallion vases are this example's lateral grips, formed as female masks, their veils drawn back, eyes almost closed, and lips parted as each twists to her left.[3] Scholars usually describe them as Cleopatra, associating the design with a vase modeled for the firm in 1769 by Jean Voyez, a noted sculptor who had worked for Wedgwood and briefly for Palmer at Church Works.[4] Voyez's design features full torsos, contorted as they struggle with a snake, which identifies them more clearly with the Egyptian queen who committed suicide by allowing an asp to bite her. Although the designs are related, Voyez's vigorously modeled, restless conception stands in sharp contrast to the restrained neoclassicism of this medallion vase. H.J.M.

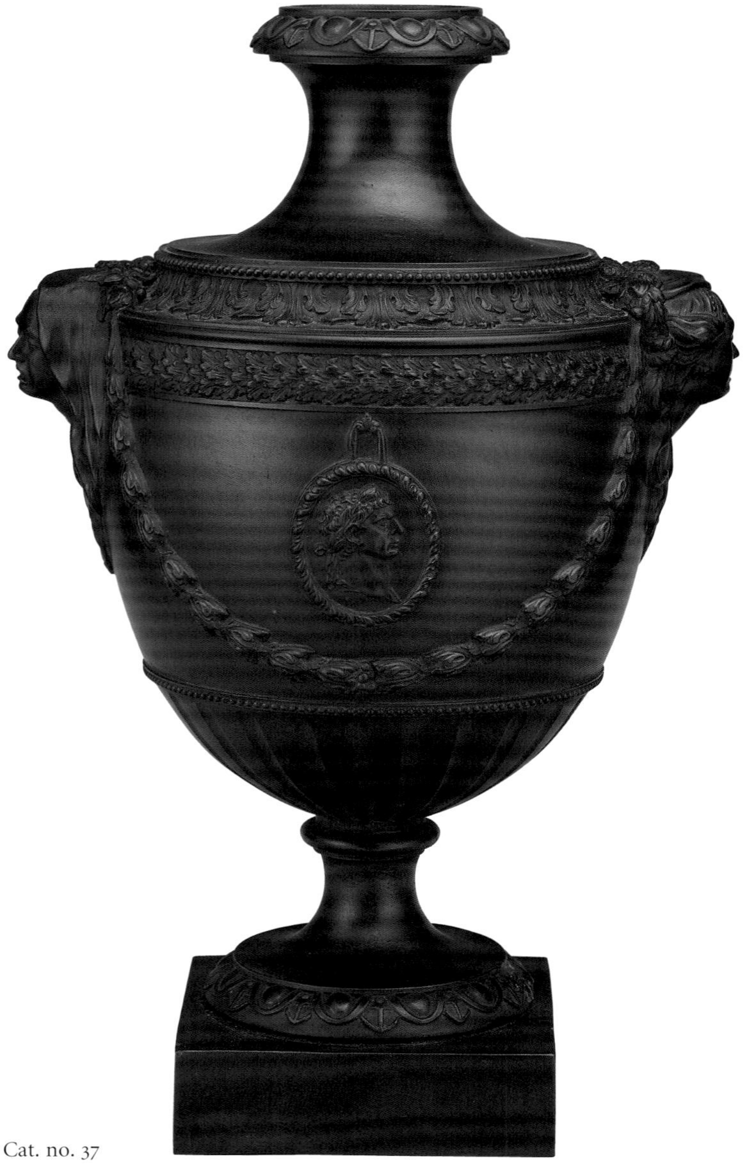

Cat. no. 37

1 Examples of this form survive with the following Church Works marks: "H Palmer Hanley"; "I. Neale Hanly"; and "Neale & Co." See for example Grant 1967: pls. 66, 68:1, 70:2, 71, figs. 1–3; Reilly 1989: vol. 1, fig. 558; and Edwards 1994: figs. 315–16, 340. The incidence of the Neale and Company mark on vases is rare; see Edwards 1994: 207, 215. The Church Works pottery manufactured several related designs, including egg-shaped medallion vases with Cleopatra-like female masks, and friezeless medallion vases with male masks or paired griffin-head/bird handles. See Grant 1967: pls. 68:2, 70:1, 70:3; and Edwards 1994: fig. 317. Wedgwood and Bentley also produced medallion vases in great numbers; see Edwards 1987: 112. William Adams of Greengates produced black basalt vases with paired female heads similar to the Palmer/Neale examples; see Reilly 1989: vol. 1, fig. 558.

2 See cat. no. 31, n. 7.

3 Such masks also appeared singly on ewers manufactured by the Church Works pottery; for an unmarked ewer of this kind attributed to the pottery, see cat. no. 38.

4 See for example Edwards 1987: 113.

38. EWER WITH FEMALE MASK

Attributed to Church Works, Hanley, Staffordshire
1778–92
Black basalt
11 ⅛ in. (28.3 cm)
The Friedsam Collection, Bequest of Michael Friedsam, 1931
32.100. 353

Long attributed to Wedgwood, this unmarked black basalt ewer was probably produced at the Church Works pottery, in Hanley, Staffordshire, one of Wedgwood's chief competitors in the production of ornamental black basalt vases.[1] Although it resembles ewers manufactured by Wedgwood (see cat. no. 31), the present piece varies in proportion and ornamentation, exhibiting several features not seen on marked Wedgwood examples.[2] For instance, while some Wedgwood versions include applied ornament at the collar and fluting near the base of the body, the foliate collar of the unmarked vessel is more elaborate than the corresponding palmette motif typical of Wedgwood examples. Similarly, the scrolling floral band on the ewer featured here varies from the relief ornament that invariably encircles Wedgwood examples: a band of laurel leaves and berries. The most obvious difference can be seen in the handles. Like Wedgwood pieces, this one features a high loop and terminates in a mask on the body, but the acanthus handle of the present vessel, with its distinctive S-curve and female mask, differs from Wedgwood's typical strap handle and satyr mask.

This unmarked ewer is actually closer to an object at the Nelson-Atkins Museum of Art, Kansas City, Missouri.[3] A smaller ewer, it features an S-curved handle, female mask, elaborate foliate collar, and band of scrolling flowers. Its mark, "I. Neale Hanly," identifies it as the product of the Hanley Church Works when it was owned by James Neale.

A London retailer, James Neale was the agent and business partner of Humphrey Palmer, the Staffordshire potter with whom Wedgwood felt a keen and bitter rivalry (see cat. no. 35). Palmer, who established the Church Works pottery sometime before 1760, was producing a high-quality black basalt clay body, suitable for ornamental wares, by the end of the decade. After business failures, the pottery was purchased by Neale in 1778. The black basalt vases Church Works produced were marked with Neale's name until 1792, when he sold the firm to his partner and factory manager, Robert Wilson.[4]

Competing for the same market, Wedgwood, Palmer, and later Neale, produced many similar wares, but rarely identical forms.[5] For this reason, we can safely attribute the Metropolitan Museum's ewer to Church Works. Dating it is more difficult,

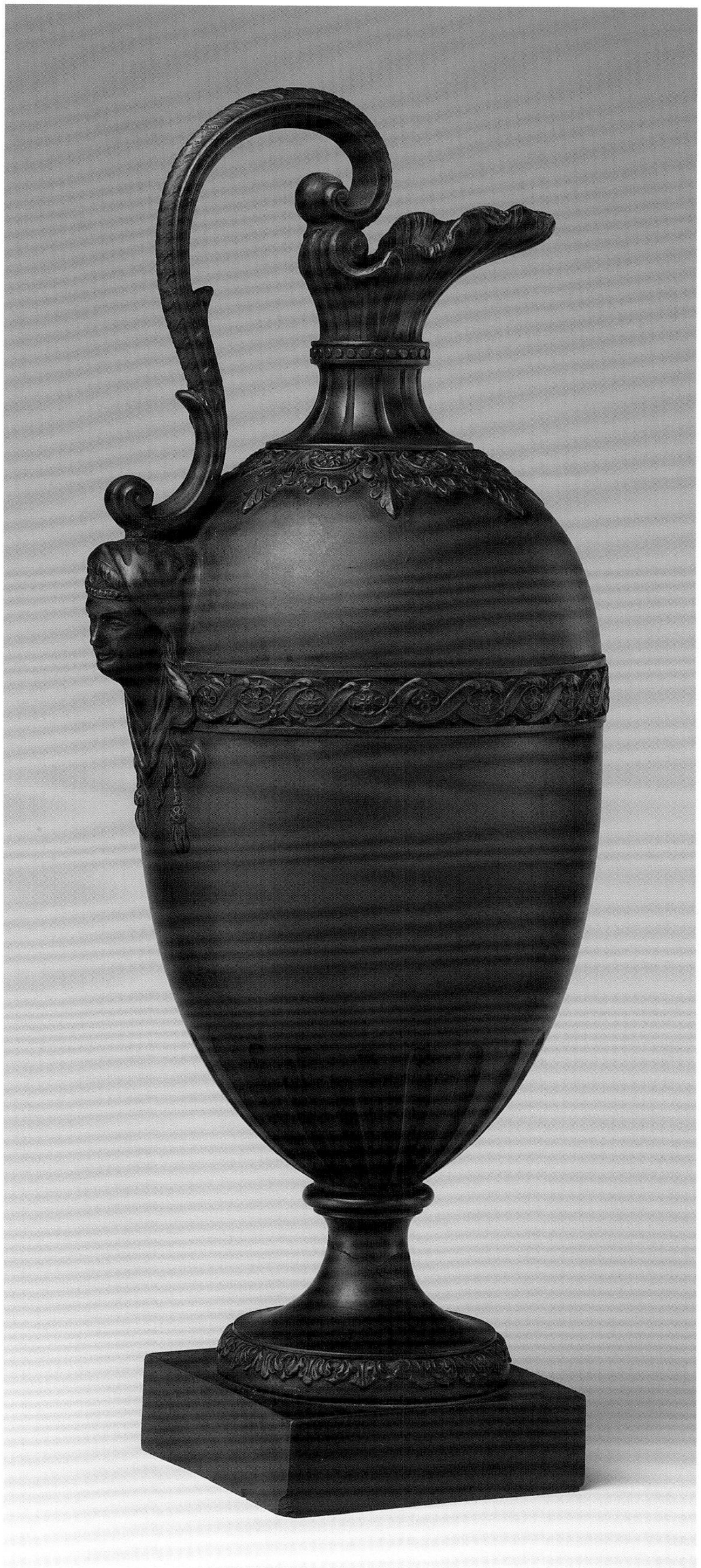

Cat. no. 38

however, since Neale is known to have produced forms introduced earlier by Palmer.[6] Indeed, the latter did not always mark his black basalt ware; of the utilitarian black ware he is known to have manufactured, no examples marked with his name survive.[7] Possibly, Neale-era wares were made at Church Works after he ceased to be involved with the pottery. It is only with caution, therefore, that one can ascribe the present ewer to Church Works during the period of Neale's stewardship. H.J.M.

1 Filler in the base obscures any marks this vase might have. It is one of eight pieces of eighteenth-century English pottery identified as Wedgwood, given to the Metropolitan Museum by the estate of Michael Friedsam in 1931; see "The Michael Friedsam Collection," *MMA Bulletin* 27, 2, special sec. 2 (Nov. 1932): 63.

2 See for example Grant 1967: pls. 7: 3, 10:2; Ross E. Taggart, *The Frank P. and Harriet C. Burnap Collection of English Pottery in the William Rockhill Nelson Gallery*, rev. ed. (Kansas City, Mo., 1967): cats. 575, 582; Reilly 1989: vol. 1, fig. 544; Edwards 1994: fig. 25; and Young 1995: cat. D41.

3 Taggart (note 2): cat. 601; Edwards 1994: fig. 318.

4 Marks on Church Works pottery of the Neale period (1778–92) include "I. Neale Hanly," "Neale & Co." and "Neale & Wilson"; see Edwards 1994: 207, 215. Rival accounts of the careers of Palmer and Neale are found in Grant 1967: 222–53; and Meteyard 1970: vol. 2, 137–38, 196–203, 254, 391, 394–95. More recent and more objective information on the Church Works pottery under Palmer and Neale can be found in Edwards 1987 and Edwards 1994: 51–54, 207–16, 217–21. Ongoing studies of archeological excavations in Hanley will undoubtedly reveal new information.

5 With the exception of some plaques and medallions, none of Palmer's wares is identical to Wedgwood designs.

6 See for example Edwards 1987: 57; and Edwards 1994: fig. 340.

7 Edwards 1994: 220–21.

39. WINE EWER, "SACRED TO BACCHUS"

Josiah Wedgwood (1730–1795), Etruria, Staffordshire
About 1800
Black basalt, with molded and applied ornament
15 3/4 in. (40 cm)
Marks: [on bottom of plinth] impressed *WEDGWOOD: A:*>; [on base] incised *236*
Gift of Martin Birnbaum, 1943
43.133.1

A decorative vessel ostensibly meant for wine, Wedgwood's ewer incorporates appropriate Dionysian motifs: festooned grapevines and leaves drape each side of the vase's body, a goat's-head mask sits on the rim of the shoulder, and a satyr embraces the neck. Regarding its theme, Josiah Wedgwood's nephew Thomas Byerley described the ewer as "Sacred to Bacchus," the wine god's Roman counterpart.[1] Similar in form to this vessel, a pendant water ewer, "Sacred to Neptune," features iconography associated with the ancient god of the sea.

The design of both vases is thought to derive from a pair of plaster vessels by Sigisbert-François Michel, exhibited at the 1774 Paris Salon. This pair, in turn, may in have been based on designs by the artist's brother, the eminent sculptor Claude Michel, better known as Clodion.[2] Ewers of a similar design appear on a list of the 1885 sale of the effects of French bronze founder Charles Crozatier. Said to be after Clodion, they are described as having necks in the form of seashells, from which, on one, a faun and, on the other, a triton appear to drink—an apt description of the Wedgwood pair.[3]

The designs made their way to Wedgwood's pottery as plaster models. In the eighteenth century, plaster casts and copies of sculpture and ornament were an extremely important means of design dissemination, second only to prints. Although they occasionally engaged artists to execute original designs, pottery owners such as Wedgwood and Humphrey Palmer, looking for ready forms to reproduce, often relied on London plaster dealers to supply them with models of both admired antiquities and newer designs, including vases.[4] One such dealer, John Flaxman, father of the renowned sculptor of the same name, sold the models for this ewer and its pendant to Wedgwood in 1775. Flaxman's plasters, which he either copied or cast from French bronze or cream ware versions of Sigisbert-François Michel's design, survive today in the collection of the Wedgwood Museum, Barlaston, Stoke-on-Trent, along with presentation drawings for the pair, also attributed to the elder Flaxman.[5]

Enduring in its appeal, this ewer was produced over many decades in various materials, including buff-colored cane ware,

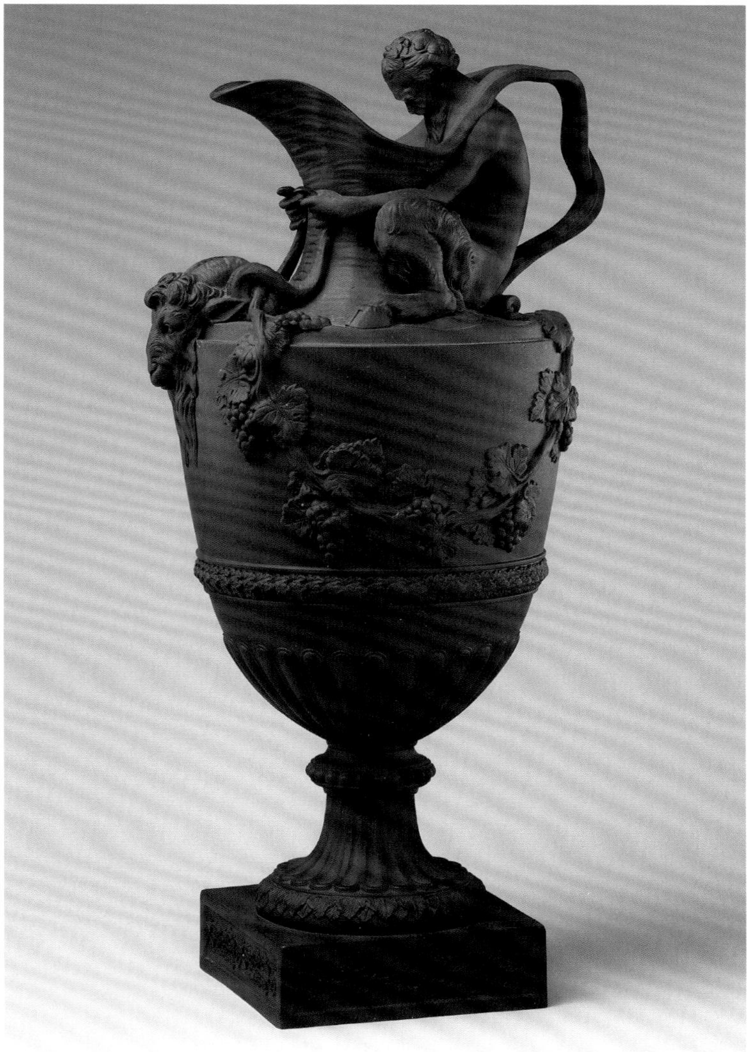

Cat. no. 39

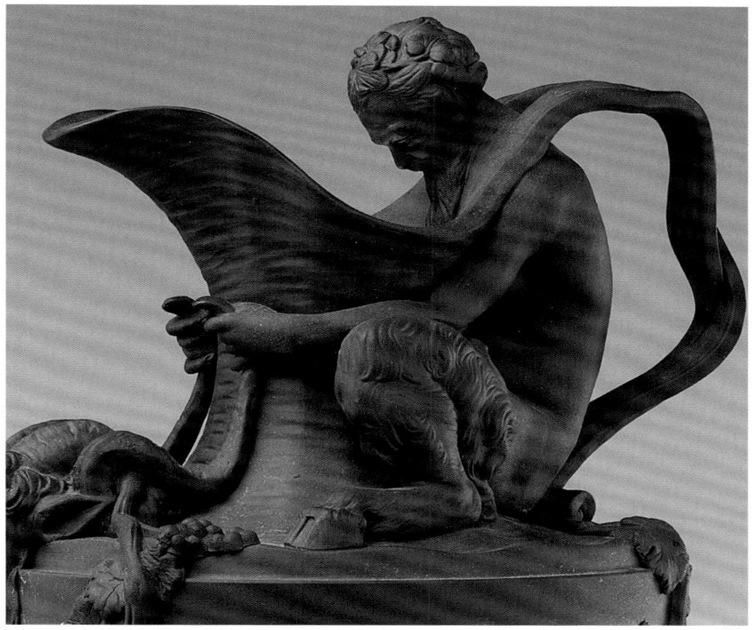

Detail of cat. no. 39

solid jasper, and jasper dip. Black basalt versions such as this one were made from the time the form was first introduced in the later years of Wedgwood's partnership with Thomas Bentley, until the twentieth century.[6] Known as number 236 in the Wedgwood Shape Books, the form was also made with several variations: in the banding, radiating ornaments on the belly, and in the choice of foot and plinth.[7] Unlike many other surviving examples, the present ewer conforms quite closely to the gadrooned design illustrated in Flaxman's presentation drawings. The mark "Wedgwood" on this ewer means that it almost certainly dates after Bentley's death in 1780, as earlier examples bear the name of the Wedgwood and Bentley partnership. H.J.M.

1 Thomas Byerley to Peter Swift, Jan. 2, 1782, Ms. 13–12452, Keele University Library, Staffordshire; cited in Reilly 1989: vol. 1, 408, fig. 556.

2 Reilly 1989: vol. 1, fig. 556; and Anne L. Poulet and Guilhem Scherf, *Clodion, 1738–1814* (Paris, 1992): 182. Another possible source were similar designs by Pierre Gouthière and François-Thomas Germain; see Pierre Verlet, *Les Bronzes dorées françaises du XVIIIe siècle* (Paris, 1987): fig. 235; Young 1995: cat. E31; and Peter Hughes, *The Wallace Collection Catalogue of Furniture* (London, 1996): vol. 3, 1411. According to Hughes, a description of stock in Gouthière's workshop from 1767 suggests that Germain produced similar vases.

3 Poulet and Scherf (note 2): 182.

4 For a detailed discussion of the London trade in plaster models, see Timothy Clifford, "The Plaster Shops of the Rococo and Neo-classical Era Britain," *Journal of the History of Collections* 4, 1 (1992): 39–66.

5 Reilly 1989: vol. 1, fig. 556; and Young 1995: 112. Wedgwood's order from Flaxman, signed as a receipt by the younger Flaxman, is in the Wedgwood Archives. See Mar. 25, 1775, Ms. Invoice L1–204, Keele University Library, Staffordshire. It is reprinted in full in Clifford (note 4): 54. Flaxman's plasters are housed with the Wedgwood Archives; the presentation drawing for the water ewer (no. 4548, Wedgwood Museum, Barlaston, Stoke-on-Trent) is illustrated and catalogued in Young 1995: cat. E31.

6 For illustrations of varied examples of this pair of ewers, see Reilly 1989: vol. 1, pl. 182, figs. 556, 994a; Adams 1992: cats. 130, 280, 298, 423, pl. 46; Edwards 1994: fig. 45; Young 1995: cats. B57, E32; Emmerson 2000: 22, fig. 22.

7 Both ewers relate to shape 236 in the Shape Number One Book; see Reilly 1989: vol. 1, fig. 994a; and Young 1995: 112. The number "236" is incised in the base of surviving examples of the form, including the Metropolitan Museum's example; see Adams 1992: cat. 280.

40. PAIR OF VASES WITH SERPENT HANDLES

Josiah Wedgwood (1730–1795), Etruria, Staffordshire
About 1787
Vitrified stoneware (jasper ware)
Each 15 ¾ in. (40 cm)
Gift of Frank K. Sturgis, 1932
32.95.8, .9

These two matching, Greek-style amphorae, with tall, curling handles, are made from solid blue jasper with molded white jasper and sprigged relief decoration. The pair is replete with neoclassical ornaments: twin serpents wrap themselves around the gracefully arched handles, pierced lancet-leaf decoration under a guilloche strap encircles the neck, a band of laurel leaves on the shoulder, flattened petals ring the vessel's base, acanthus-leaf tips surround the rim of the foot, and fret-work panels adorn each side of the square plinth.[1]

Such motifs contrast with the more baroque-style, figural scenes featured on each of the two sides of the vases: on one, the goddess Venus in a chariot drawn by swans; and, on the other, cupids attending the swans, taking off their harnesses and removing the vehicle. These crisp bas-relief designs are listed in the Wedgwood Ornament Catalogue of 1787 as subject number 245; they can be found as well on factory plaques and other similarly shaped vases.[2]

A composition by the principal painter at the court of Louis XIV, Charles Le Brun, has been suggested as the inspiration for the mythological vignettes on these pieces, but the exact source has not been determined.[3] Le Brun himself may have adapted a painting by Raphael of Venus in her chariot for a tapestry design or another decorative purpose.[4] Such transfers of ancient iconography over the course of several classical revivals were common among the practitioners of neoclassicism. In this case, models of the Renaissance and baroque periods illustrate how eighteenth-century designers working in the style mined all possible sources for ideas.

While the jasper body was new in 1787, the classical style of such pieces as those featured here was not unfamiliar to Wedgwood's designers and modelers. A similar cream ware model, with surface marbling and gilding on a sprigged leaf ornament, but without mythological scenes, was made as early as 1772.[5] Related versions of the amphora shape, featuring different mythological subjects, were produced by the factory in cream ware, black basalt, green jasper, black jasper dip, and later green jasper dip.[6] For these clay bodies, Wedgwood offered an array of variations in decoration: serpent-entwined or plain handles, a number of choices for the central subject, interchangeable ornamental motifs and molded bands. By making available different sizes and types of clay bodies, and a variety of decorative schemes, Wedgwood assured the success of his vases, as he presented products to suit every fashionable taste and permitted his customers greater freedom than they had enjoyed previously in decorating their homes.

The delicate "Chinese" fretwork applied to the present vases' plinths also appear on other Wedgwood vessels, and on bough pots in other shapes.[7] The simple design may be based on a frieze from plate 1 of Giovanni Battista Piranesi's 1765 *Della introduzione e del Progresso delle Belle Arti*. The pattern was copied in Pierre d'Hancarville's illustrated volumes of Sir William Hamilton's vase collection, with which Wedgwood was quite familiar.[8]

The fact that Wedgwood also featured the snake-handled amphora vase design in an illustration in his factory catalogue of 1787 further demonstrates the potter's deliberate attempt to popularize such antique forms and create an audience for his designs.[9] Operating in a fiercely competitive market and without the benefit of royal or state sponsorship, such as Sèvres or other continental porcelain manufactories enjoyed, Wedgwood had to rely on innovative marketing strategies as much as on new clay bodies, technology, and fashionable forms to garner a greater customer base. His mix-and-match choices, illustrated catalogues, and company-owned showrooms, frequented by aristocrats and other members of genteel society, were remarkable and daring tactics for this time, although they may seem commonplace today. Not surprisingly, this approach ultimately contributed to the widespread commercial success of Wedgwood's neoclassical wares.[10] C.H.

1 The shape of this vase is listed as no. 266 in the factory's Shape Number One Book and is first recorded in the factory Oven Books on Sept. 6, 1783, as "266, 4 [blue & white jasper vases] snaked round handles 17 ins."; black versions are recorded on May 8, 1784. See Emmerson 2000: 23, fig. 24. The snake handles are based on late sixteenth-century Italian majolica; see *Wedgwood Trio*, ed. Harry M. Buten (Merion, Pa., 1967): 203–04, pl. 23; and Grant 1967: 71, fig. 3. A pair of such vases, of about 1786, are marked "H," which may indicate that the modeler was William Hackwood; see Adams 1992: 119.

2 Robin Reilly, *Wedgwood Jasper* (New York, 1972): 235; and Adams 1992: cat. 281, pls. 98–99. The same Venus chariot reliefs are illustrated on a smaller, twelve-inch black dip jasper vase of about 1790, featuring scrolled white jasper handles rather than snakes; see Reilly 1994: 54, pl. 39. The piece, one of three that comprise a garniture, is also marked "H" on the plinth.

3 Reilly 1994: 235. Adams 1992: cat. 281, pls. 98–99, also attributed the design featured on plaques to Le Brun.

4 George Leland Hunter, *The Practical Book of Tapestries* (Philadelphia/London: 1925): 150.

5 See Wolf Mankowitz, *Wedgwood* (London, 1953): fig. 48.

6 See *A Catalogue of a Collection of Plaques, Medallions, Vases, Figures, & c., in Coloured Jasper and Basaltes: Produced by Josiah Wedgwood...at Etruria, in the*

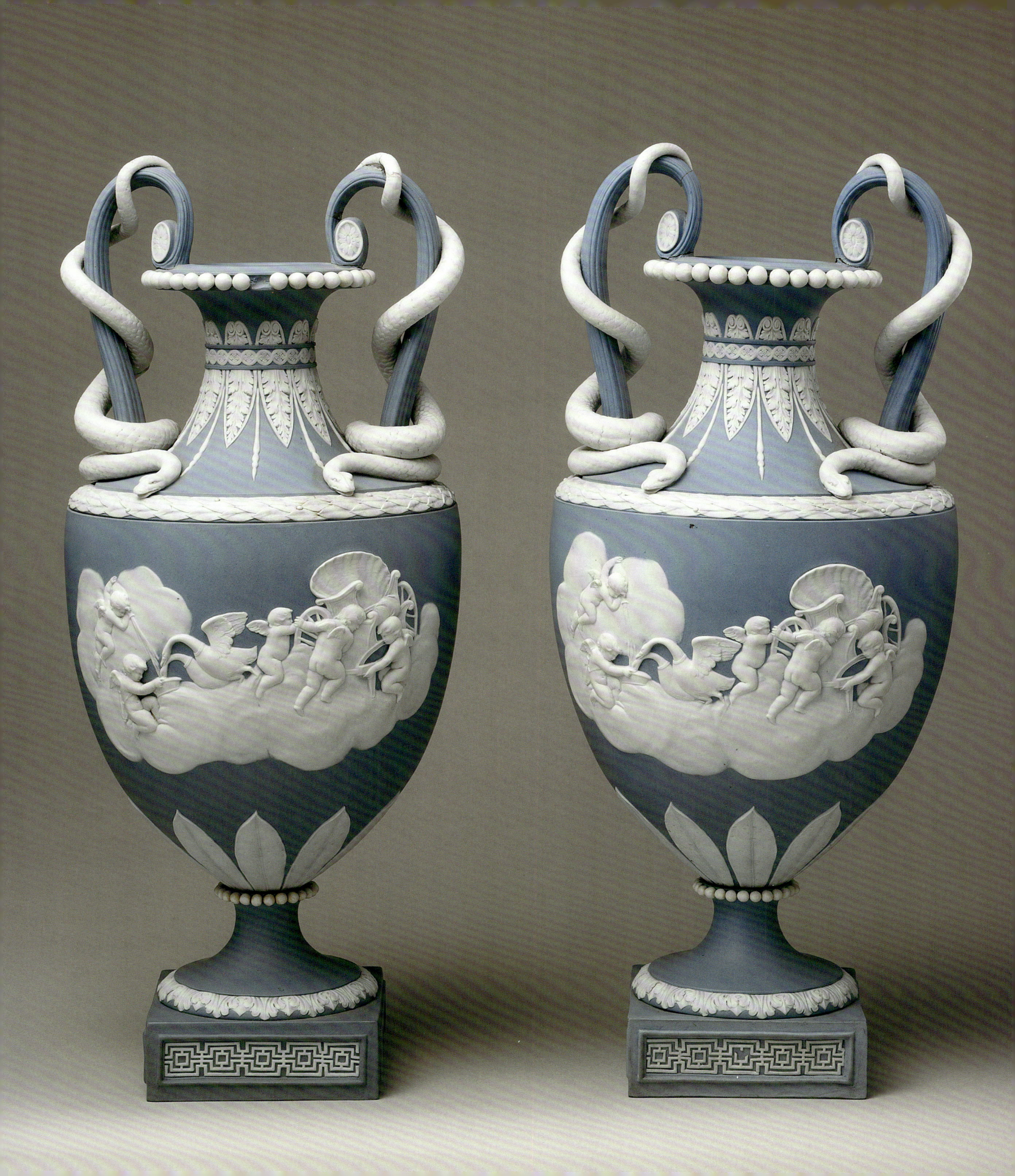

County of Stafford, 1760–1795... (Edinburgh, 1901): fig. 136; *Early Wedgwood Pottery Exhibited at 34 Wigmore Street London W. I* (Barlaston, Stoke-on Trent, 1951): 2 (fig. 4), 25 (fig. 69); Frederick Rathbone, *Wedgwood by Rathbone: A Reprint in its Entirety with an added index of Old Wedgwood by Frederick Rathbone...*, ed. Henry M. Buten (Merion, Penn., c. 1968): pls. 12 , 15; and Reilly 1989: vol. 1, 650, fig. 991. Other Wedgwood examples in the same shape are illustrated in Adams 1992: 118–19, pls. 98–99. A similarly shaped, green dip jasper vase, of about 1790, features a "Sacrifice to Cupid."; see Reilly 1994: 202, pl. 275.

7 Identical versions of this plinth survive in the Lady Lever Art Gallery, Liverpool (LL 1373 and LL 1374), and the Wedgwood Museum, Barlaston, Stoke-on-Trent, Staffordshire. For an illustration of the latter, see David Buten, *18th-Century Wedgwood: A Guide for Collectors and Connoisseurs* (London, 1980): 138, fig. 129; and Reilly 1994: 35, fig. 20. The same fret design is found on a hexagonal plinth on a green and white jasper dip bough pot of about 1786, now in Nottingham Castle. See Nancy H. Ramage, "Piranesi's Decorative Friezes: A Source for Neoclassical Border Patterns," *Ars Ceramica* 8 (1991): 18; and Reilly 1994: pl. 54.

8 For Piranesi's sources, see Ramage (note 7): 14–15. For the fret design in Hamilton/d'Hancarville 1767–74 , see vol. 1, pl. 88; vol. 2, French title page, pls. 41, 84; and vol. 3, pls. 56, 130.

9 Reilly 1994: 35, fig. 20.

10 For a detailed discussion of Wedgwood's strategies, see Neil McKendrick, "Josiah Wedgwood and the Commercialization of Potteries," in idem, *The Birth of a Consumer Society: The Commercialization of Eighteenth-Century England* (London, 1982): 34–99.

41. VASE WITH CUPID FRIEZE

Josiah Wedgwood (1730–1795), Etruria, Staffordshire; probably modeled by Henry Webber (1754–1826) or Giuseppe Angelini (1742–1811)
About 1790–1800
Vitrified stoneware (dipped jasper ware)
14 in. (35.6 cm)
Mark: *WEDGWOOD*
Rogers Fund, 1910
10.101.2

The sprigged, or applied, ornamentation on this wide-mouthed vase—anthemion band, bearded masks, cupids, laurel sprays, swags, and medallions—illustrates the manner in which Josiah Wedgwood successfully adapted classical figurative and architectural motifs for his vase designs in jasper ware. He repeatedly copied classically inspired mythological scenes such as this, using wax models created by his factory for jasper ware plaques and tablets.[1]

The vase depicts a procession of dancing cupids, or *erotes*, playing instruments and carrying torches below a row of bearded masks. The same scene appears on a 1795 Wedgwood jasper dip vase (Birmingham [Alabama] Museum of Art) in a similar shape; it has a lid decorated with white jasper foliage and berries.[2] The molded ledge on the inside of the present example indicates that

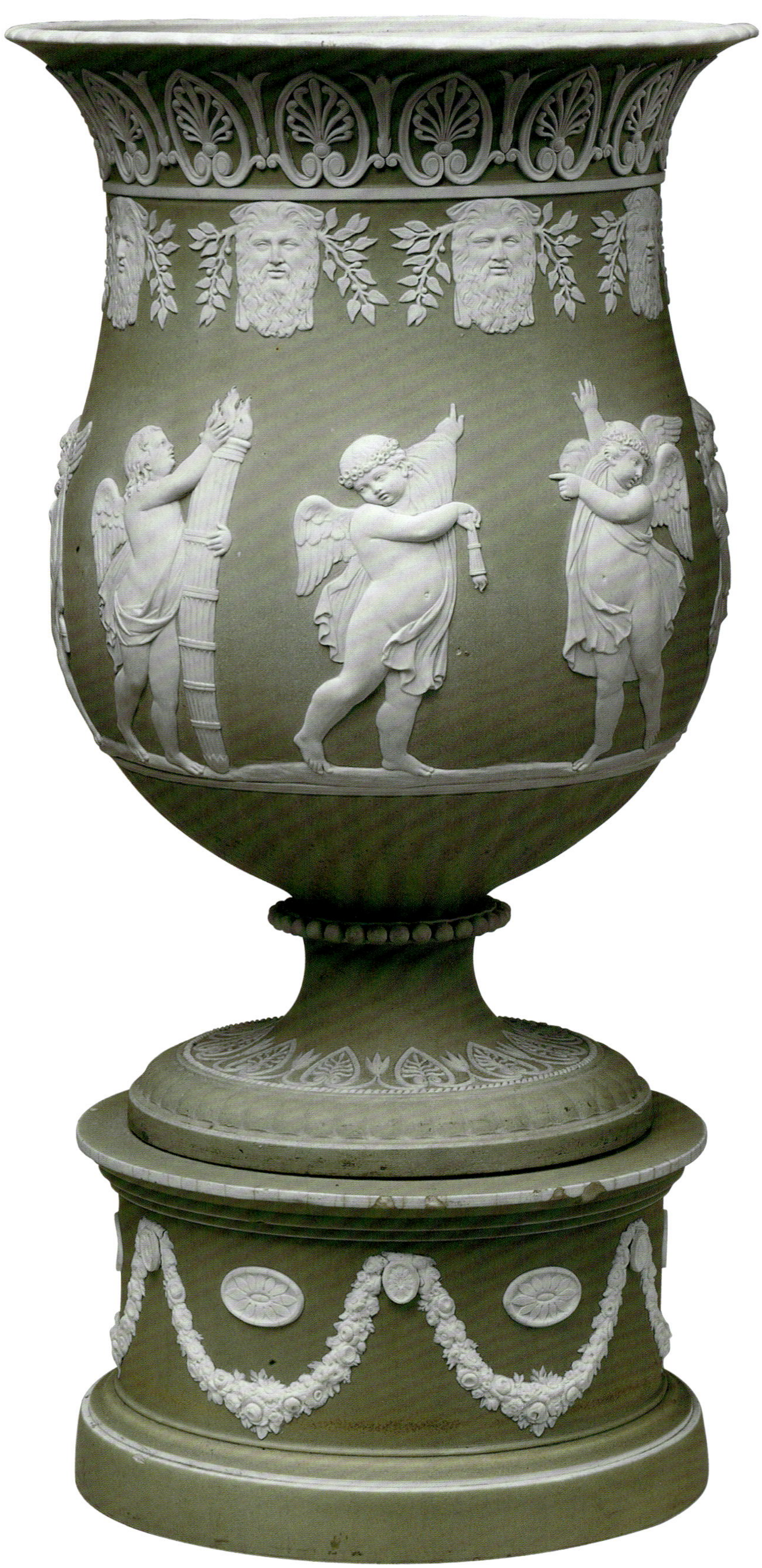

Cat. no. 41

originally it may have had a lid as well. The sixth version of the 1787 Wedgwood Factory Ornamental Catalogue, under the bas-relief category, lists the scene of cupids in procession as design 259, modeled by Henry Webber, head of the modeling studio at Wedgwood from 1782 to 1794.[3]

The composition was adapted from a 1787 wax version that replicated an ancient frieze of a line of cupids playing musical instruments on an octagonal cinerary urn for D. Lucellus Felix, in a tomb in Via Appia, Rome.[4] Some scholars have interpreted the marching figures as representing the hope for life after death and identified them as "several Geniuses representing the Pleasures of the Elysian Fields," a scene modeled in wax in 1787 by Giuseppe Angelini, a sculptor who worked in Rome for Wedgwood.[5] Wedgwood did not limit this particular composition to vases of this shape. The same design is depicted on a rare lilac and white jasper bough pot from 1790 (Lady Lever Art Gallery, Liverpool), and also appears without the masked border on plaques from the 1820s.[6]

Although classified in the general category of jasper ware, this vessel, like that in the Birmingham Museum collection, was made in white jasper and dipped in a colored solution rather than fully stained. The factory frequently used this cost-effective technique, which had been invented there in 1777, because the dip process required less of the expensive cobalt pigment, which had to be imported from the continental Europe.[7]

As early as 1771, Wedgwood was trying to develop a ceramic material, such as jasper ware, that was not reflective, took matte coloration in a range of colors, and did not require glazing, which might blur crisply molded neoclassical reliefs.[8] The result of Wedgwood's technical acumen was a new type of fine-bodied stoneware, to which a special component, barium sulphate (commonly known at the time as "cawk") was added. Wedgwood chose to keep his new formula a secret, due to stiff competition among Staffordshire potters, who also sought profitable developments in ceramic processes. Therefore, when asking his partner, Thomas Bentley, to purchase this ingredient, Wedgwood demanded that it be sent furtively: "I dare not have it the *nearest way*, nor undisguis'd."[9]

Although Wedgwood's jasper ware was in production from 1776 on, the factory vigorously marketed it only after Bentley's death in 1780.[10] Consequently, the first jasper ware vases made at the factory were probably held back from the 1781 stock sale of the Wedgwood and Bentley partnership, and were not displayed in London until 1782.[11] Wedgwood's development of the solid and dipped jasper ware formula, as well as his frequent use of classical motifs and themes on such wares, illustrates his continuing search for better and more popular ceramic materials in order to remain competitive in the growing eighteenth-century pottery market.
E.S./J.L.M.

1 Reilly 1994: 196.
2 Reilly 1994: 212, pl. 300; and Adams 1992: 119 (pl. 100), 240 (pl. 282). A jasper ware medallion with the "Sacrifice to Hymen," probably modeled by John Flaxman, is in the British Museum, London. Unlike the present example, it depicts a group of putti near a sacrificial altar, rather than in a procession; see Reilly 1994: 139, pls. 135B, 382. Another medallion, from 1775–80, also shares the name; see Adams 1992: 97 (pl. 73), 359, cat. 911.
3 Reilly 1994: 386. Both Flaxman and Webber modeled "Sacrifice to Hymen" scenes. For the attribution to Webber, see Adams 1992: 119, cat. 282, pl. 100. The factory sent Webber to Rome in 1787 to make drawings of classical prototypes to be used as models for ceramics: see Reilly and Savage 1980: 459.
4 Reilly 1994: 179. The classical source is discussed in Carol Macht, *Classical Wedgwood Designs* (New York, 1957): 61, pl. 36.
5 Reilly 1994: 178–79, pls. 225, 225a. See also Adams 1992: 119, cat. 282. While the wax model has often been attributed to Flaxman based on its quality, according to a Wedgwood invoice, Angelini was paid twelve *zecchini* for the delivery of the first set of five wax models, one described as "57. Several Genius representing the Pleasures of the Elysian Fields, in games, dancing & banqueting." The manufactory received these models from its Rome studio in March 1790. See Robin Emmerson, "Wax Models by Pacetti and Angelini in the Lady Lever Art Gallery," *Ars Ceramica* 12 (1995): 30.
6 Reilly 1994: 178.
7 Finer and Savage 1965: 171. Wedgwood wrote Bentley on April 13, 1777, that some vases are "color'd with the Cobalt @36/-Per lb which being too dear to mix with the clay…we have wash'd them over"; see Reilly 1994: 84.
8 Reilly 1994: 69–71. Reilly declared the material to be the "most significant innovation in ceramic history since the Chinese invention of porcelain" (69).
9 Finer and Savage 1965: 131, 171; Reilly and Savage 1980: 51. "Cawk" was referred to in Wedgwood's formulary as "no. 74," or *radix jasper* (jasper root).
10 Young 1995: 109. The Wedgwood and Bentley Ornamental Catalogue of 1774 lists 93 tablets, plaques, and medallions made from biscuit and basalt: the 1777 factory catalogue includes more than double that number, and the 1787 catalogue has 275. See Reilly 1994: 129.
11 Young 1995: 109.

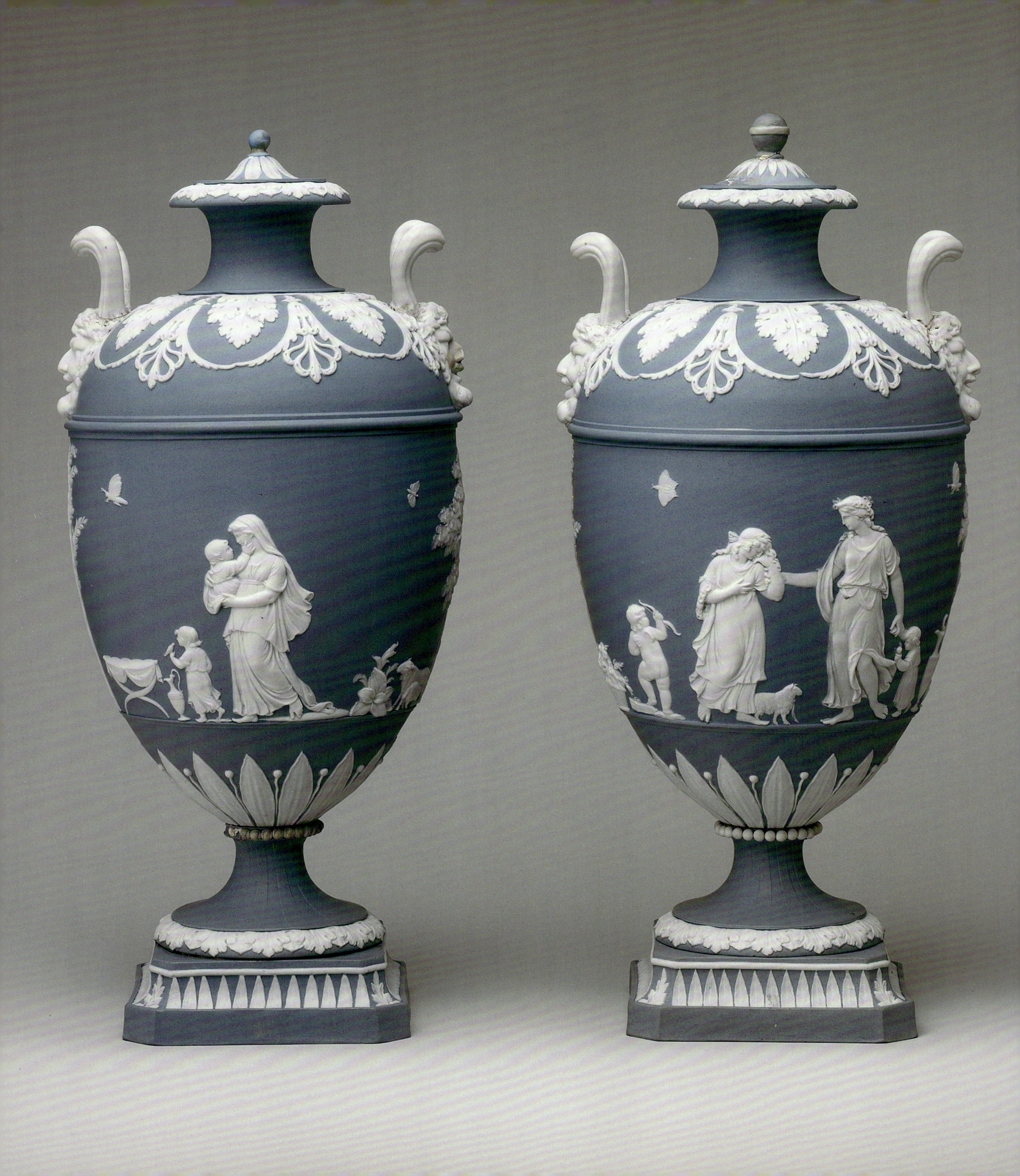

42. PAIR OF VASES WITH COVERS

Josiah Wedgwood (1730–1795), Etruria, Staffordshire; 32.95.12: designed by John Flaxman (1755–1826); 32.95.13: possibly modeled by William Hackwood (1757–1839) after designs attributed to Lady Elizabeth Templetown (1747–1823)
About 1790–1800
Vitrified stoneware (jasper ware)
Each 11 ½ in. (29.2 cm)
Gift of Frank K. Sturgis, 1932
32.95.12–.13

The four ornamental narratives on this pair of solid jasper vases were created by two different designers. They demonstrate the late eighteenth-century trend for infusing the neoclassical style with elements of sentimentality or moralizing that reflect certain Enlightenment ideas.[1] The scenes of "Maternal Love" and "Childhood" on the vase at the left were designed by John Flaxman, a sculptor who created many works for Wedgwood. The designs on the other vase, "Charlotte at the Tomb" and "Offering to Peace," have been attributed to Lady Elizabeth Templetown, an amateur artist who is recorded as having provided Wedgwood with fourteen designs in the 1780s.[2]

As documented by a factory invoice, William Hackwood, one of the primary modelers in Wedgwood's firm, often used "beautiful cut Indian paper" templates supplied by Templetown to model ornament for sprigging.[3] The mournful image of Charlotte was inspired by Johann Wolfgang von Goethe's successful novel *The Sorrows of Young Werther* (1774). The adaptation of this motif into a vase design illustrates the factory's attempts to incorporate contemporary literary taste into its classically derived repertory. The modern, virtuous heroine Charlotte, the object of Werther's tragic love, is presented in ancient garb. Thus, in such an image, Wedgwood both classicized a popular theme and modernized elements derived from antiquity.

These covered vases illustrate how Wedgwood's designs reflect changes in taste and the increased consumption of material goods by the upper classes in the eighteenth century. His London showroom was designed to meet the expectations of female customers: "Six or eight at least such services are absolutely necessary to *do the needful* with the Ladys in the neatest, genteelest & best method...these articles may, every few days, be so alter'd, revers'd, & transform'd as to render the whole a new scene, even to the same Company."[4] Cultivating this kind of "experience in Female Taste" explains in part Wedgwood's association with Lady Templetown, whose impressive social connections helped the enterprising potter sell his wares.[5] By seeking the approval and artistic involvement of other socially influential women, including Lady Diana Beauclerk, daughter of the second Duke of Marlborough; Miss Emma Crewe, "the daughter of one of the most celebrated beauties of her time"; and Whig hostess Frances Anne Crewe, Wedgwood was able to link his products to the prestige of aristocracy.[6] While the ceramist garnered a few designs and more than a few admirers from his female connections, most importantly he obtained social acceptance of his jasper ware vases at the highest levels of society. E.S.

1. Reilly 1989: vol. 1, 604–05; and Reilly 1994: 155–56. During the last ten years of Wedgwood's life, the factory began to produce more designs that combined neoclassical motifs with domestic scenes, reflecting the growing influence of romanticism.
2. Young: 1995: 41. Elizabeth was the daughter of Shuckburgh Boughton of Poston Hereford; she married Clotworth Upton, who became Baron Templetown in 1776. The attribution of "Charlotte" is based on a note Wedgwood wrote on his 1787 catalogue. Templetown's "Offering to Peace" is also featured on an oval jasper plaque; see Reilly 1994: 157, pl. 176.
3. Although almost all of Templetown's designs for Wedgwood were modeled by Hackwood, including two of the four on the pair featured here, this invoice is dated April 30, 1785, two years before Templetown created these two scenes; see Reilly 1994: 157.
4. Josiah Wedgwood to Thomas Bentley, May 31, 1767, Wedgwood Ms. E. 18149–25, Keele University Library, Staffordshire; quoted in Nancy F. Koehn, *Brand New* (Boston, 2001): 30.
5. Josiah Wedgwood to Thomas Bentley, Jan. 1768, Wedgwood Ms. E 18183–25, quoted in Reilly 1989: 55.
6. Young 1995: 65.

43. DICED WARE CASSOLETTE VASE

Josiah Wedgwood (1730–1795), Etruria, Staffordshire
About 1790
Vitrified stoneware (jasper dip)
11 ½ in. (29.2 cm)
Gift of Frank K. Sturgis, 1932
32.95.4

A variation on the form of the bell krater with a lid, this vase is decorated with a novel type of surface pattern. Within an overall classical shape, the interior geometric patterns on the surface create an almost rhythmic tension with the vase's form and outline. Its perforated lid indicates that it is a cassolette, a type of vessel meant to contain fragrant potpourri (originally, a liquid) for masking unpleasant odors.[1] Such wares were manufactured in pairs or multiples; the Metropolitan Museum owns the companion to this vase as well.[2]

This piece is an example of Wedgwood's brilliant experimentation with new ceramic bodies and production methods to enliven his classical forms. The so-called diced ware resulted from his development of the jasper dip technique. Creating the precise patterns required expert handling; an engine-run lathe was used to cut regular notches out of an unfired, leather-hard vase, thereby removing one layer of solid jasper dip in certain areas to expose a second color underneath. Despite Wedgwood's claim in his Common Place Book that he introduced lathe work to the English ceramic industry in 1763, Staffordshire potters John Philips and David Elers had employed lathes about 1700.[3] Nonetheless, Wedgwood's use of the tool distinguished itself from all previous attempts by English potters because of the way he applied it to integrate form, color, pattern, and style.[4]

Wedgwood's first important use of lathes occurred with his cream-colored earthenware, or "Queens ware," of the early 1760s. In 1767 he toured Matthew Boulton's factory in Soho, near Birmingham. and subsequently wrote to inquire about the rose-engine-turning lathe he had seen there.[5] It is not known whether Wedgwood procured it from Boulton, but the use of engine-turned dicing at the factory from 1785 on allowed the ceramist to construct more complex surface ornamentation in his dipped jasper ware vases. In 1786 the pottery began to add applied motifs, often rosettes or quatrefoils, in a third color, which further complicated the checkered patterns and produced more variations. In the present example, a third color, sage green, accents the rhythmic pattern of blue and white.
E.S.

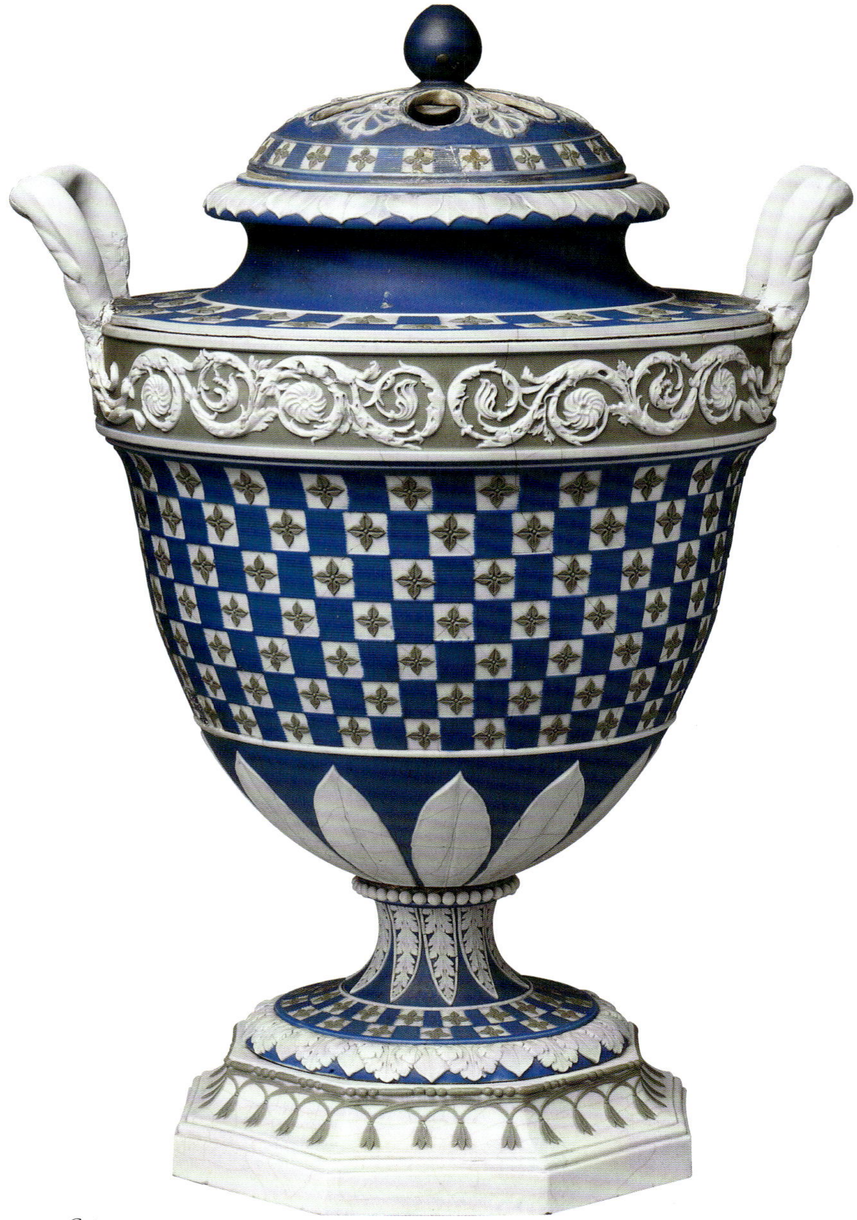

Cat. no. 43

1 The vase was designed to come apart in five pieces, which permitted its interior to be cleaned easily; see Reilly 1994: 214. A similar pair of cassolette vases, from around 1790, is in the Birmingham (Alabama) Museum of Art; see Adams 1992: 214, pl. 307. See also Reilly 1989: vol. 1, 662, pl. 1023.
2 32.95.5.
3 Reilly and Savage 1980: 255; Young 1995: 23; and J. V. G. Mallet, "Engine-turning on Chelsea Porcelain, with Considerations on Its Previous Use at Meissen and Vincennes/Sèvres," *English Ceramic Circle Transactions* 17, 3 (2001): 421.
4 Evidence provided by Jacques-Louis Brolliet suggests that Nicolas Sprimont used engine-turning at the Chelsea factory before 1763. Engine-turned plates, cups, and saucers dating from 1756 to 1758 have been identified; see Mallet (note 3): 422.
5 Finer and Savage 1965: 53; and Reilly and Savage 1980: 74. Pots were mounted eccentrically and carved as they turned in select areas to create patterns. A guide, called a rosette, was applied to the surface; see Reilly and Savage 1980: 150–51.

44. DICED WARE URN WITH COVER

Josiah Wedgwood (1730–1795), Etruria, Staffordshire
1785–1840
Vitrified stoneware (jasper dip)
21 ¼ in. (54 cm)
Gift of Frank K. Sturgis, 1932
32.95.1

Josiah Wedgwood was motivated to search for new ceramic formulas such as jasper ware in part by the complex combination of colors used in Adam-style neoclassical interiors. Pastel yellows, pinks, lilacs, sage greens, and tawny browns crisply intersect on the ceilings Adam designed for 20 St. James Square, London. In Adam's designs for Mellerstain Castle, built in 1770, the blue walls and ceilings, compartmentalized with stucco, foreshadow Wedgwood's jasper ware plaques and vases. Both the ceramist and architect were drawn to the way white bas-relief figures emerge from a colored field.

Jasper ware took color well; Wedgwood described its "white body, succeptible of being colour'd & which shall polish itself in burning Bisket."[1] The factory continued to develop new shades, including brown, green, yellow, and lilac dip and solid jasper as variations to the popular blue and white.[2] With the advent of engine-turned dicing and sprigged details in contrasting colors, Wedgwood's emulation of Adam's style became more intricate, as the present example demonstrates. This lidded urn, in the shape of a calyx krater, indulges in the dizzying patterns that dicing permitted. Despite its classical shape and structure, the piece exhibits an eclecticism that defies the logic of antique prototypes.

Adam described this effect as an improvement on the antique, for it added novelty and variety.[3] Similarly, about his diced ware, Wedgwood stated, "I have endeavoured to preserve the stile and spirit or if you please the elegant simplicity of the antique forms...but not with absolute servility."[4] Indeed, the quatrefoils in this diced ware vessel do not derive from ancient Greek or Roman sources, although the vase's other applied ornamental devices—including the applied palmettes and flower on the base—are more overtly classical. E.S.

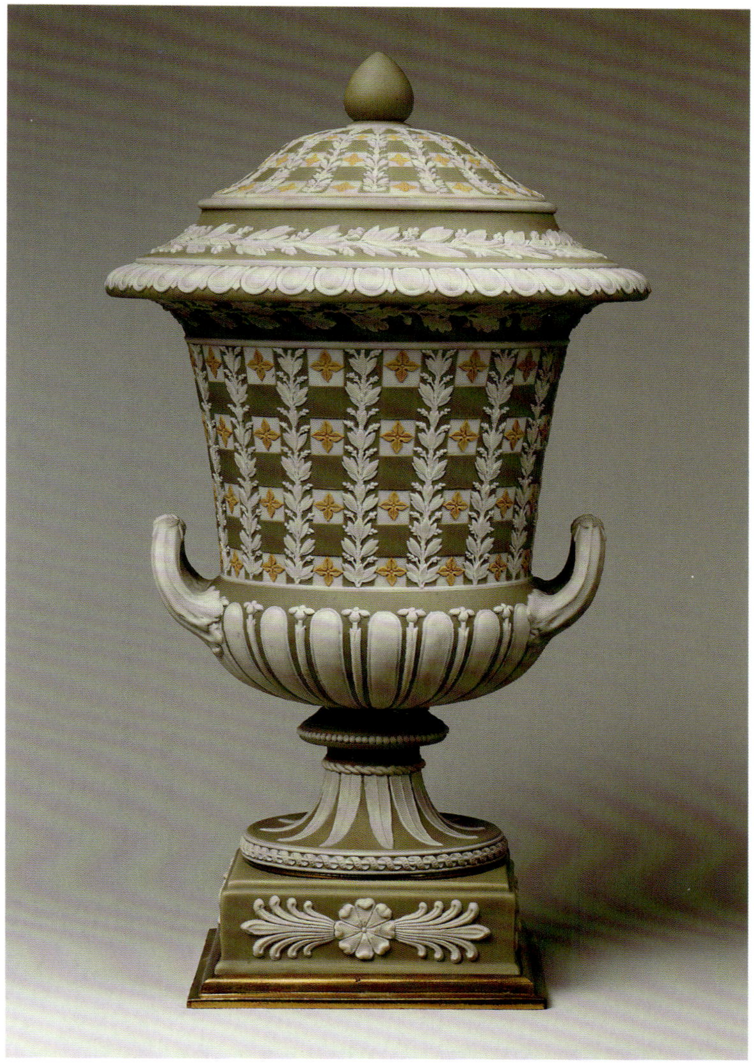

Cat. no. 44

1 Josiah Wedgwood to Thomas Bentley, Jan. 13, 1771, quoted in Reilly 1994: 69. Interestingly, when the potter sought the praise of "Capability" Brown in the early 1780s, the architect rejected his jasper ware because he preferred the tones of pale marble to the potent color of Wedgwood's product.
2 Reilly 1994: 86.
3 Reilly and Savage 1980: 112.
4 Steve Parissien, *Adam Style* (London, 1992): 147.

45. PAIR OF PERFUME BURNERS, "SPHINX VASES"

Matthew Boulton (1728–1809) and John Fothergill (d. 1782),
Soho Manufactory, New Birmingham
About 1770–72
Fluorspar vases with gilt-brass mounts, marble, tortoiseshell, and wood
13 in. (33 cm)
Gift of Irwin Untermyer, 1964
64.101.1633–34

Matthew Boulton's Soho Manufactory produced a number of objects that exploited the commercial opportunities created by the "antique taste" and the resulting mania for vases.[1] Boulton was the first English manufacturer to engage in large-scale speculative production of ormolu, as the gilt-bronze ornaments in imitation of French models were known.[2] The mainstay of Boulton's ormolu business was the mounting of vases, often in quasi-utilitarian forms such as perfume burners and candelabra.[3] Boulton called this pair "sphinx vases," after

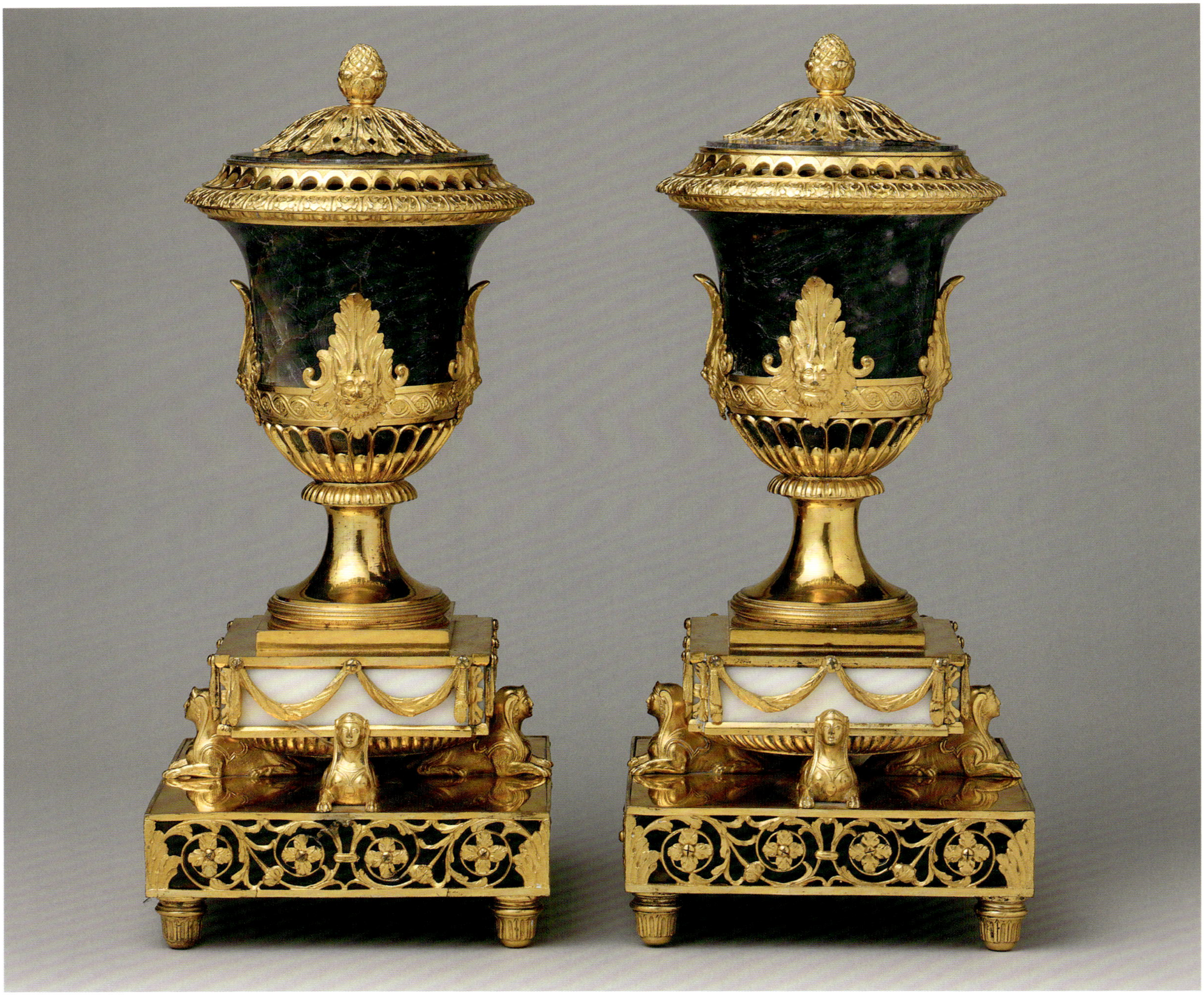

Cat. no. 45

the unusual supports carrying the vessels' raised plinths. The pair assumes a standard urn form, based on the Greek calyx krater.[4] William Chambers, architect to King George III, may have provided Boulton with the designs for both the sphinxes and grotesque masks on the body of the vases. Chambers was among the first to reject the rococo style and return to antique models for design inspiration. In the spring of 1770, Boulton wrote to his wife that the architect "made me a present of some valuable, usefull and acceptable modells."[5] A similar sphinx appears in *Chambers's Franco-Italian Sketchbook*.[6]

The basic vase form could be endlessly varied by combining component metal mounts and other materials to create different, ostensibly unique, assemblages. Boulton was able to extend the number of designs he offered by using a variety of finishes and materials.[7] Like Josiah Wedgwood, he marketed his vases to aristocratic patrons, the most important of whom were members of the royal family. A nearly identical pair of sphinx vases in the royal collection at Windsor Castle, together with a clock case and two "King's" vases, were likely part of a garniture ordered by Queen Charlotte in March 1770.[8] The upper plinths of the more finely finished Windsor vases are faced with panels of brown- and gold-lacquered glass, while the lower plinths are veneered with tortoiseshell. The present pair also features tortoiseshell-veneered lower plinths, but the bases are marble.[9]

The perfume burners incorporate vases of the mineral fluorspar, known as blue john, which are mounted in gilt brass, a cheaper material than the traditional French gilt bronze. The mounts were generally cast, but long bands of moldings were rolled mechanically in the flattening mills, then stamped with or hammered in steel dies. In 1770 Boulton recorded the purchase of six "sphinx" blue john vase bodies from Robert Bradbury of Castleton, Derbyshire, probably referring to the shape used for this design.[10]

Capitalizing on his royal commission from Queen Charlotte, Boulton held sales at Christies and Ansell's in Pall Mall in 1770, 1771, and 1772. Vases comprised most of the nearly four hundred objects in the second sale, and the material most frequently used for them was blue john, including a dozen "in the antique taste *radix amethysti* [blue john] and *or moulu*, lined with silver and perferated for essence, supported by four sphinxes upon an ornamental base of ebony."[11] The perfume burners featured here are not lined with silver, but rather with gilt copper. At least sixteen other sphinx vases survive, which exhibit slight variations in their design or in combinations of mounts and materials. The model continued to be mentioned in Boulton's records through 1777. The years 1770 to 1772 represented the peak of the Soho ormolu venture, which ultimately proved to be a financial failure. As Boulton's friend James Keir said, ormolu vases were "too expensive for general demand, and therefore not a proper object of wholesale manufacture."[12] J.L.

1 Perfume burners were considered "classical" due to the generally held belief that burning incense was a quintessential practice among the ancients. Joseph-Marie Vien's early neoclassical painting *The Virtuous Athenian Maiden* (1763; Musée du Louvre, Paris) depicts a priestess offering incense at a tripod burner, a type popularized in England by James Stuart and Robert Adam. Several objects featured here use the form or incorporate it as a decorative motif (see cat. nos. 70, 72).

2 In 1768 Wedgwood wrote his partner, Thomas Bentley, "Mr. Boulton tells me I sho'd be surprised to know w't a trade has lately been made out of Vases at Paris. The artists have even come over to London, picked up all the old whimsical ugly things they could meet with, carried them to Paris, where they have mounted and ornamented them with metal, and sold them to the virtuosi of every nation, and particularly to Millords d'Anglaise, for the greatest rarities"; see Wedgwood to Bentley, Mar. 15, 1768, cited in Meteyard 1970: vol. 2, 77.

3 Boulton started ormolu production in 1768; see Goodison 1974: 23, 143.

4 New York, Metropolitan Museum of Art, *Highlights of the Untermyer Collection of English and Continental Decorative Arts* (1977): 183–84.

5 See Boulton to his wife, Anne Robinson, Mar. 6, 1770, quoted in Goodison 1974: 51–53, 163–65.

6 The sketchbook is now in the Victoria and Albert Museum, London. Boulton purchased Chamber's published works the same year.

7 Goodison 1974: 23, 65, 67.

8 Boulton recorded that "the Queen sent for me into her bed chamber, shewed me her chyney piece and asked my opinion how many vases it would take to furnish it, for she says all the china shall be taken away"; Boulton to his wife, Anne Robinson, [Mar. 3, 1770], quoted in Goodison 1974: 31–32, 85, 163–64.

9 New York (note 4): 184.

10 Fluorspar is a material unique to England. In 1768 Boulton wrote his friend John Whitehurst, "I have found a use for Blew John which will consume some quantity of it. I mean that sort which is proper for turning into vases"; Goodison 1974: 29–30. In 1769 Boulton purchased fourteen tons of blue john. A drawing in his Master Catalogues, now at the Birmingham (England) Reference Library, appears to represent the stone shape required for the sphinx vase; see Goodison 1974: pls. 161s, 77.

11 Catalogue from the archives of Christie, Manson and Wood, London, quoted in Goodison 1974: 34, 36, 136, 163.

12 James Keir, "Memorandums for the Memoir of M. Boulton," Dec. 3, 1809, Matthew Boulton Papers, Birmingham (England) Reference Library, Box Boulton M, item 112, sheet 2, quoted in Goodison 1974: 42–44.

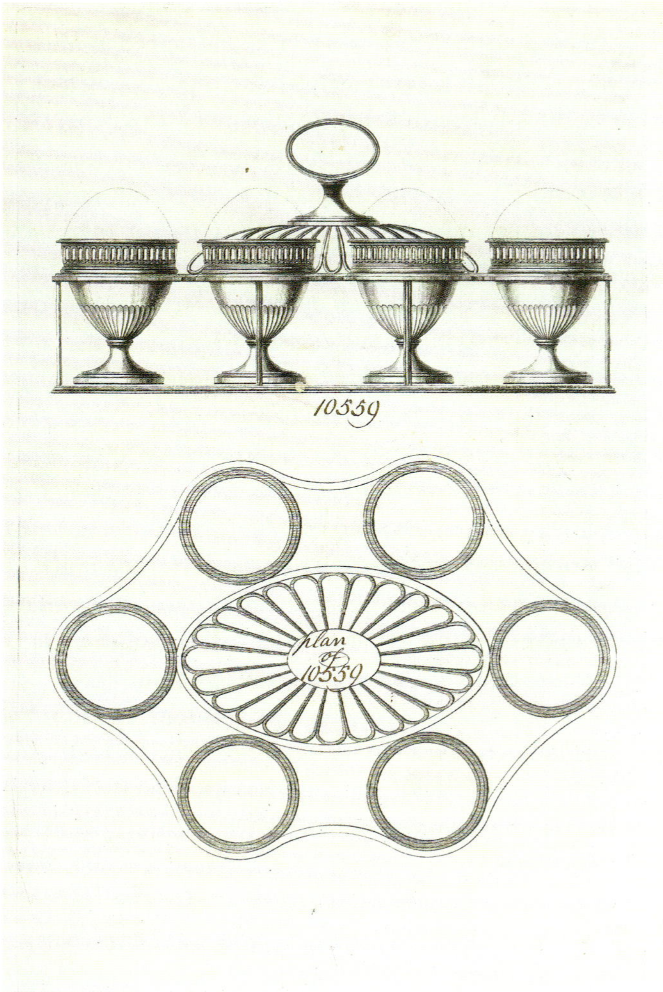

Cat. no. 48

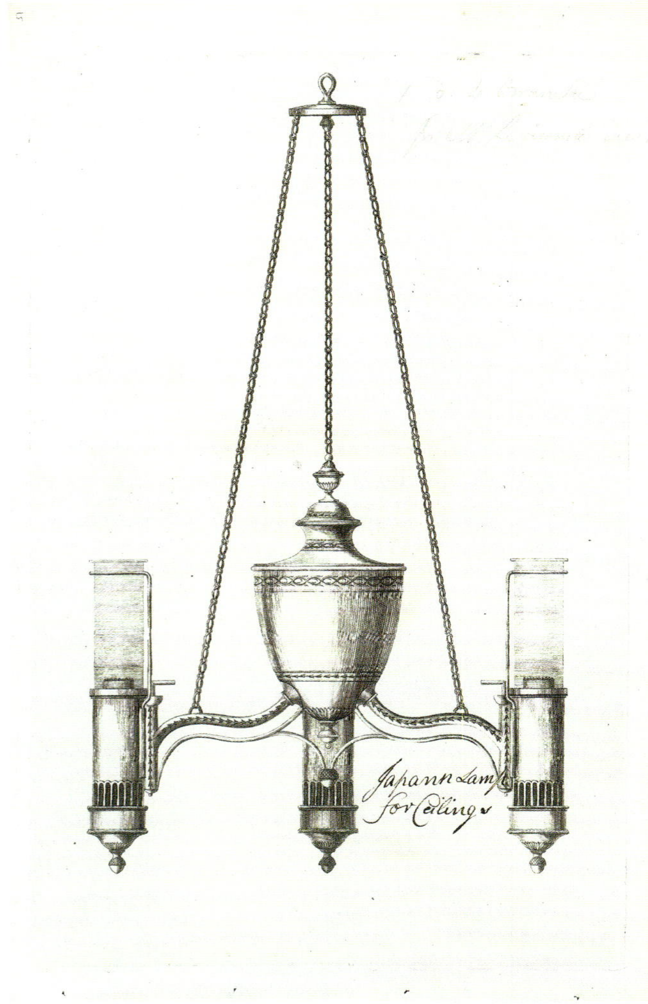

Cat. no. 49

version "which the bourgeoisie likes best." King Louis XV ate eggs every Sunday and was renowned for the dexterity with which he could knock the small end off an egg with single stroke. The breeding of poultry became something of an aristocratic fad, part of a general interest in scientific and agricultural improvements.[15]

49. ARGAND LAMP (JAPANN LAMP FOR CEILINGS)

54.569.2 (51)

The inclusion in this catalogue of several designs for a type of lamp invented by François-Pierre-Amié Argand in 1783 provides an approximate date for the publication. The Argand lamp contained a hollow wick, which was placed inside a narrow glass chimney. It admitted air into the center of the oil-fueled flame, which further increased combustion to create an unprecedented intense light, largely free from smoke. To provide fuel at a steady rate, increase efficiency, and maintain a consistent level of brightness, Argand utilized a fountain-feed principle, locating the oil reservoir to one side and connecting it to the burner by an arm. The level of illumination could be adjusted by a mechanism that raised or lowered the wick.[16] As shown here, several burners could be served by a single reservoir.

A vase serves as the lamp's central motif, embellished only with simple, running husk moldings. The form remained standard for oil reservoirs until the mid-nineteenth century, evoking one of the functions of vases in antiquity. Boulton's design repeats the vase shape in finials above and below the reservoir. The pierced arcades at the base of the burners, similar to those at the edges of the egg cups (cat. no. 48), serve as ornamentation and as openings that admit the oxygen necessary for combustion. The *Catalogue of Silverware* includes other examples of Argand table lamps and wall brackets, all utilizing the vase-shaped reservoir.[17]

Discouraged by lack of interest in his invention in France, Argand traveled to England in 1783. He was introduced to Boulton by a Mr. Parker, a mutual acquaintance who seems to have been manufacturing the lamps, though not in the quantities sufficient to meet the demand its inventor expected.[18] On March 12, 1784, Argand patented his lamp (no. 1425) and on May 1 announced that Boulton would manufacture it. No doubt the clear superiority of the lamp, "so constructed to produce neither smoke nor smell and to give considerable more light than any lamp hitherto known," led both Boulton and Argand to expect a huge commercial success.[19] In fact, the very ease of manufacture and a universal demand for the new lamps led to a flood of unauthorized versions.[20]

In 1786 Argand's patent was invalidated, due to a claim that the invention had been introduced in England a few days before the date he had patented it. "Oppressed with grief," he returned to France, where he set up his own lamp manufactory.[21] Boulton nevertheless continued to manufacture Argand lamps without the protection and exclusive rights of a valid patent.[22] J.L.

1 Matthew Boulton to James Adams, Oct. 1, 1770; cited in Goodison 1974: 98. Throughout his career, Boulton considered opening a London showroom but, perhaps because of his constant financial difficulties, never did. This increased his reliance on sketches and, later, on prints to sell his designs. Another printed catalogue at the Victoria and Albert Museum, London (E2060–1952, location M63V), includes a similar range of designs.

2 W. A. Seaby and R. J. Hetherington, "The Matthew Boulton Pattern Books, Part I," *Apollo* 51 (Feb. 1950): 48. The earliest datable object from the Birmingham catalogues is a 1767 fluorspar candelabrum made for Queen Charlotte. Designs reproduced in the Metropolitan's trade catalogue appear in the first Birmingham volume and probably date before 1795; see idem, "The Matthew Boulton Pattern Books, Part II," *Apollo* 51 (Mar. 1950): 79.

3 As early as 1774, Boulton was working with Francis and John Eginton to produce "mechanical paintings," reproductions of oil paintings whose exact method of execution (probably involving aquatint) remains unclear; see Robinson and Keith R. Thompson, "Matthew Boulton's Mechanical Paintings," *Burlington Magazine* 112 (Aug. 1970): 497–507; and Goodison 1974: 60, 68, 184, n. 111. In 1780 Boulton began manufacturing a roller printer that had been patented by James Watt (Feb. 14, 1780, no. 1244); his firm used a large version extensively. Because the process reversed the original, lettering and numbers had to be inserted by hand after printing. Hundreds of drawings marked "reverse" are in the Boulton and Watt Collection at the Birmingham (England) Reference Library; see H. W. Dickinson, *Matthew Boulton* (Cambridge, Eng., 1937): 107–09.

4 Boulton to Elizabeth Montagu, Jan. 16, 1772; see Goodison 1974: 47. Montagu was a celebrated London intellectual and hostess of the famed "blue-stocking" literary gatherings; see Elizabeth Montagu, *The Queen of the Blue Stockings* (London, 1906).

5 A similar Boulton helmet-shaped creamer of about 1785 survives, but it lacks fluting; see Eric Delieb, *The Great Silver Manufactory, Matthew Boulton and the Birmingham Silversmiths, 1760–1790* (London, 1971): illus. 85.

6 James Keir, "Memorandums for the Memoir of M. Boulton," Dec. 3, 1809, Matthew Boulton Papers, Birmingham (England) Reference Library, Box Boulton M, item 112, sheet 2, quoted in Kenneth Quickenden, "The Planning of Boulton and Fothergill's Silver Business," in Michael Hallet, ed., *Silver and Jewellery: Production and Consumption since 1750* (Birmingham, Eng., 1995): 25.

7 Peter B. Brown, *In Praise of Hot Liquors: The Study of Chocolate, Coffee and Tea-Drinking 1600–1850* (York, 1995): 52, 55, 73, 87–90.

8 Delieb (note 5): 73–74.

9 Adam used the ram's-head motif frequently in a large service dating around 1771–74, which he designed for Sir Watkyn Williams-Wynn; see Michael Snodin, "Putting Adam in Context," in Hallet (note 6): 13. Designs of about 1775 for the Duke of Roxburghe also exhibit the basic "spool-top" vase shape and ram's heads; see Adam and Adam 1773: vol. 1, no. 1, pl. 2 (details for the gateway at Syon House).

10 Boulton owned treatises by William Chambers and publications by James Stuart and Nicholas Revett, as well as Robert Adam. In 1768 Boulton paid a subscription fee for a volume of vase designs and in 1771 asked a friend to procure a new book on vases being published by the bookseller Darby; see Goodison 1974: 49–50, 54–56, n. 56.

11 Goodison 1974: 49–50.

12 Delieb (note 5): illus. 132.

13 A similar Boulton egg caddy (hallmarked 1789) is in the City Art Gallery, Birmingham. The egg cups are identical to those featured here, but omit the gadrooned base. The mixing and matching of various options and variations are typical of Boulton's designs. See Seaby and Hetherington Mar. 1950 (note 2): 81; and Delieb (note 5): illus. 96.

14 Reay Tannahill, *Food in History* (New York, 1989): 238–39.

15 Maguelonne Toussaint-Samat, *History of Food*. trans. Anthea Bell (Cambridge, Mass., 1994): 359–60.

16 Roger W. Moss, *Lighting for Historic Buildings* (New York, 1988): 75–76.

17 The handwritten inscription on this design, "Japann Lamp for Ceilings," may derive from the technique of "japanning" (using layers of painting, gilding, and varnishes in imitation of Asian lacquer), which was often applied to tinned goods. That Boulton and Argand both worked in "japanning" is perhaps indicated by the former's documented production of "japanned" wares and Argand's employment of twenty-eight to thirty tin workers in his later factory; see Delieb (note 5): 110.

18 Ibid.: 109.

19 Brian Bowers, *Lengthening the Day: A History of Lighting Technology* (Oxford/New York, 1998): 27–29.

20 British manufacturers evidently took up the production of the Argand lamp enthusiastically. In *The Diary of Sophie von La Roche* (1786), the author, a German novelist, described the shop windows in Oxford Street, London, where she admired "a stall with Argand lamps situated in a corner house and forming a really dazzling spectacle. Every variety of lamp, crystal, lacquer and metal ones, silver and brass in every possible shape"; quoted in Neil McKendrick, *Birth of a Consumer Society: The Commercialization of Eighteenth-Century England* (Bloomington, Ind., 1985): 79.

21 Boulton to James Watt, Feb. 24, 1786; quoted in Dickinson (note 3): 128.

22 Delieb (note 5): 108–09.

Neoclassicism and the New Aesthetic in Silver

Stefanie Walker

Prized since antiquity, silver was an ideal medium for neoclassical vase designers and craftsmen to convey some of their main artistic goals. They could shape it into symmetrical forms with precise outlines and mirror-smooth surfaces; they could bend handles into elegant loops, while adding reflective beading, gadroons, and fluting to create shimmering areas of regularized ornament. Not surprisingly, the rarity and expense of silver continued to allure artists and patrons alike.

Throughout the 1770s in continental Europe, a typical, conservative approach to silver design prevailed among its makers. As they had in the earlier decades of the century, artists continued to demonstrate their technical prowess and inventiveness by creating elaborately decorated items. Sporting a vase-shaped finial, Jean-Nicolas Roettier's tureen (cat. no. 50) exhibits the new, solid proportions and strict symmetry of early neoclassicism, the *goût grec*, but is covered with ornamentation. The decoration of the wine cooler by Ignaz Josef Würth (cat. no. 51), with its multiple mythological references, is equally dense. On the other hand, the small mustard pot by Marc-Étienne Janety (cat. no. 52), from the early 1780s, combines a classicizing urn shape with figures and ornament along a symmetrical axis in an expression of the mature phase of neoclassicism, known as the *goût étrusque.*

Enhanced production methods, developed in England and France, helped to establish a new aesthetic in silver, one that focused on the contrast between expanses of shiny surface and the glitter of detailed ornament. An increased use of the lathe allowed a silver disk to be spun over a wooden core, a quicker and more precise method of raising the basic shape of a vessel than hammering it entirely by hand. Individual motifs, such as masks or rosettes, were produced in series of identical casts, while thinner decorative bands could be stamped or rolled. The English silver featured here exemplifies the growing use of these practices and the variety of functions the evolving vase shape could take on.

By the first decades of the nineteenth century, during the so-called Empire period, neoclassical silver reached its ultimate expression under the influence of designers Charles Percier and Jean-Guillaume Moitte. The focus on the tensile curves of an object's outline resulted in a preference for smooth, ovoid, and wide bowl shapes, with ornament often restricted to intricate bands or free-floating medallions. Figures, when they occur, usually assume calm poses and display intentionally standardized, almost expressionless features.

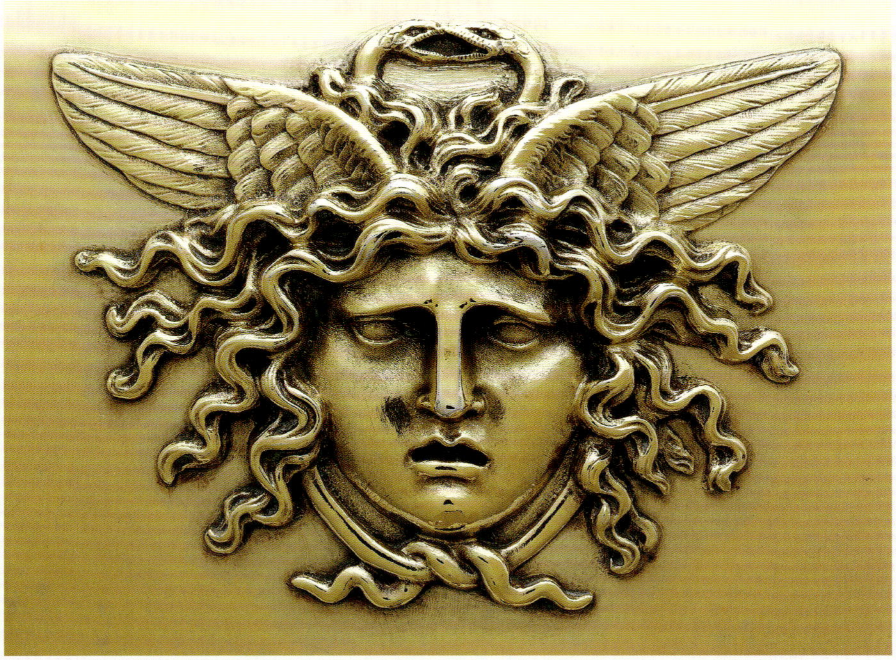

Detail of cat. no. 61

50. TUREEN WITH COVER

Jacques-Nicolas Roettiers (1736–1788)
1775–76
Silver
12 ¼ × 15 ½ × 9 ⅞ in. (31.1 × 39.4 × 25.1 cm)
Marks: [inside bottom, inside cover, on bottom of liner] Fleur-de-lis, two grains de remède, *J N R*, a sheaf of wheat (maker's mark); [inside bottom, inside cover, on bottom of liner] crowned *M* (warden's mark 1775–76); [inside bottom, inside cover, on bottom of liner] crowned *A* (charge mark, 1774–80); [on top rim of liner] monkey's head (discharge mark, used between 1774–80)
Gift of Dorothy C. Livingston in loving memory of Robert R. Livingston, 1986
1986.320

This tureen, in which new ornamental elements have been grafted onto a traditional shape, exemplifies the transitional phase of early neoclassicism, called *goût grec*.[1] Silversmiths had to be aware of new fashions, while at the same time respecting the conservative tastes of their patrons. The vessel's oval shape was prevalent since the seventeenth century, but its maker, Jacques-Nicolas Roettiers, transformed it into a new kind of object by adding a battery of classical motifs. Heavy gadrooning decorates the cover, while fluting, gadrooning's concave counterpart, is applied to the rest of the vessel. The tureen's imbricated, volute legs are applied with acanthus leaves half-way down the sides, and a frieze of rosette-filled, guilloche ornament forms the rim. Handles are rendered as elaborately wrought oak and acorn branches tied with ribbons.

The piece features a cover with an extravagant finial in the shape of an urn—a favorite *goût grec* motif and the most interesting neoclassical ornament on the tureen.[2] The finial is draped with heavy oak-leaf festoons, and lancet leaves sheath the base. A ring of pearled beads encircles the waist, the center of which bears a small, square frame containing a four-leaf blossom. K.C.

[1] See Clare Le Corbeiller, "Craftsmen and Elegance," *Apollo* 106 (1977): 306–401; and idem, "Grace and Favor," *MMA Bulletin* 27 (1969): 289–98, for extensive entries on this tureen and its relationship to Roettier's most important commission, a service for the Russian Count Orlov. The Metropolitan Museum possesses a matching tureen by Roettiers (1976.357.1).

[2] Another tureen, described in an inventory dating from Dec. 31, 1775, was made for the daughters of Louis XV. It had a frieze and handles decorated with olive leaves, and a finial in a shape of a vase imbricated with fish scales and wrapped with rose garlands; see Le Corbeiller, "Craftsman and Elegance" (note 1): 396–97.

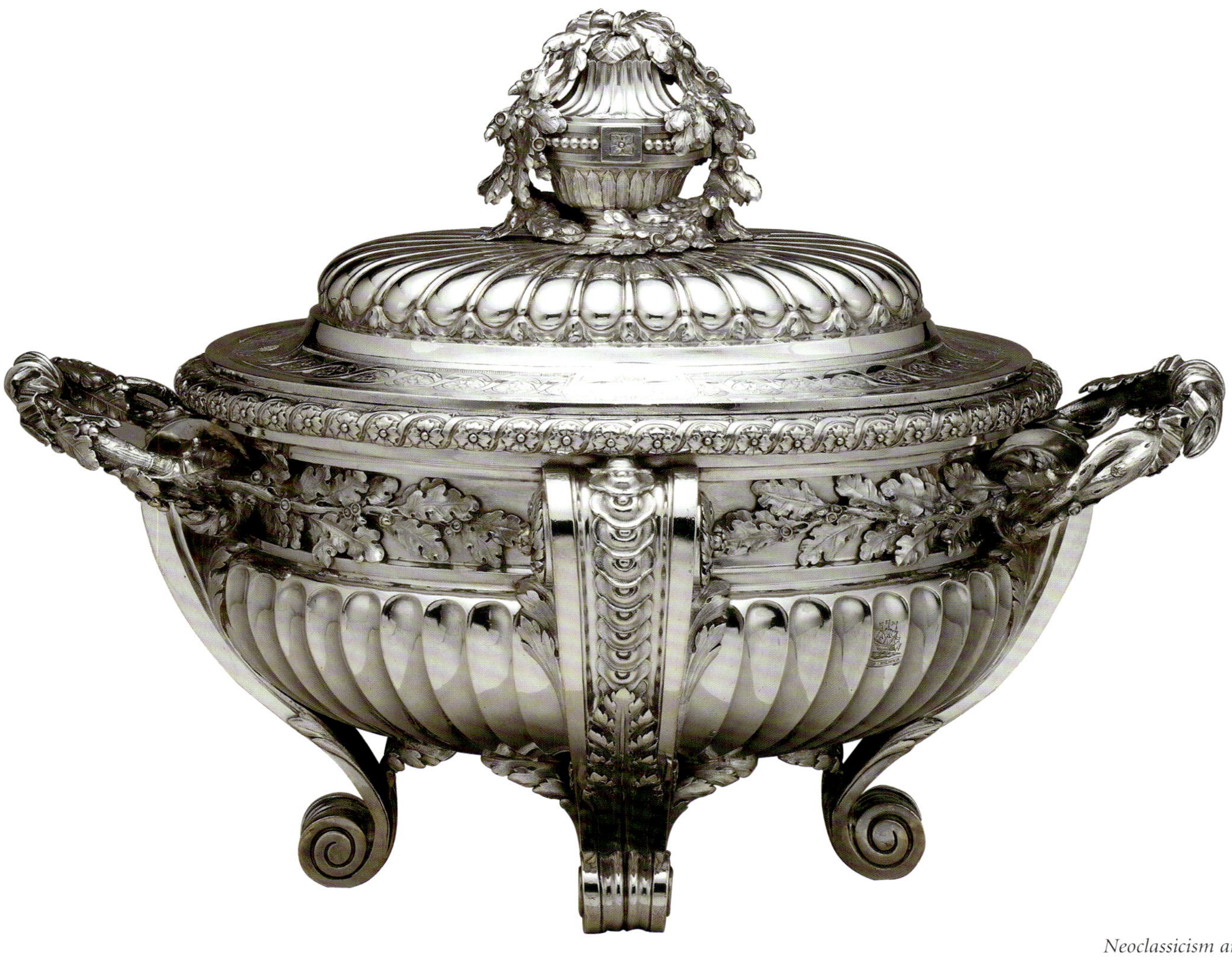

Cat. no. 50

51. WINE COOLER

Ignaz Josef Würth (act. 1769 on, died 1792)
1781
Silver
11 5/16 in. (28.9 cm)
Marks: [on rim of foot] *IIW* in an oval (master's mark); *13* over an oval shield with a cross, surrounded by *1, 7, 8,* 1 (Vienna assay and year mark)
Purchase, The Anna-Maria and Stephen Kellen Foundation Gift, 2002
2002.265.1

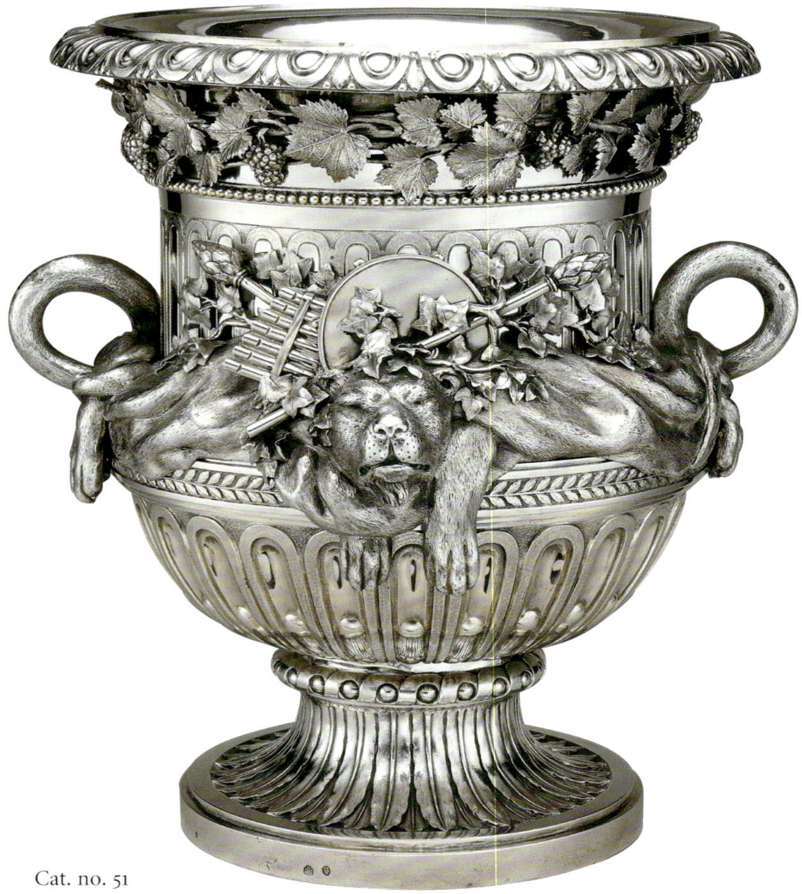

Cat. no. 51

The antique krater vase form, with its wide mouth, cylindrical waist, circular spreading foot, and stepped base, was adapted by eighteenth-century designers as a wine cooler, holding one bottle and enough ice to keep Bacchus's favorite drink at the desired temperature. By the sixteenth and seventeenth centuries, krater-shaped vessels figured as elements in silver buffets, serving as richly decorated show items without utilitarian purpose.[1] However, with the development of large, ostentatious silver and porcelain table services at the European courts during the late baroque era and especially the neoclassical period, this type of vase-shaped wine cooler became a popular contemporary manifestation of ancient vessel design.

The dramatic, bold outline of the krater form provided an appropriate new paradigm for neoclassical ornamental details and patterns. This wine cooler and its pair (also in the Metropolitan Museum) belonged to a set of eight that formed part of the now-dispersed so-called Second Teschen, or Polish Service, made for Duke Albert Casimir of Saxony-Teschen and his consort, Archduchess Maria-Christina of Austria, sister of Queen Marie-Antoinette, and daughter of Empress Maria-Theresia.[2] The set included round and oval tureens, cloches, sauceboats, candelabra, candlesticks, dozens of plates, and various other kinds of tableware exemplifying the splendor of royal dining during the ancien régime.[3] The overall style of this piece reveals the influence of French neoclassical designs, and reflects a strong appreciation of contemporary French art and culture of its owners. Grandson of Augustus the Strong, Duke Albert had inherited the "collector's bug" of his ancestors, but not the means to realize such a costly passion. Nonetheless, with the help of the substantial dowry of his wife, who shared her husband's interests, Albert assembled one of the great art collections of his time (see also cat. no. 30).

The vigorous design of this wine cooler, the sparkling play of textures it affords, and the daring juxtaposition of classical elements with whimsical sculptural details embody the freshness of Viennese neoclassicism, in contrast to the more restrained outlines of high-style French examples.[4] The malleable, light-reflective quality of the precious metal, exploited to the maximum, and the superb technical accomplishment demonstrate an experienced workshop guided by a master of European importance, the imperial court goldsmith Ignaz Josef Würth.[5] The supreme quality of the service's items is confirmed by the enormous amount of silver expended. The weight of just two wine coolers totals 383 ounces, or 11.911 kilograms of the valuable material.

Duke Albert and his wife were appointed joint governors of the Austrian Netherlands in 1780; the coolers bear the year mark for 1781. Baroque imagery often associated such positions of power with Hercules, who wore the fur of the Nemean Lion as a symbol of great strength. Lion furs draped around the wine cooler's body refer to the mythic hero, but also teasingly evoke the insulation of its liquid contents. Lion tails form the unusual handles. The applied trophies include crossed *thyrsoi*, tambourines, reed pipes, and ivy as well as grapevines. While primarily signaling the alcoholic contents of the bottles held in the cooler, these motifs also symbolize Bacchus, the god of wine and erotic ecstasy. The witty decoration of the wine cooler, visually the most accomplished item of the service, represents the triumph of the pleasures of Bacchic life over the Herculean symbol of worldly power. W.K.

1 *Versailles et les tables royals en Europe* (Paris, 1993): 23, 86, cat. 4.
2 Ibid.: cats. 205, 210; and Christie's New York, sale cat., Apr. 19, 2002, lot 74: 313–15.
3 2002. 265.2. For more on this pair, see Wolfram Koeppe, "Pair of Wine Coolers," *Recent Acquisitions, MMA Bulletin* 61 (Fall 2003): 26. For the Second Teschen Service, see Edmund Wilhelm Braun, *Das Tafelsilber des Herzogs Albert von Sachsen-Teschen...* (Vienna, 1910): 7, pl. 4. The name of the Polish Service reflects an unspecified family tradition of the dukes of Teschen in the nineteenth century and cannot be traced to the time of Albert Casimir. For the First Saxony-Teschen Service, made in Vienna around 1748, see Wolfram Koeppe, *Die Lemmers-Danforth-Sammlung Wetzlar* (Heidelberg, 1992): 484–86, no. GO23, illus. P.450 (with further literature).
4 Braun (note 3): pls. 7–8, fig. c; and Christie's New York, sale cat., Apr. 11, 1995: lot 119. See also Jean-Luc Martinez and Anne Dion-Tenenbaum, "L'athénienne," *D'Après l'antique* (Paris, 2000): 336–53.
5 Waltraud Neuwirth, *Wiener Silber/ Namens-und Firmenpunzen 1781–1866* (Vienna, 2002): 183 (mark no. P1274–1279). Würth descended from a long line of silversmiths active in Vienna. He became a master in 1770 and from 1779 to 1782 produced his most notable work, a silver table service for Duke Albert. See *Allgemeines Lexikon der bildenden Künstler...von Ulrich Thieme und Felix Becker*, vol. 35–36 (Munich, 1992).

52. MUSTARD POT

Marc-Étienne Janety (1739–1820)
1783
Silver, cobalt-blue glass
4 3/4 × 3 1/2 × 2 1/2 in. (12 × 9 × 6.4 cm)
Marks: [underside of the base] *M E J*, a weight, and two grains of tolerance, surmounted by crowned fleur-de-lis (master's mark); [underside of the base] *U* (year mark for 1783–84); [underside of the base] interlaced *L*'s (mark of tax administration of 1782–89); [outside of the base and on top of lid] female head (discharge mark used between 1782–86); [outside of the base] nineteenth-century control mark
Collection of Paul Micio, Paris

The shape of this mustard pot (*moutardier*), like that of Josiah Wedgwood's ceramic version of the Borghese Vase (cat. no. 6), is based on the Greek krater. The master silversmith Marc-Étienne Janety replaced the low handles of the original with more elegant, elongated handles, bifurcated at the top. Two dancing bacchantes, also reminiscent of figures on the Borghese Vase, lift an oval cartouche, which was originally engraved with the owner's coat of arms. Even though the ancient marble original had not yet arrived at the Louvre, Janety could well have seen engravings of the famous vase and was, in all likelihood, familiar with reproductions of it on display in the gardens of Versailles from the late seventeenth century on.[1] The hinged top of the mustard pot has a small cutout to allow for a mustard spoon. Openwork decoration allows the blue of the original glass liner to shimmer through the metal and adds to the delicacy of the object, which manages nonetheless to retain some of the monumentality of its large-scale antecedent.

Vessels of similar design are also found among English silver models. During the second half of the eighteenth century, the English created a type of mustard pot shaped like a tankard with one vertical handle, a hinged top with a small notch to allow for a mustard spoon, and a pierced silver frame enclosing a blue-glass liner. The liner protected the silver from the corrosive effects of the vinegar in the mustard.[2] This model was common throughout Europe; in France the silver *moutardier* acquired openwork decoration of pilasters, garlands, and other neoclassical motifs.[3]

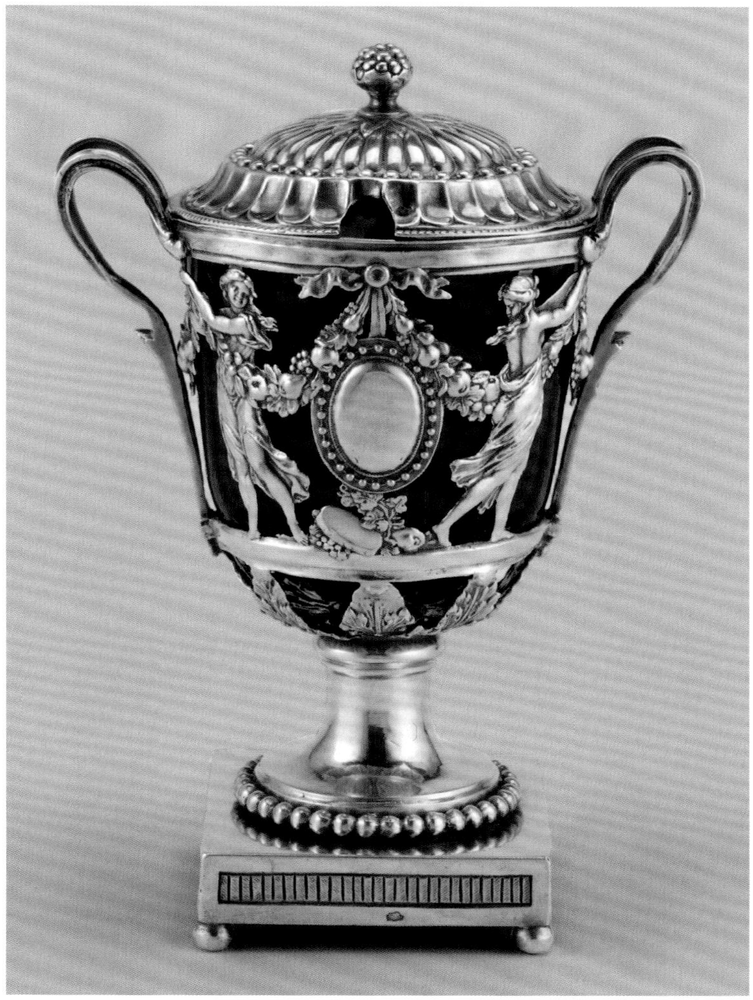

Cat. no. 52

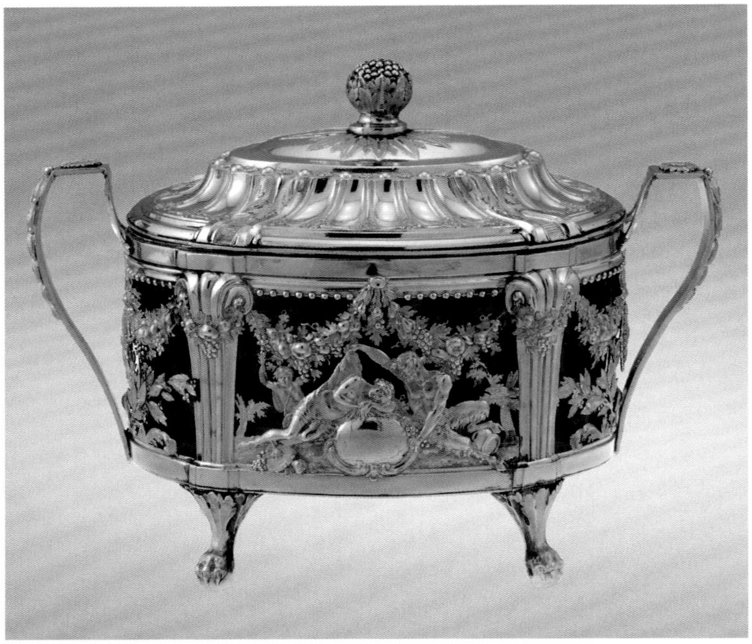

Fig. 1: *Sugar Bowl*, 1786 Marc-Étienne Janety. Platinum and blue-glass liner; 5 ¼ × 7 in. (13.3 × 17.8 cm). Purchase, Gift of Dr. and Mrs. A. L. Garbat Starr, and Julian A. Garbat, by exchange, and Harris Brisbane Dick Fund, by exchange (1974), 1974.164.

Janety's shop was located in Paris, on the rue de l'Arbre Sec, where he was established until 1793. He was the first European goldsmith to work in platinum, which has an extremely high melting point; he made both domestic wares and scientific instruments in this non-tarnishing metal.[4] The only known surviving example of Janety's work in platinum is a 1786 sugar bowl, also in the Metropolitan Museum (Fig. 1). P.M.

1 Sophie Baratte et al., *Les Bronzes de la couronne* (Paris, 1999): 147–48. Oechslin 1982: cat. 71, illustrates an etching of the Borghese Vase from Charles Errard's *Recueil de divers vases antiques*, from about 1651, which was widely available and served as a primary study tool.

2 The silver mustard pot did not become a standard element of table services until the end of the seventeenth century, when the sauce form of the mustard—made from mixing ground mustard seed and vinegar—began to replace the powder that had been served since the fourteenth century. In France the *moutardier* often took the shape of a keg, as was the case in the silver services of both Louis XIV and Madame de Pompadour; see Véronique Alemany-Dessaint, *Orfèverie français* (Paris, 1988): 146.

3 In Paris the sale of mustard was regulated by the corporation of distillers of vinegar, mustard, and spirits, with a membership of 188 masters at the end of the eighteenth century. According to the *Almanach Dauphin* of 1777, the shop of a distiller named Capitaine carried "more than thirty kinds of mustard, notably those in powder form and from England"; see Michel Beurdeley, *Les Petits Métiers de la France d'autrefois* (Paris, 1992): 149.

4 Henry Nocq, *Le Poinçon de Paris*, vol. 2 (Paris, 1927): 353.

53. DESIGN FOR A URN WITH PORTRAIT MEDALLION AND GARLANDS

Pierre-Élisabeth de Fontanieu (about 1730–1784)
1770
Etching, from *Collection de Vases*
Image: 13 ⅛ × 8 ¾ in. (332 × 222 mm)
The Elisha Whittelsey Collection, The Elisha Whittelsey Fund, 1949
49.69. 6 (37)

This two-part design for an urn indicates the degree to which, by 1770, neoclassicism had matured and gained acceptance in France. The vessel's balanced form employs a measured range of ornament between the spherical finial and the plain, square plinth. At the center of the body is a portrait of Louis XVI, crowned with the laurel wreath of Roman emperors. The medallion's oval frame is tied with a bow and decked with floral festoons overlaying the vase's channel-cut mid-portion. Beneath the band of ribbon-tied reeds, flattened acanthus leaves decorate the rounded lower section.

Each of the twenty-six designs for vases in this volume is elaborated in two plates, one with ornamentation and the other showing only the outline.[1] Thus, the precise and elaborate rendering of the decorated urn is reduced on the next page to a contour indicating basic components of its shape, which could be turned on a lathe (Fig. 1). On a practical level, the unusual pairing of prints was meant to instruct designers on how to work with classical forms in metal, ceramic, or even wood.

On a more personal level, the underlying motivation for this volume reveals the passion of Pierre-Élisabeth de Fontanieu, who in 1767 took his father's place as Administrator-General of the Royal Wardrobe, for ornamental lathe-turning. This high-skill hobby called for a sophisticated set-up that, at a minimum, required a treadle-operated lathe with an overhead drum and a cross-slide capable of holding the various tools and chucks in place to work the wood. Several lathes are listed among Fontanieu's tools in the inventory of his private workshop, which was purchased upon his death by Louis XVI. He was also an enthusiast of lathe-turning, continuing a longstanding tradition of this princely hobby.[2] Fonantieu's apparently genuine enthusiasm for lathe-turning matched the king's—what better way to please and impress the monarch than with such a book? His florid dedication in the front of the volume confirms this intention.[3]

It was unusual for a court official to produce such a body of work. Whether Fontanieu, who seems not to have been trained as an artist, produced the original designs or etchings himself remains unclear. He did, however, relish his position and took particular interest in his role, participating for example in the

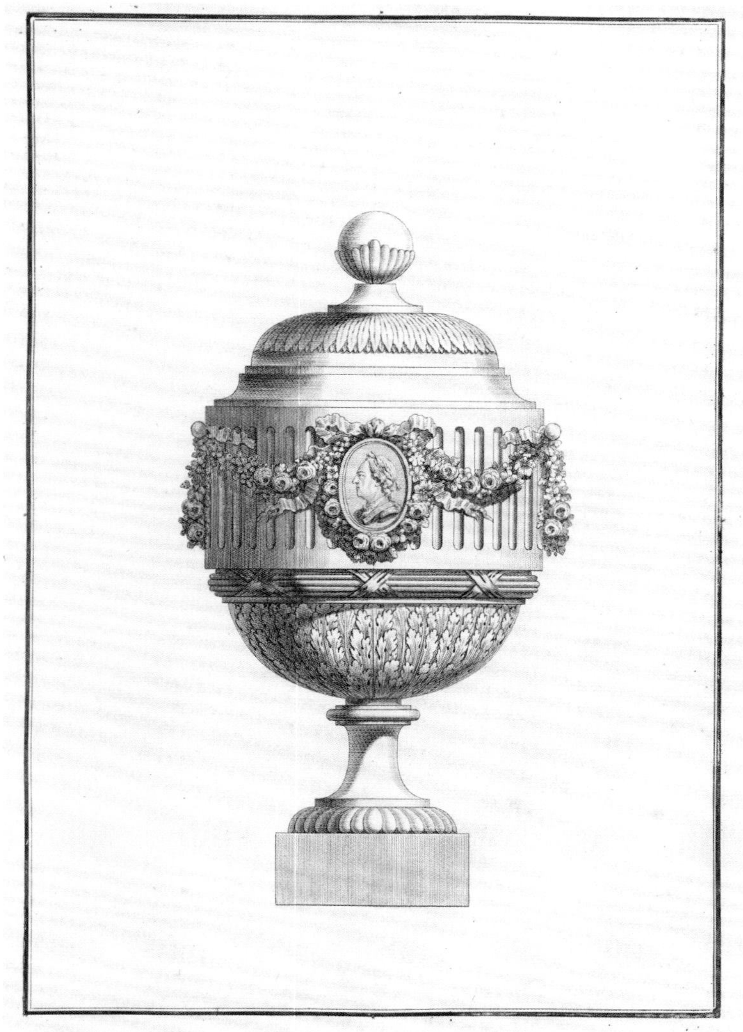

Cat. no. 53

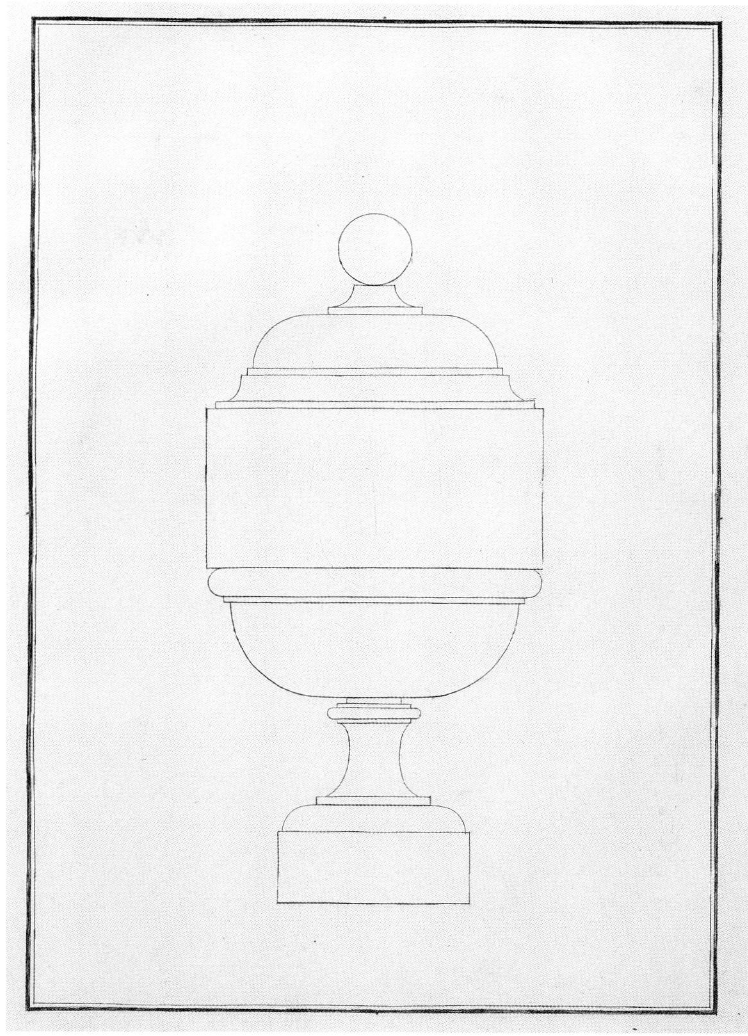

Fig. 1: *Outline Drawing* for cat. no. 53. Image: 12 ⅞ × 8 ⅝ in. (327 × 219 mm). The Elisha Whittelsey Collection, The Elisha Whittelsey Fund (1949), 49.69.6 (38).

design process of the famous cabinetmaker Jean-Henri Riesener and directly supervising the conversion of royal furnishings from the rococo to the neoclassical style. The *Collection de Vases* had an immediate impact: the Sèvres factory produced at least two designs from the volume that could be formed in molds and trimmed on the lathe.[4] C.H.

[1] Not including the dedication page, the *Collection de Vases* contains forty-six plates. The first forty comprise twenty vase designs in pairs—the fully ornamented version, followed by an outlined one. Six of these pairs illustrate two designs per plate. Six additional plates show "pièces de tour" (objects for lathe-turning), plinths and columns with elaborate neoclassical ornament.

[2] Eriksen 1974: 180.

[3] The dedication translates:

> Sire, the goodwill with which you kindly considered the works for [lathe-] turning that I have had the honor of presenting to you, in addition to the permission that you granted to dedicate this new collection to you, has encouraged me to lay it before your eyes. I hope this collection will please you and demonstrate the zeal that animates throughout. [From] the most respectful and most humble of your servants, Fontanieu (Sire, Les bontés avec lesquelles Vôtre Majesté a bien voulu regarder les Ouvrages de Tour que j'ai eu l'honneur de lui présenter, ainsi que la Permission que vous m'avez accordée de vous offrir cette nouvelle Collection, m'a encouragé à l'exposer souz vos yeux; J'aurai rempli tous mes voeux si ce Recueil peut vous plaire et vous prouver le zele qui anime sans cesse. Le plus Respectueux et le plus soummis de vos sujets. Fontanieu).

The design of the dedication plate subtly integrates some of the lathe-turner's tools, such as the end of a pair of calipers, in the elaborate trophies that ornament the neoclassical frame enclosing the text.

[4] Eriksen 1974: fig. 288.

54. TEA URN WITH LION AND SATYR MASKS

John Carter II (act. 1768–77)
About 1773–74
Silver, ivory
20 ¾ in. (52.7 cm)
Marks: [inside section of urn into which heated iron is inserted and on base] *IC* in rectangular reserve (master's mark); [inside section of urn into which heated iron is inserted and on base] *S* (year mark for 1773–74); [inside section of urn into which heated iron is inserted and on base] crowned leopard's head (mark for London assay office); [inside section of urn into which heated iron is inserted and on base] lion passant (assay mark for sterling silver)
Rogers Fund, 1911
11.28

A new addition to the tea service, the classically inspired tea urn appeared in large numbers around 1770. Resting on low, square stands, the aptly named vessels were generally plainly ornamented, as suited the new taste in metalwork.[1] The decorative scheme here, however, is more robust than most. Unlike slightly earlier models that show greater expanses of unadorned surfaces, this example displays an array of bold sculptural ornament. Gadroons follow the sharp rise of the body from the shoulder to the narrowing on neck, body, and pedestal foot. A pair each of lion and satyr masks alternate just below the shoulder and are linked by swags of drapery, which cross the forehead of the satyrs, pass behind their pointed ears, and attach to the lions' heads. The delicate coat of arms, inscribed with an ermine and pair of lions to the left of the central satyr mask, was added later for a Philadelphian, General Philemon Dickinson, or his daughter.[2]

Despite such attention to detail, the tea urn's unusually severe shape has been noted for its lack of playfulness in comparison to other neoclassical examples.[3] Indeed, the long neck with its domed cover relates to Roman sepulchral urns. Drawings and prints of these sober forms by Pietro Santi Bartoli appeared as early as the second half of the seventeenth century, and such illustrations enjoyed popularity during the eighteenth.[4] Most certainly the nobility of the form and not its original function interested the designer of this urn.

The very name of this object—tea urn—can be misleading, as tea was never made or stored in this particular vessel. Early examples such as this merely held and kept hot the water that was needed for brewing tea. One could remove the domed cover, with its acanthus-leaf finial, in order to fill the urn with hot water, which was kept warm by an iron bar inserted into a tube inside the vessel.[5] A turn of the spigot, originally in ivory on this urn, provided a flow of water over loose china leaves in a teapot.[6] The urn's spool-shaped foot and square plinth provided the necessary height to place a pot below the tap. Resting prominently on a stand apart from the tea

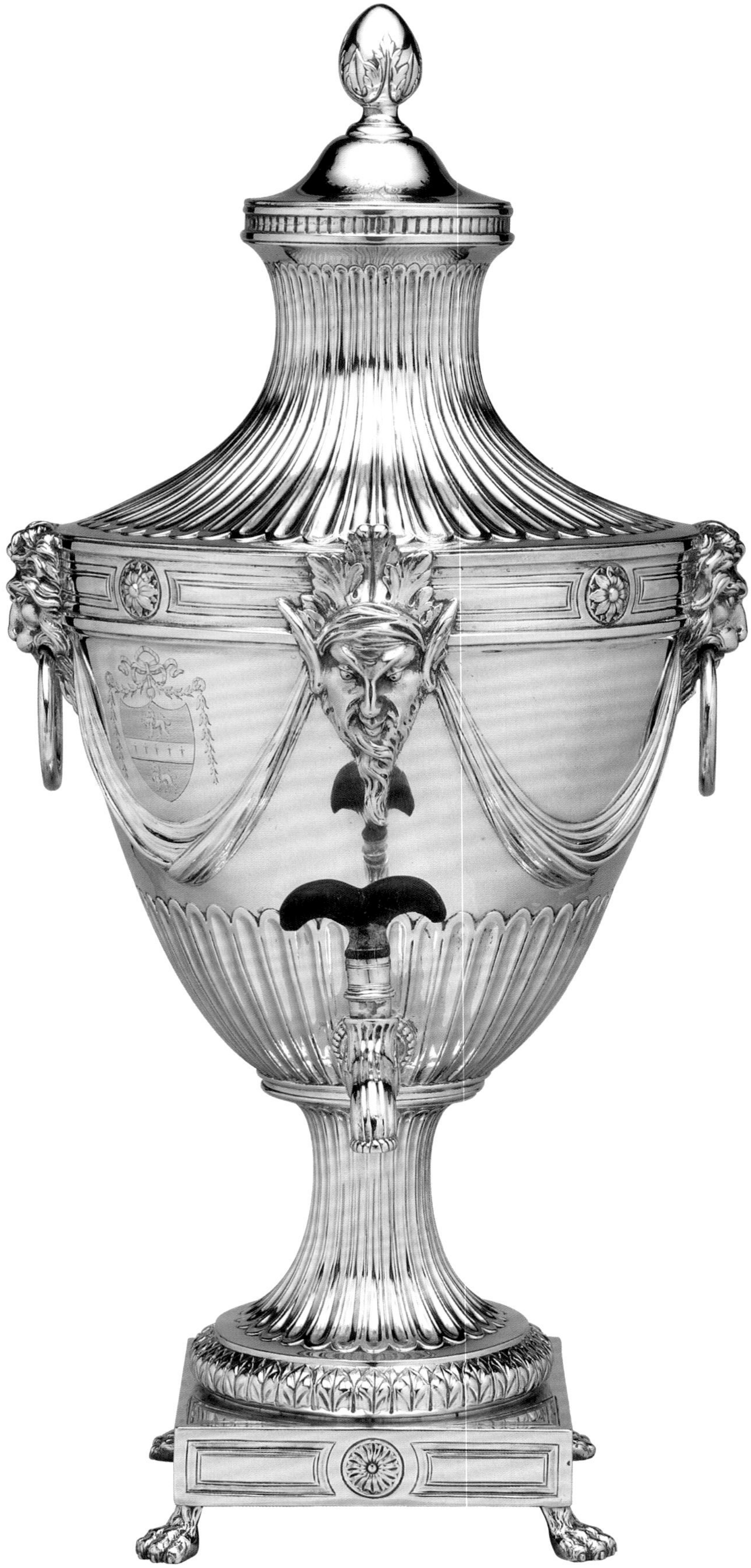

Cat. no. 54

table, the form was far more convenient than the kettle it replaced. The internal heating element also made for a more elegant design solution than external charcoal burners, but by the 1790s spirit lamps were again used.[7] The change to a hotter heating mechanism in the spirit lamp led to tea being brewed directly in the urn.[8]

The master goldsmith John Carter II was well versed in the neoclassical style. He is best known for executing salvers and candlesticks, the latter designed by Robert Adam in the shape of Corinthian columns, a standard conceit of neoclassical silver.[9] Whether the silversmith procured the design for the present urn or followed his own scheme is not revealed by the marks, but the boldly applied neoclassical elements bear no resemblance to tea urns in Adam's pattern books.[10] As such, Carter's vessel reveals an alternate interpretation of the progression of neoclassical form and ornament in eighteenth-century English silver. C.H.

1 G. Bernard Hughes, "The Georgian Vogue of the Tea Urn," *Country Life* 132 (Oct. 25, 1962): 1026.
2 Curator Jessie McNab confirmed the attribution of the coat of arms via facsimile on Nov. 21, 2002.
3 Jessie McNab Dennis, *English Silver* (New York, 1970): 78.
4 Oeschlin 1982: 85, fig. 4. For Carlo Antonini's 1772 engraving of a marble urn on an antique altar, see idem: 88, fig. 74.
5 For a tea urn from about 1770, see Hughes (note 1): fig. 1.
6 China tea did not require boiling water for brewing, as did teas from India.
7 Clayton 1985: 424.
8 Hughes (note 1): 1027.
9 He supplied candlesticks to the firm of Wakelin and Parker as early as 1769. Rowe 1965: 37, pls. 10–11a; and Michael Snodin, "Putting Adam into Context," in idem, *Silver and Jewellery: Production and Consumption since 1750* (Birmingham, Eng., 1995): 16. For more on Carter, see Grimwade 1982.
10 *Jackson's Silver and Gold Marks of England, Scotland and Ireland*, ed. Ian Pickford, 3rd ed. (1905, reprint London, 1989): 215.

55. HOT WATER JUG WITH LEATHER HANDLE ON STAND

Matthew Boulton (1728–1809) and John Fothergill (d. 1782), Soho Manufactory, near Birmingham; designed by James Wyatt (1748–1813)
1775–76
Silver, leather
Jug: 14 ½ in. (36.8 cm)
Stand: 5 ⅛ in. (13.2 cm)
Marks: [on foot of jug] *IF* (master's mark for John Fothergill); *MB* (master's mark); *C* (year mark for 1775–76) leopard passant (assay mark for sterling silver); anchor (mark for Birmingham assay office)
Jug: Metropolitan Museum of Art Voluntary Anniversary Gift, 1993
1993.73
Stand: Gift of Madame Liliana Teruzzi, 1966
66.172.5

James Wyatt's design for this jug evokes, but does not duplicate, the classical wine ewer, or *oenochoe*. The nuanced profile of the handle and the faceted neck exemplify English neoclassicism of the 1770s.[1] Like Robert Adam before him, Wyatt had spent several years studying ancient art and

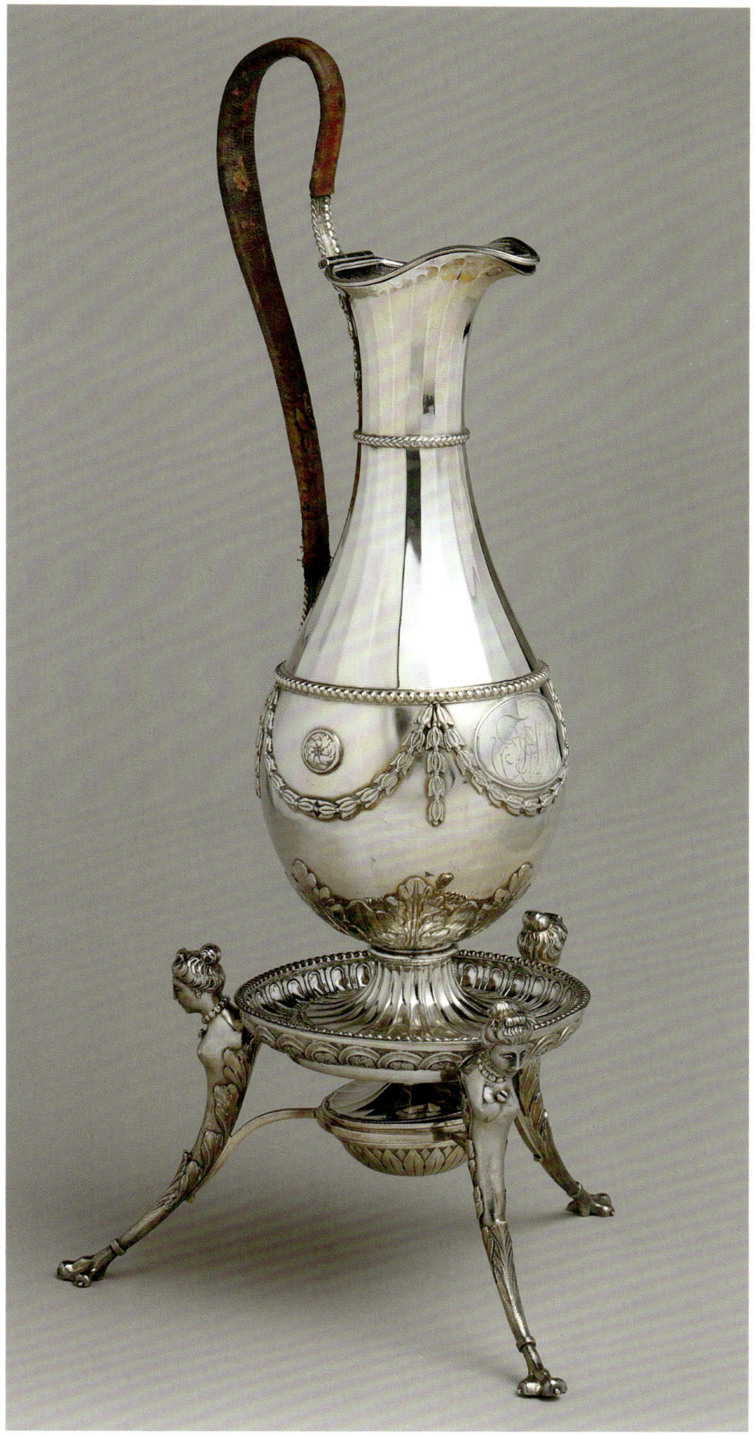

Cat. no. 55

architecture in Italy. When he returned to England in 1768, the Adam style was in full force. Wyatt received many architectural commissions, and like Adam developed accessories that matched his classically styled interiors. Matthew Boulton's silver works in Soho, near Birmingham, produced designs by both architects.

While he unabashedly emulated Adam's motifs, Wyatt's designs are lighter in spirit, reflecting a taste for expanses of smooth, reflective surface between areas of ornament, as seen in this elegant piece.[2] Its basic form, without ornament or stand, appears in a Boulton and Fothergill pattern book, and a closely related drawing appears in an album of Wyatt's, once in the collection of the Vicomte de Noailles.[3] The drawing shows the same fluted foot, but the leaf festoons and alternating bay and acanthus leaves on the lower body of the Metropolitan Museum's example more closely resemble the ornament of a Wyatt design for a candlestick.[4]

Most of Wyatt's silver designs were commissioned by architectural clients, but the nature of this piece's relationship to Thomas Brettell, the owner identified by the monogram on the large, central oval plaque is not known.[5] The particulars of Boulton and Wyatt's interaction likewise remain obscure, but it seems clear that the manufacturer did seek designs from the aspiring architect (whose brother was one of his agents), and sold the resulting products independently.[6] The Soho manufacturer recognized that favor among London society would boost sales of lesser lines among England's growing middle class. Eager to promote himself as a man of taste, he used the new neoclassical style to enhance the reputation of his city's fledgling silver industry.[7]

Boulton was also quick to realize the practicality of the new aesthetic. Unlike rococo silver, which depended on handcrafted surface decoration, neoclassical silver called for flawless surfaces and repetitive, precise ornament that was easily and accurately produced with new machines, such as the roller mill and die cutter, both of which he owned.[8]

While Boulton and his partner, John Fothergill, embraced modern technology to manufacture most of their products, this jug was apparently never mass-produced.[9] Although a piece such as this required some mechanized processes, the amount of hand finishing involved would have been costly. Initially, the Soho factory employed such production methods for more exclusive goods, but it dropped silver-making in favor of steam-engine manufacture within a decade.[10] Wyatt too had moved on to bigger commissions, and he evidently stopped designing silver. C.H.

1 Jessie McNab, *Recent Acquisitions: A Selection, 1992–1993, MMA Bulletin* 51, 2 (1993): 41.
2 See for example Frances Fergusson, "Wyatt Silver," *Burlington Magazine* 106 (May 1960): 750 (fig. 8), 754 (fig. 50).
3 Ewer no. 237 on the page of designs from the first pattern book (nos. 234–42), reproduced in W.A. Seaby and R. J. Hetherington, "The Matthew Boulton Pattern Books, Part I," *Apollo* (Feb. 1950): 48, fig. 2. For the jug and stand in the Noailles collection, see Fergusson (note 2): 753, fig. 53.
4 Fergusson (note 2): 750, fig. 49.
5 The curatorial files in the Metropolitan Museum's Department of European Sculpture and Decorative Arts indicate that descendents of Thomas Brettell sent this piece to auction.
6 Fergusson (note 2): 752.
7 Birmingham had not been a center for silver makers before 1770, when Boulton successfully lobbied for a local essay office.
8 Kenneth Quickenden, "The Planning of Boulton and Fothergill's Silver Business," in Michael Hallet, ed., *Silver and Jewellery: Production and Consumption since 1750* (Birmingham, Eng., 1995): 21–28.
9 Other examples do exist, including an identical model with a tripod stand, from 1779, in the Nelson Atkins Museum of Art, Kansas City, Missouri (F95–3a–c).
10 Fergusson (note 2): 752.

56. CHOCOLATE POT WITH OVAL CARTOUCHE

John Wakelin and William Taylor (partnership act. 1777–92)
1779
Silver, wood
Marks: *IW*, a trefoil above (master's mark for John Wakelin); *d* (date mark); lion passant (assay mark for sterling silver); crowned leopard's head (mark for London assay office)
12 ⅜ in. (31.5 cm)
Anonymous Gift, 1976
1976.424.1

Evoking the shape of the *oenochoe*, or classical wine ewer, this chocolate pot offers no distinctive attributes that would identify its purpose. The combination of a beaklike spout, hinged cover, arching loop handle in wood, silver, or ivory with the elongated vase shape and pedestal foot made a stylish solution that suited coffee as well as chocolate.[1] While such pots could be replete with neoclassical ornament, the decoration of this example is restrained. Besides a small acorn finial, only the molded gadrooned edge on the spout, shoulder, and foot break up the smooth expanse of silver. These simple details seem to contrast with a more delicate engraved oval tied with a lover's knot on the front of the vessel. Most neoclassical silver favored shield-shaped cartouches and festoons engraved in the bright-cut method. This oval may have been a later addition, possibly to soften the strong form for use in an Edwardian lady's parlor.[2]

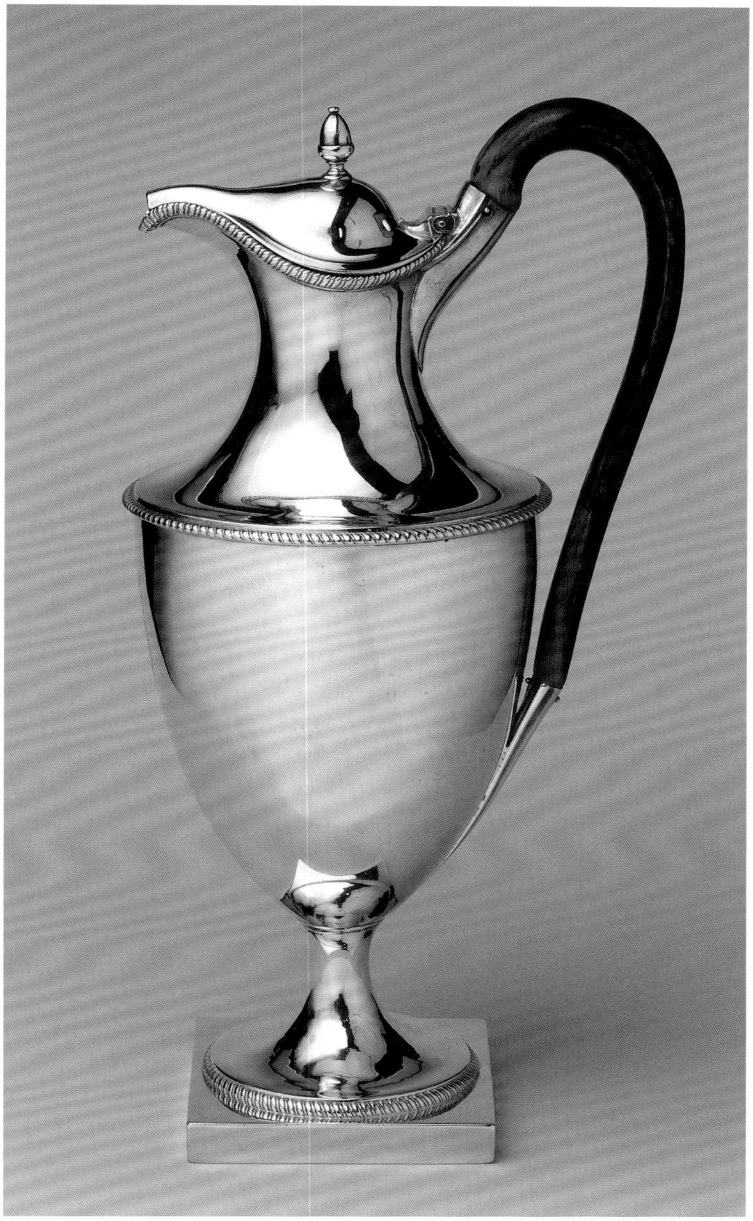

Cat. no. 56

For the majority of silver holloware made in late eighteenth-century England, the new aesthetic meant that the form should follow the geometry prescribed by classical antecedents. With the chocolate pot, such rigid adherence to neoclassical principles often meant a disregard for functional innovation. Since the introduction of chocolate to London society around 1660, the luxury beverage had become increasingly popular and the demand for vessels geared to its preparation grew as well.[3] Some of the earliest known pots position the handle on the side at a right angle to the spout.[4] The 1700s saw further improvements: a hole in the lid to let steam escape while stirring a beater without losing excessive heat, a long spout for allowing the mixed liquid to pour from the bottom (chocolate sediment, unlike that of coffee or tea, was to be savored), a heavy foot for insulating the table, and often a burner as well.[5] Except for the thick, square base under the flared foot, these characteristics disappeared under Robert Adam's influence and made neoclassical chocolate pots virtually indistinguishable from those for coffee or hot water.[6]

A double-duty vase shape such as this could be spun on the lathe, making it economical to produce as well as aesthetically pleasing. John Wakelin and William Taylor, who were registered as platemakers, created this piece three years into their partnership.[7] Wakelin, who was the son of a prominent London goldsmith, subsequently joined with Robert Garrard, founder of the famous London firm that bore his name.[8] Aspiring eignteenth-century silversmiths such as Wakelin seemed to prefer classical shapes as much for their adaptability as for their fashionable form. C.H.

1 See neoclassical coffee pots by William Holmes and Nicholas Dumée, in Vanessa Brett, *Sotheby's Directory of Silver, 1600–1940* (London, 1984): 220, figs. 971–72.

2 A nineteenth-century retailer could have added the engraving in an effort to update the piece to modern tastes. I am grateful to Jessie McNab this suggestion.

3 G. Bernard Hughes, "Silver Pots for Chocolate," *Country Life* 20 (Oct. 1960): 856.

4 Edward Wendem, "Silver Chocolate Pots," *Antique Collector* 17 (Sept.–Oct. 1946): 167–73, fig. 1.

5 See examples in ibid: 167–73, figs. 2–6, 9–10.

6 Hughes (note 3): 857.

7 Grimwade 1982: 691.

8 Christopher Lever, *Goldsmiths and Silversmiths of England* (London, 1975): 147, 149.

57. CREAM JUG WITH ENGRAVED CARTOUCHE AND FESTOONS

Probably Thomas Streetin (act. 1791–after 1820)
1792–93
Silver
5 ½ in. (14 cm)
Marks: *T S* in rectangle (master's mark); *r* (London year mark for 1792–93); crowned leopard's head (mark for London assay office); lion passant (mark for sterling silver); king's head (mark for silver tax)
Rogers Fund, 1913
13.42.54

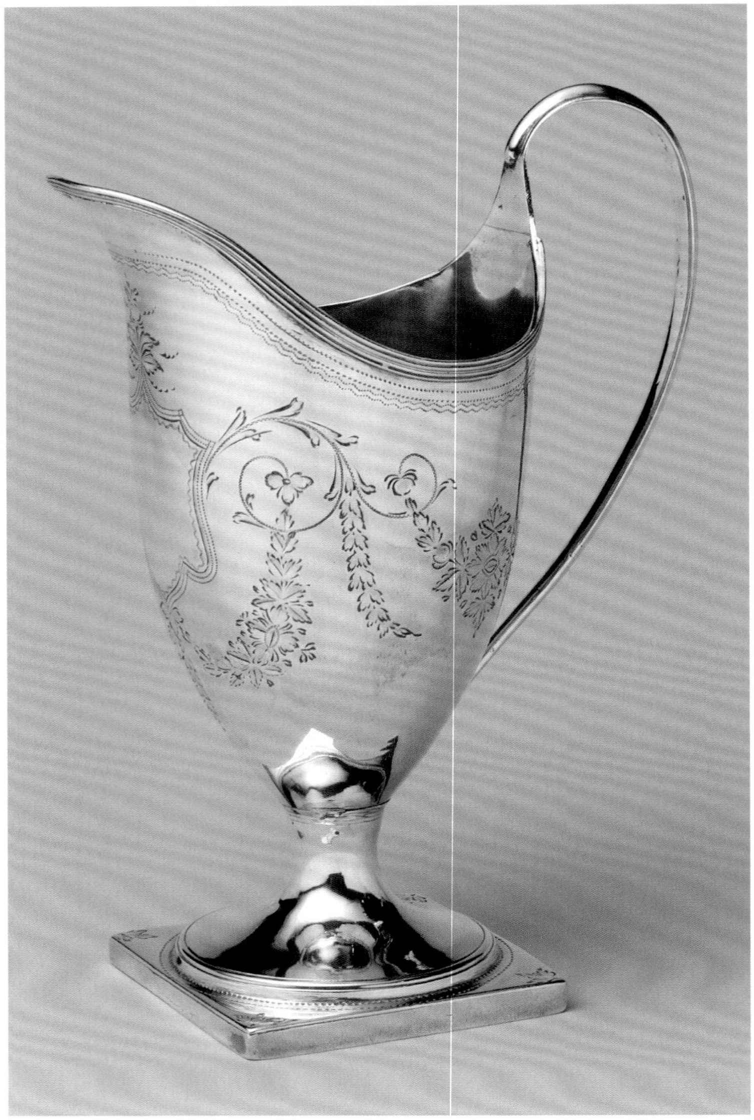

Cat. no. 57

The marks on the vessel suggest that it was made in 1792 or 1793 by a little-known London silversmith named Thomas Streetin.[1] Adhering only partly to the outlines of the classical vase, this cream jug derives from another antique form: the upturned helmet of a warrior. Helmet-shaped jugs and ewers appeared in England during the seventeenth century and, by the eighteenth, bore the marks of such notable silversmiths as Paul Crespin, Paul de Lamerie, and David Willaume.[2] Unlike its baroque antecedent, generally a squat cylinder attached to a hemispherical bowl, this diminutive cream jug possesses a subtly curved, unbroken profile that is nearly elliptical. The rim rises gently to form the spout and then more sharply to a peak on the opposite side, where the narrow loop handle attaches. The balanced composition combines the baroque helmet-ewer form with more attenuated, classically derived amphora-like shapes.

Equally understated is the jug's molded decoration: reeding along the rim and a rope pattern along the edge of the circular foot. In contrast, the decoration on the body is much richer: bright-cut engraving envelops the body with floral festoons and a large cartouche, roughly shaped like a shield, under the spout.[3] Bright cutting, so called for its sparkly brilliance, involved lifting out short slivers of silver with a sharp, beveled chisel to create an energetic yet delicate pattern.[4] Requiring less time to execute than more technically demanding methods of engraving with precise, long lines, the technique allowed decorators easily to add embellishments like the rows of tiny cuts directly under the reeded rim in this piece. The overall playful effect lightens an otherwise somber form.

This cream jug was most likely part of a set, along with a coffee or teapot, sugar bowl, tongs, and hot-water urn. Cream jugs similar to this one were a mainstay of the tea table from the 1770s until at least the first quarter of the nineteenth century. C.H.

1 Streetin served his apprenticeship to the silver-spoon maker John Lambe between 1783 and 1790, when he entered his first mark as platemaker. He took on his son Henry as an apprentice in 1816. Because Streetin's registered mark has a point between the initials, lacking here, there is some doubt about this attribution.
2 Christopher Lever, *Goldsmiths and Silversmiths of England* (London, 1975): fig. 45; and Vanessa Brett, *Sotheby's Directory of Silver, 1600–1940* (London, 1984): 175 (figs. 711–12), 188 (fig. 782).
3 The cartouche was a common motif for receiving a coat of arms or family crest; apparently, this one was never used.
4 Harold Newman, *An Illustrated Dictionary of Silverware* (London, 1987): 49.

58. DESK SET WITH MINIATURE URNS ON TRIPODS

John Plimmer (act. 1771–1803)
About 1802
Silver
Width 13 $^{3}/_{16}$ in. (33.1 cm)
Marks: [underside of frame, urns, their covers, and tray] *I•P* in rectangle (master's mark); [underside of frame, urns, and their covers] *G* in square with canted upper corners and shaped lower edge (year mark for 1802–03); [underside of frame and urns] leopard's head (mark for London assay office); [underside of frame, urns, their covers, and tray] lion passant (mark for sterling silver); [underside of frame, urns, and their covers] profile of king's head (mark for silver tax)
Anonymous Gift, 1976
1976.424.2

As this group of unusual tripod urns demonstrates, desk and library sets, and inkstands were among the most adaptable objects for interpretation in the neoclassical style. Made as early as 1610, the form was originally called a "standish," shorthand for the "standing dish" that held the many writing accessories of persons of rank.[1] The more common term, inkstand, was in use by the time this example was made, but only for sets featuring the newer, less expensive and easier-to-clean flint glass containers.[2] Silver, however, continued to be employed, as it denoted its owner's status. Here, the malleable precious metal allowed the faceted bowl of each pot to be seamlessly soldered onto a narrow pedestal and placed into the fitted tripod stand.

Only the variation in the flat covers reveals the function of each urn. Of the middle two, from left to right, a plain top covers

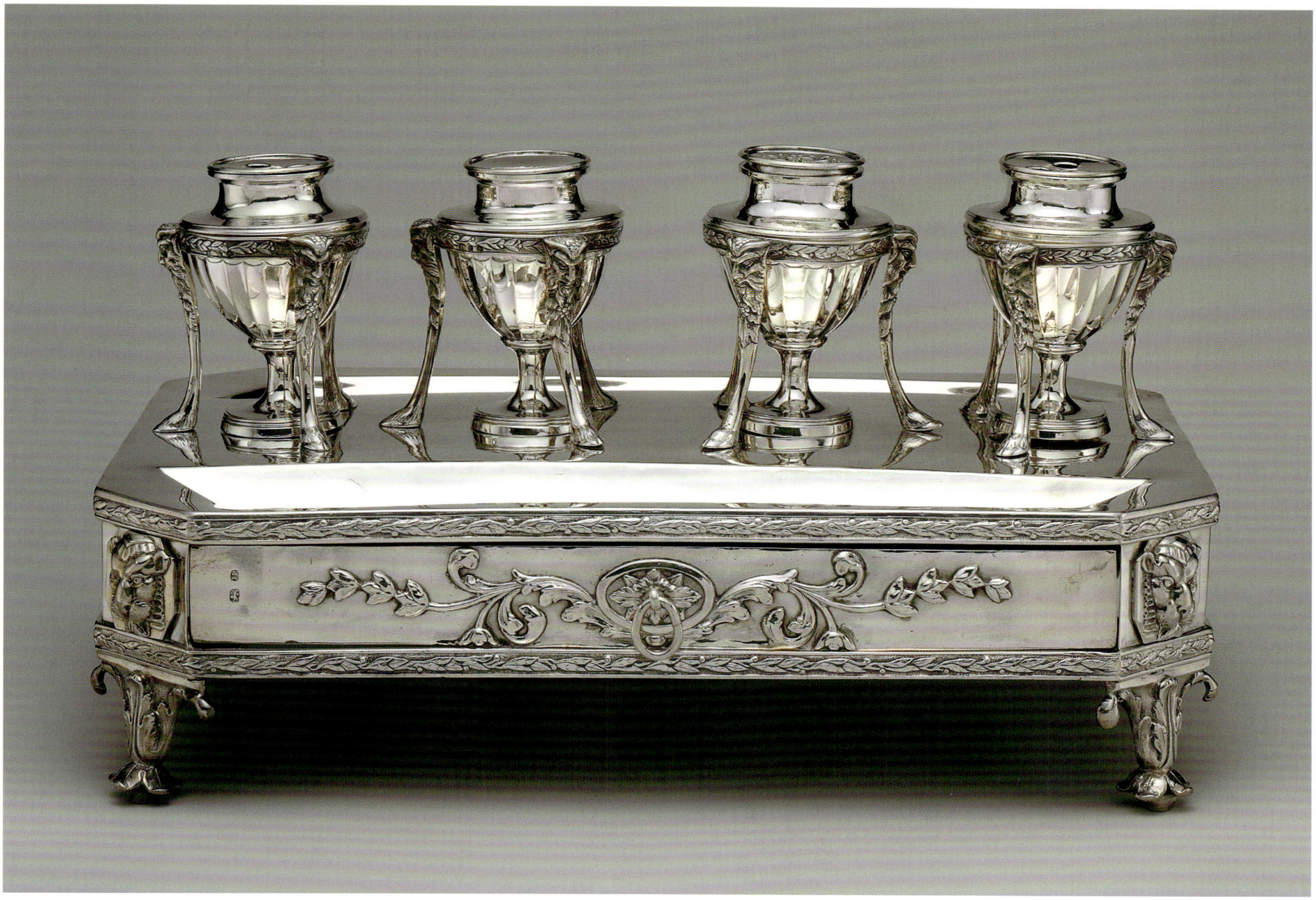

Cat. no. 58

the container for wafers—wax disks used for receiving the seal of the owner—while the ornately cut top belongs to the vessel that would have dispensed pounce, a fine powder of gum sandarac used for blotting ink. Covering the containers at either end are lids with a trio of holes, large enough to hold pens or quills (during this period, the latter were still made from goose feathers precisely cut with a penknife). One of these vessels held ink, a ground pigment made with gum tragacanth, while the other stored the cleaner of preference, tiny balls of lead shot.[3] Rounded channels on either side of the tray and the shallow drawer below housed other writing implements: the newly invented steel-nibbed pens, pencils, penknives, rulers, erasers, and dog-tooth sharpeners.

Typical of the later neoclassical style, the decoration here intermingles animalistic and floral elements with expanses of smooth surfaces. The animal-like legs of the tripod stands feature molded satyr masks and hoofed feet, while the more sedate laurel leaf and berry band on the ring corresponds to the molded banding that reinforces the edges of the rectangular base. On the fascia, foliate scrolls of acanthus leaves unfurl from a central oval, attached by bolts to a drop ring. The silversmith also appropriated Egyptian motifs (see cat. no. 66), more unusual in England than purely Greek- or Roman-inspired English silver. Full lion faces, with pharaonic headdresses fit for a pharaoh, ornament the four canted corners of the tray just above its short legs, embellished with acanthus leaves, a device that would become common in the future Regency style.

According to his registered mark (entered in 1771), John Plimmer was a smallworker, a silversmith focused on specialty items not associated with dining.[4] Freed from having to concentrate on large matching sets, this class of silversmith could be more daring in the execution of designs. Here, Plimmer displayed full understanding of the history of the form, as well as the use of neoclassical ornament. The rectangular base is reminiscent of an earlier type, the so-called treasury inkstands produced a century before, as well as casket shapes from the 1760s.[5] Freestanding vase-shaped containers were not entirely new. In the 1720s, the Huguenot silversmith Paul de Lamerie used florid vase forms on his elaborate rococo creations; as early as the 1770s, the silversmiths Edward Wakelin and John Parker were decorating neoclassical vases.[6] Plimmer took this idea a step farther with his tripod stands, a device that secures the vases while also allowing them to be moved around.[7] Plimmer's desk set offers a highly literal adaptation of the neoclassical vase, especially when compared to other unusual neoclassical designs for inkstands, particularly the sphere-shaped variety, those with canoe-shaped trays, and later Regency models. C.H.

1 Jonathan Stone, "English Inkstands in Silver and Sheffield Plate," *Antiques* 90 (Sept. 1966): 342.
2 G. Bernard Hughes, "Georgian Silver Inkstands," *Country Life* 127 (Feb. 1957): 230; and Stone (note 1): 342.
3 Hughes (note 2): 230.
4 Grimwade 1982: 628. See also *Jackson's Silver and Gold Marks of England, Scotland and Ireland*, ed. Ian Pickford, 3rd ed. (1905, reprint Woodbridge, Eng., 1989): 227.
5 Stone (note 1): 334; and Clayton 1985: 219.
6 See the Paul de Lamerie inkstand in Stone (note 1): 334, fig. 4; and a 1771 inkstand by Edward Wakelin and John Parker in Rowe 1965: pl. 31.
7 Similar classical stands are found in a desk set using glass containers with silver mounts, made almost a decade earlier for example by W. Pitts and J. Preedy; see Clayton 1985: 218, fig. 313.

59. HOT WATER POT WITH TRIPOD STAND AND SPIRIT LAMP

Paul Storr (1771–1844)
1807–08
Silver, wood
11 3/4 in. (29.8 cm)
Marks: *PS* (maker's mark); *M* (year mark for 1807–08);
lion passant (mark for sterling silver); sovereign's head (dirty mark)
Gift of Dr. and Mrs. John C. Weber, 1979
1979.495.1

As another solution to the "vexing problem of replenishing the teapot with boiling water, without ringing for a servant," this hot-water pot by Paul Storr, made toward the end of the period examined here, seems altogether different from the tea urn by John Carter II (cat. no. 54) realized some thirty years earlier.[1] Yet, when one focuses on their basic structure, both vessels reveal a similar classical urn shape. Unlike his predecessor, Storr exposed the heating element and chose to embellish the vessel only minimally. The sparse ornamentation includes a molded band break that encircles the neck and a gadrooned band on the shoulder. The latter is repeated on the circular frame of the tripod—also a neoclassical form—and the urn-shaped spirit lamp it carries. Storr relied upon classical sources for other ornamental details as well, such as the fruit-and-leaf finial, pendant rings, reeded legs, and leonine paw feet.

This hot-water pot was part of a set consisting of a teapot, sugar bowl, creamer, and waste bowl; but it also may have been sold separately.[2] Such vessels seem to have been stock items Storr produced through much of the first decade of the nineteenth century. Other examples reveal a variety of available sizes and handles, easily exchanged in the volute-socket on the body of the vessels.[3]

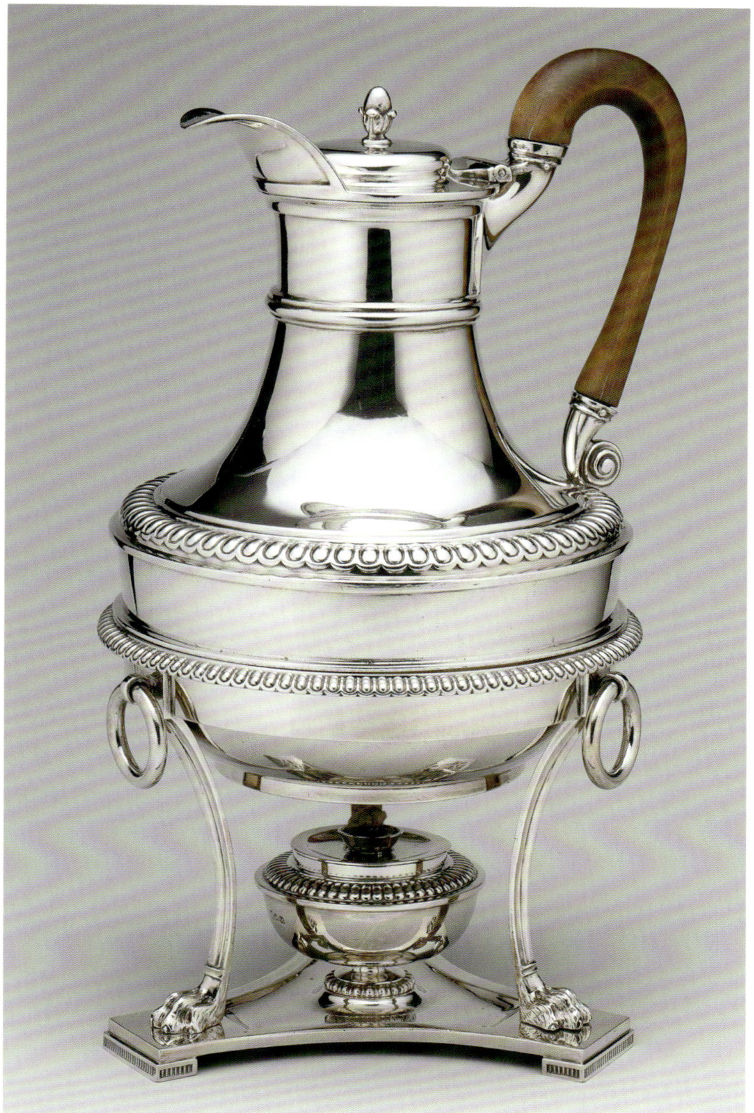

Cat. no. 59

Storr received a traditional apprenticeship training, a practice that was no longer obligatory by the early nineteenth century. In 1792, under William Frisbee, he produced his first piece, a Greek-styled vase, a fitting debut for a silversmith born during the heyday of the Adam style.[4] Storr's early work, like this hot-water pot, moves beyond this vein, although its assured forms and spare decoration do not yet suggest the influence of the French Empire style apparent in later pieces. The silversmith worked independently at the start of his career, but joined Rundell, Bridge, and Rundell shortly after he made this piece. They were appointed goldsmiths to the Prince Regent, and much of their output reflected their royal client's taste for heavier, more decorative pieces. Storr's designs for the firm reflect this emphasis on ornament. C.H.

1 Jessie McNab, "Paul Storr Hot Water Pot, Stand and Spirit Lamp" in *Notable Acquisitions, 1979–1980 MAA Bulletin* (1980): 33.
2 For a similar hot-water pot with a tea set, see N. M. Penzer, *Paul Storr, 1771–1844: Silversmith and Goldsmith* (New York, 1954): 143, pl. 32; and McNab (note 1).
3 McNab (note 1) stated that "the earliest recorded example" dated to 1801. For an example of the same form, with carved loop handles, see Penzer (note 2).
4 Penzer (note 2): 19.

60. TUREEN FOR WILLIAM MURDOCH

Probably William Stroude (act. 1788–1823)
1808–09
12 ¾ in. (32.4 cm)
Marks: [rim of cover and rim of foot] *W S* in a rectangle (maker's mark); *N* (year mark); [top of cover and rim of foot] crowned leopard's head (mark of London assay office); [top and rim of cover and rim of foot] lion passant (assay mark for sterling silver); [top and rim of cover and rim of foot] king's head (duty mark)
Bequest of Alphonso T. Clearwater, 1933
33.120.1

This imposing tureen, more than many earlier neoclassical interpretations of the form featured here, emphasizes its austere profile by relying on a modicum of ornamental detail. Creating an oval bowl deeper than one finds in earlier Adamesque designs for similar objects, the sides of this tureen rise steeply and terminate in a cabled edge on the shoulder that corresponds to the molding on the periphery of the round base of the pedestal foot. Narrow molded strap handles, sprouting on either end from attenuated acanthus leaves near the top of the foot, sweep upward widely and break in sharp right angles to reattach at the top of the foot. Identical handles appear to have occurred first on an elliptical silver tureen over a dozen years earlier, but by 1800 were a fairly common device and possibly indicate the influence of French ceramic designs for geometric handles.[1] The half-domed lid, capped with a pair of confronted foliate C-scrolls, extends to the rim of the inset shoulder. This would make serving soup difficult without a liner, which, if the tureen originally had one, has long since disappeared.[2]

The popularity of soup notwithstanding, the wide expanse of this deep bowl was clearly intended for a more lofty purpose.[3] The lengthy inscription indicates that it was presented to William Murdoch, an Englishman, by the American Supreme Court Chief Justice John Marshall, along with his two business partners, his brother James and one R. Colston. A date mark located on the rim of the foot gives the year of production as 1808–09. Seen on both sides of the cover, the crest of the Murdoch family,

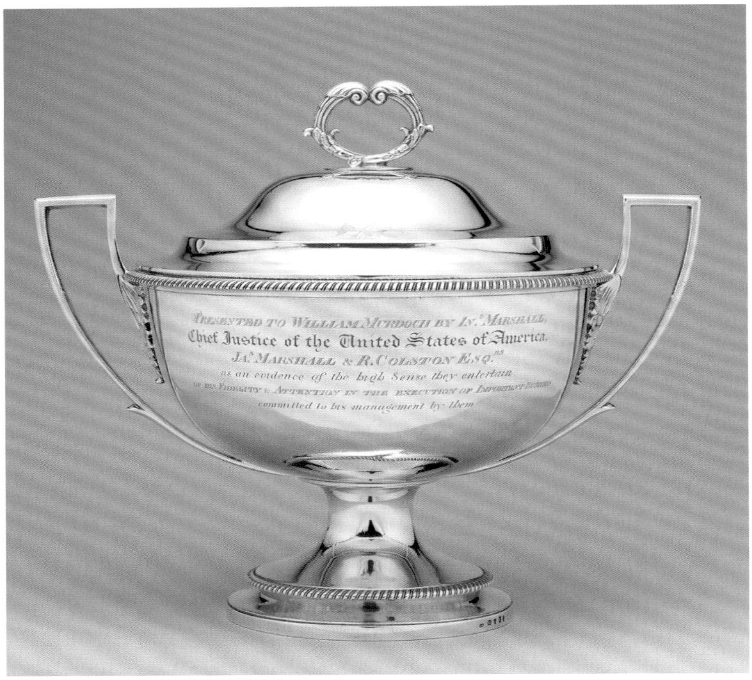

Cat. no. 60

"a raven rising sable, with an arrow through the breast gules," is the only other engraved decoration.[4]

Like many vessels, tureens were well suited for wholesale adoption of the classical vase or urn shape. A French import in regular use in England by the 1720s, the covered form played a significant role in disseminating the formal restraint of the neoclassical style as much as it once had in promoting the exuberant rococo.[5] It mattered little that few Roman and Greek antecedents in metal existed; inventive silversmiths, eager to appropriate the new manner, looked to various other media for inspiration.[6]

Archeological treasures like the stout marble vase excavated in 1770 from Pantanello, near Tivoli, and immortalized in etchings by Giovanni Battista Piranesi inspired some to adapt such forms to practical use.[7] The Warwick Vase (see cat. no. 2) inspired a brief vogue for circular tureens.[8] Ideas for classical soup tureens appear in Michelangelo Pergolesi's *Designs for Ornaments*, first issued in 1777 (see cat. no. 81).[9] In the 1770s, Mathew Boulton and John Fothergill carried out more than a dozen designs of vase-shaped covered oval dishes in the Adam style with any number of variations: round pedestal foot or four claw feet; ornamental garlands, fluting, or plain sides; upswept handles or pendant rings.[10] By the end of the century, the taste was for more restrained decoration, such as gadrooning.[11] Monumental forms like this 117-ounce covered vessel proved over time to be more practical as a presentation piece than as a serving dish.[12] C.H.

1 Robert Makepeace also used them in a tureen of circa 1795–96; see Gerald Taylor, "Silver and Plate," *The Late Georgian Period, 1760–1810* (London, 1956): 74, fig. 1; and *The Campbell Museum Collection*. 2nd ed. (Camden, N. J., 1972): fig. 11.
2 Liners made of silver were often melted down.
3 The inscription reads: *PRESENTED TO WILLIAM MURDOCH BY IN. MARSHALL / Chief Justice of the United States of America / Jas MARSHALL & R. COLSTON ESQrs. / As evidence of the high sense they entertain / OF HIS FIDELITY OF ATTENTION IN THE EXECUTION OF / IMPORTANT BUSINESS / Committed to his management by them.* This tureen probably never served "that green turtle soup of which the Chief Justice John Marshall of the Supreme Court of the United States was so preeminently a connoisseur," as described in the *MMA Bulletin* 26 (Apr. 1931): 103–04. Marshall was a presenter, not the owner.
4 Ibid.: 104.
5 Judith Banister, "A Century of Tureens," *Antique Dealer and Collectors Guide* 21 (Dec. 1966): 68.
6 Rowe 1965: 41.
7 Oechslin 1982: 86–87, fig. 75.
8 Clayton 1985: 364.
9 Rowe 1965: 65, pl. 46.
10 See a page from Boulton and Fothergill's pattern book showing "sauce tureens," in the Birmingham (England) Reference Library, in Rowe 1965: pl. 57.
11 Clayton 1985: 364.
12 *MMA Bulletin* (note 3): 104.

61. HOT-WATER URN

Jean-Baptiste-Claude Odiot (1763–1850)
1798–1809
Silver gilt, bronze gilt, ebony
20 ½ × 8 in. (52.1 × 20.3 cm)
Marks: [inside of cover, below lip of urn on band with rosette motifs, and on central portion of bottom of lamp] *J B C O*, a bellows, in a vertical lozenge (maker's mark); [inside of cover, on neck, on side of square base, and on central portion of bottom of lamp] cock, flanked by *A* and *1*, in vertical lozenge (maker's mark); [inside of cover and on rim of foot] male head flanked by *8* and *5* in an oval (Paris large excise mark, 1798–1809); [on central portion of bottom of lamp] male head, flanked by *8* and *5*, inverted and reversed, in a circle (Paris medium excise mark, 1798–1809)
Rogers Fund, 1934
34.12.2

A close analysis of this magnificent hot-water urn provides insights into how the silversmith Jean-Baptiste-Claude Odiot harmonized taste and technology. This item was an essential element in the ritual of drinking tea and hot chocolate in eighteenth-century European dining rooms and salons; however, only the most lavish examples were made of silver. Thus, Odiot's design, which assumes the fashionable form of a classical amphora with short handles, would have been the epitome of conspicuous consumption and display.

The urn is highly burnished, one of the stylistic traits that distinguishes Odiot from his prominent competitor Martin-Guillaume Biennais (see cat. 62), who liked to contrast polished and matte surfaces. Odiot often applied cast ornament, with detailed sculptural modeling, to smooth, shiny backgrounds.[1] He typically employed mythological figures, like the medusa mask on this example, rather than floral decoration that Biennais favored.[2] Much of the ornamentation, such as the border around the neck and the leaves around the bottom of the urn and top of the foot, is engraved. The medusa mask, handles, and spout, however, were attached separately. The vessel rests on a classicizing platform set with a silver-gilt lamp and laurel border. Heat and steam rose through an ingeniously conceived central hollow tube and escaped through a hole near the cover.

Odiot was deeply influenced by the architect-designers Charles Percier and Pierre-François-Léonard Fontaine. Their grand, sculptural manner, known as the Empire style, which they also applied to metalwork, prompted Odiot to develop his cast-ornament technique. Nonetheless, the juxtaposition of applied, volumetric elements and the sleek surface of this neoclassical urn achieves a subtle balance that exemplifies Odiot's own vision of antiquity, design, and artistry. K.C.

1 Rather than employing such traditional methods as chasing, engraving, and repoussé, Odiot pierced a hole into the silver for each attachment. Each was equipped with a threaded bolt that was then screwed onto the object and secured with nuts. This allowed the silversmith to burnish the vessel before applying any ornament. For more on Odiot's processes, see Clare Le Corbeiller, "The Construction of Some Empire Silver," *MMA Journal* 16 (1981): 195–98.

2 Frégnac 1966: 266.

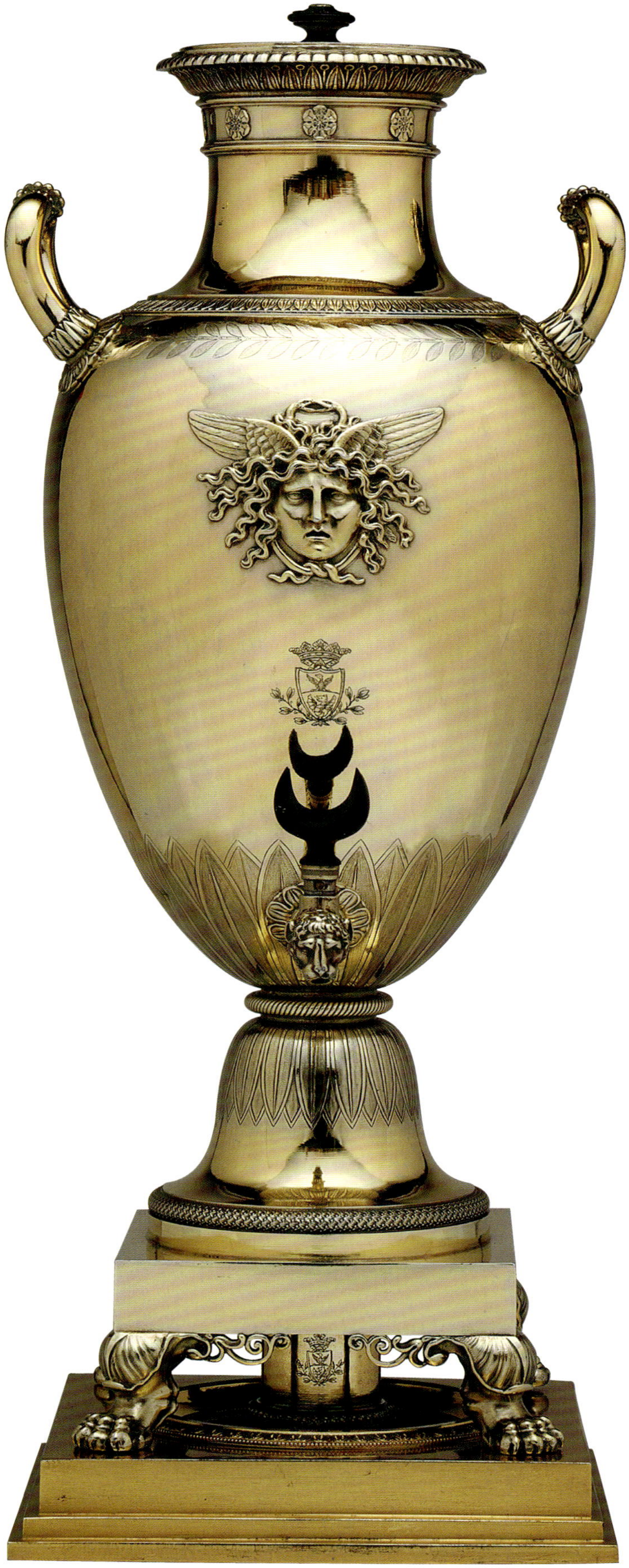

Cat. no. 61

62. COFFEEPOT

Martin-Guillaume Biennais (1764–1843)
1797–1809
Silver gilt, ebony
13 1/8 in. (33.3 cm)
Marks: [inside cover and on each leg] *B*, 2 pellets, a monkey above, in a vertical lozenge (maker's mark); [inside cover and on each leg] head of Apollo with *1*, in an oval (unofficial quality mark for first standard silver, 1794–97); [on neck and inset rim of cover] cock, flanked by *A* and *1*, in a vertical octagon (Paris mark for first standard silver, 1798–1809); [on neck and inset rim of cover] male head, flanked by *8* and *5*, inverted and reversed, in a circle (Paris medium excise mark, 1798–1809); [on rim of lip] ear (Paris medium census mark, 1809–1819)
Rogers Fund, 1934
34.12.1

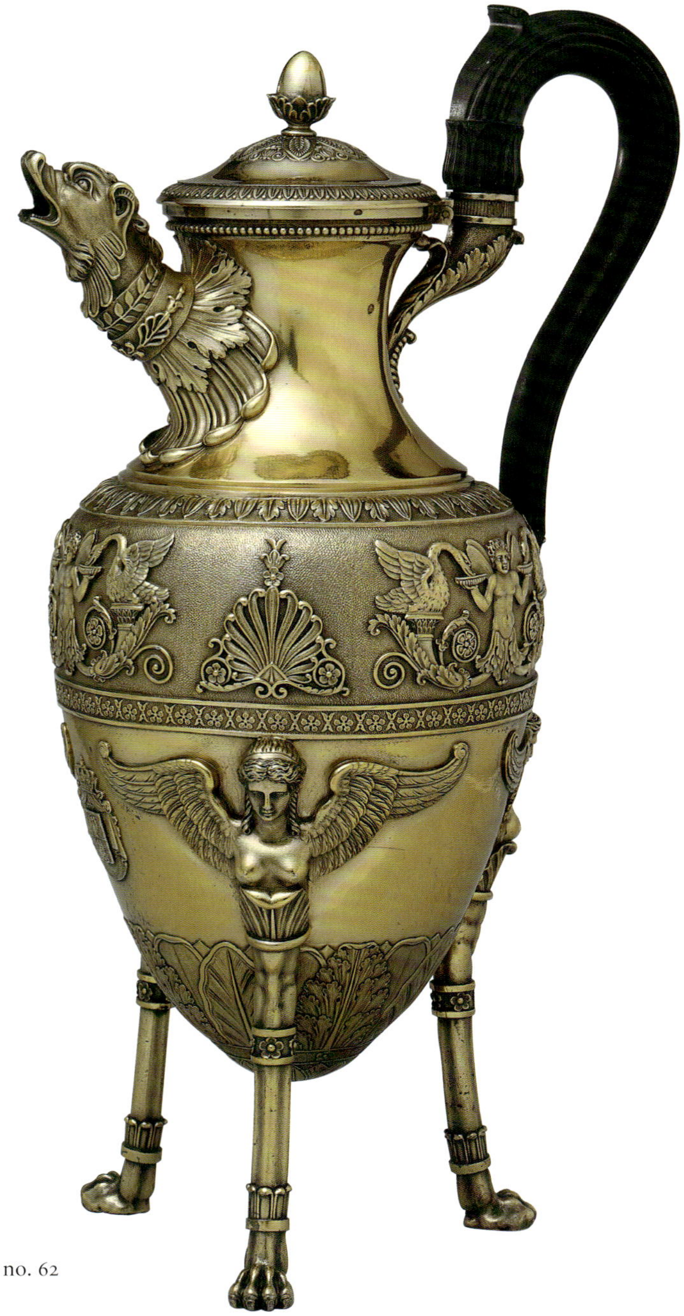

Cat. no. 62

Martin-Guillaume Biennais, one of the most prominent and successful goldsmiths of the late eighteenth and early nineteenth centuries, was not in fact trained in metalwork. He was a *marchand ébéniste*, a specialist in small cabinetwork; nevertheless, his designs for silver captured the changing aesthetic in art and design during the Napoleonic era and led to his distinguished and prolific career in this field.

This coffeepot illustrates how Biennais utilized the decorative vocabulary of the early Empire aesthetic. Its form derives from an ancient amphora, but, because of the piece's function, he substituted for the traditional pair of handles a tall, curving ebony one and a fanciful spout, shaped as a gargoyle-like, open-mouthed head (Fig. 1).

The design and ornamentation of this object reflect the inspiration of architect-designers Charles Percier and Pierre François-Léonard Fontaine, who dominated European neoclassicism in the early nineteenth century. Relying heavily on design motifs and ornaments from their influential publication *Recueil des decorations intérieurs*, which first appeared in 1801, Biennais, as well as other silversmiths, such as Jean-Baptiste-Claude Odiot (see cat. no. 61) and Henry Auguste (see cat. nos. 64–67), came to represent official Empire taste in silver design. An early twentieth-century publication, a folio of loose prints, *Dessins inédits d'orfèvrerie par Percier*, illustrates designs said to have been made explicitly for Biennais. They are certainly very close to actual works known by him, and perhaps directly inspired the ornamentation on this coffeepot.[1]

The ovoid shape of the coffeepot is divided into three main areas: an unpolished ground in the middle zone and upper and lower sections of burnished gilt silver. Resembling a delicate ancient bas-relief, the middle register comprises a central frieze: palmettes alternating with winged grotesque half-figures feeding pairs of swans. At the base, acanthus and lancet leaves radiate upward from a central rosette. Three legs with lion-paw feet, each terminating at the top by winged female busts, support the body. The design of the legs is similar to one on plate 15 of *Dessins inédits*, and the central relief is related to a motif in plate 10 (Figs. 2–3). Biennais knew how to use this stock of neoclassical ornament to his best advantage. He employed the basic ovoid shape numerous times and only slightly altered its decoration.[2] His contemporaries praised this conservative approach to form and his disciplined use of a set vocabulary of decorative motifs. Critical of "the incredible mania for change," Percier and Fontaine argued that ornament should be used sparingly so as not to corrupt the basic integrity of form.[3]

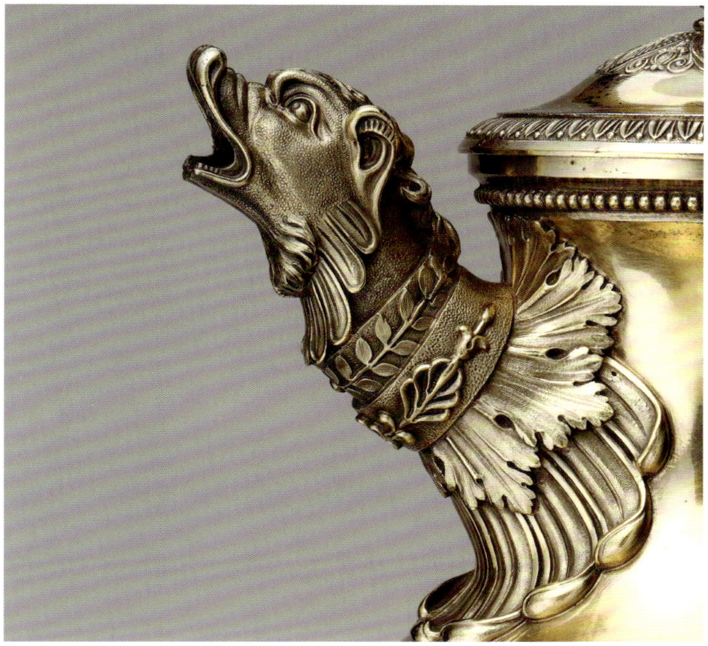

Fig. 1: Detail of the spout of cat. no. 62.

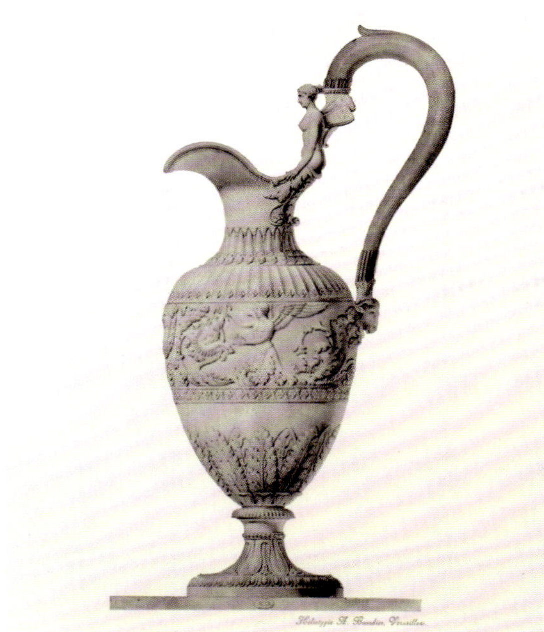

Fig. 2: *Design for a Coffee Pot*, about 1800–10. Attributed to Charles Percier. Lithograph; plate 10 from Egon Hessling, ed., *Dessins inédits d'orfèvrerie de Percier*... (Paris, 1911).

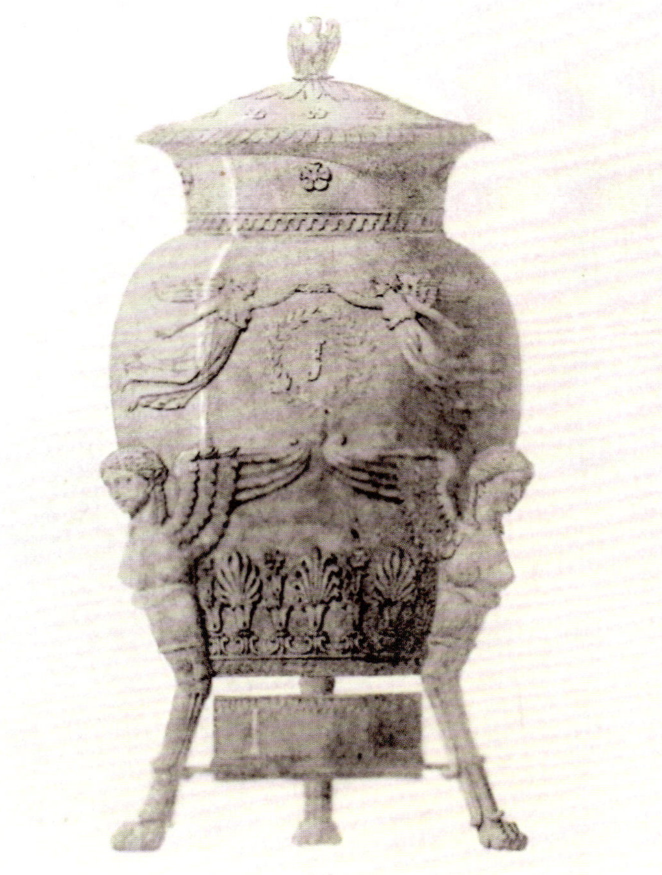

Fig. 3: *Design for a Hot Water Urn*, about 1800–10. Attributed to Charles Percier. Lithograph; plate 15 from Egon Hessling, ed., *Dessins inédits d'orfèvrerie de Percier*... (Paris, 1911).

Biennais may not have been involved in the design, or even the execution, of his silver; that, he left to his draftsmen and assistants. It would, however, be a mistake to view this coffeepot simply as a production piece. Its carefully calculated design plays polished surfaces against matte areas and emphasizes elaborate repoussé and cast decoration. Biennais did, nonetheless, organize his workshop to produce large quantities of silver that frequently reveal a standardization of shape and ornament. His numerous and prestigious commissions in France, as well as abroad, demonstrate that his clients seem not to have cared, or even to be conscious of, such distinctions. K.C.

1 Egon Hessling, ed., *Dessins d'orfèvrerie de Percier: conservés à la bibliothèque de l'Union centrale des arts décoratifs de Paris* (Paris, 1911). In his introduction, Hessling stated clearly that most of the prints in this album of twenty plates reproduce designs in a collection now in the Bibliothèque du Musée des Arts Décoratifs, Paris. I am grateful to Paul Micio for this reference.

2 See for example the coffeepot from Empress Maria Feodorovna's service, in Frank Davis, *French Silver: 1450–1825* (New York, 1970): fig. 74. See also the coffeepot made in 1810 for Napoleon and Empress Marie-Louise, now in the Musée du Louvre, Paris (OA 9537); Anne Dion-Tenenbaum, *L'Orfèvre de Napoléon. Martin-Guillaume Biennais* (Paris, 2003): 69–70, no. 37.

3 J. E. Poole, "A Napoleonic Silver-Gilt Service by Martin-Guillaume Biennais," *Burlington Magazine* 119 (1977): 392.

63. DESIGNS FOR SILVER

Jean-Guillaume Moitte (1746–1810)
Late 1780s
Pen and black ink, black and gray wash
7 15/16 × 12 7/8 in. (202 × 327 mm)
Inscribed: [upper register, left, in black ink] *Trépied*; [center left, in black ink] *Sceau à demie bouteille*; [center right, in black ink] *Sceau à glace*; [right, in black ink] *Vase à fruit*; [verso, in brown ink] *No. 77*
The Elisha Whittelsey Collection, The Elisha Whittelsey Fund, 1978
1978.638.1

The sculptor Jean-Guillaume Moitte's style of neoclassical ornamentation reached its apogee in his designs for the goldsmith Henri Auguste (see cat. nos. 64–67). Their collaboration was so extensive that contemporaries claimed Moitte and his assistants produced more than one thousand drawings for Auguste.[1] Like the present example, many of these depict variations of ancient vases and urns with simple, restrained outlines that are embellished with distinctive classical ornament.

This sheet displays designs for a variety of objects: two candelabra, two wine coolers, and four footed bowls. Decorative motifs on these objects include classically inspired rams supporting bowls, a caryatid, female masks, handles in the form of satyrs with snake hair, and friezes of swags, rosettes, and intertwining snakes, all typical of Auguste's designs. These motifs derive from antique prototypes found in Roman architecture, sculpture, and painted decoration that Moitte studied during his stay in Rome in 1771–73.[2]

Because of its modest size, this sheet may have served as a template at the beginning of the design process to help realize larger-scale drawings for actual production. It may also have served as a visual record or part of an inventory of current models. Most important is that this drawing provides an index of the forms, motifs, and overall style of the silver Auguste produced. For example, the wine cooler with mascaron handles in the top row, second from the left, is similar to Auguste's design for a wine cooler with variant handles (cat. no. 64). A gilt-silver

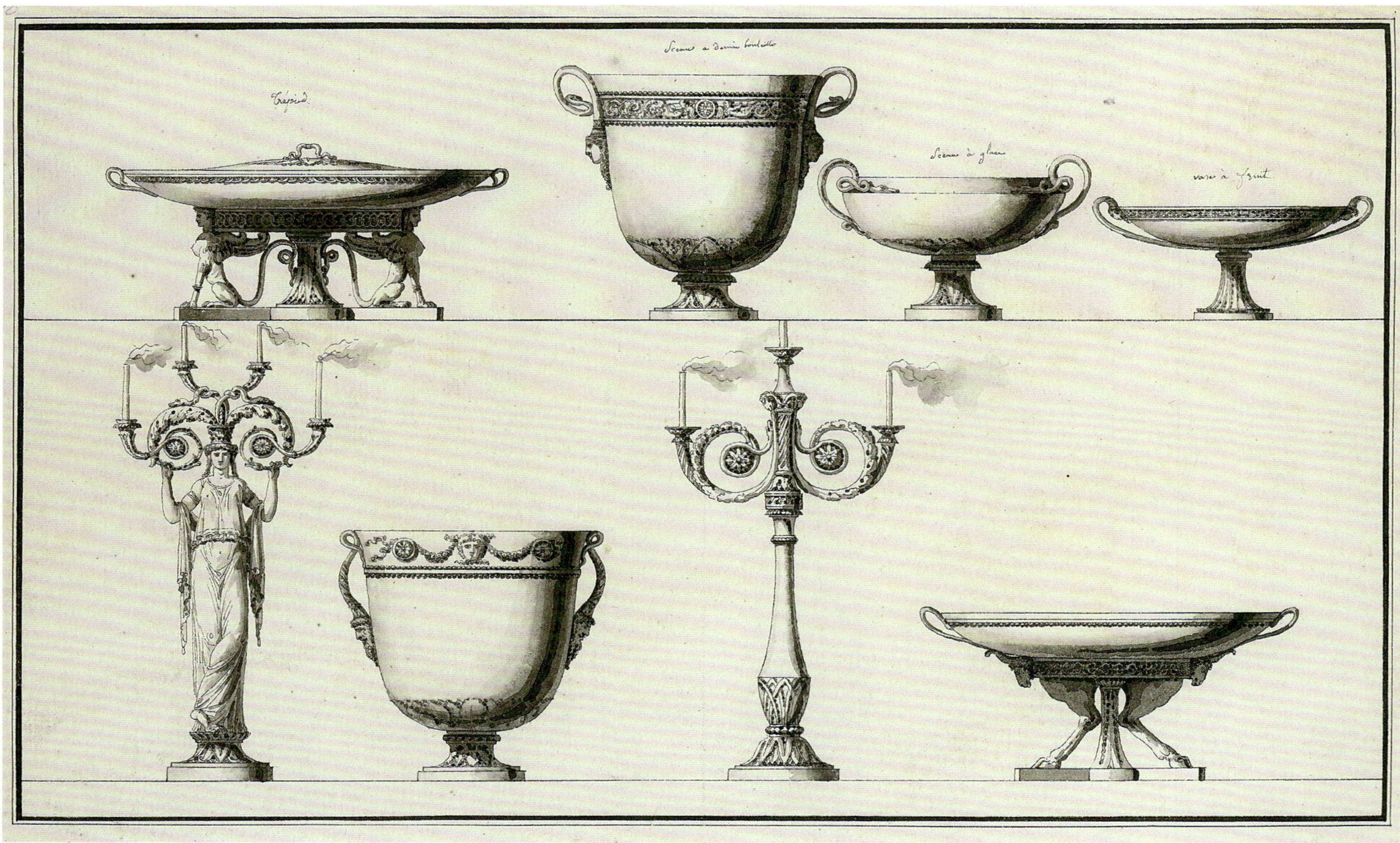

Cat. no. 63

basin, designed by Auguste and once owned by the renowned collector of antiquities William Beckford, is closely related to the design of the low-bodied container labeled as a fruit bowl, first from the right, in the top row of this sheet.[3]

Although some of Moitte's drawings were sold after his death, a large number remained in the Auguste workshop. After Auguste went bankrupt in 1809, his rival, Jean-Baptiste-Claude Odiot (see cat. no. 61), acquired the contents of his workshop. While the Metropolitan Museum owns some of these drawings (see cat. nos. 64–66), all identified by the presence of Odiot stamps, this drawing and *Design for a Tureen with Swan Handles* (cat. no. 67), which bears Auguste's signature, do not have the stamps found on many others; thus, their provenance is uncertain.[4]

This unsigned drawing shows such close stylistic affinities with secure works by Moitte that one can attribute it to him, but with a cautionary note.[5] Numerous drawings from Auguste's workshop were not signed because they were executed for the shop, not by Moitte, but by Auguste's other draftsmen.[6] Therefore, it is often difficult to distinguish between designs in Moitte's hand and copies or adaptations made by the Auguste workshop.[7]

Moitte's austere yet inventive classicism of the late 1780s presaged the beginning of the so-called Empire style by almost a decade and remained immensely popular throughout his lifetime.[8] At the artist's funeral in 1810, the critic Antoine-Chrysostome Quatremère de Quincy praised the consummate sculptor-designer as follows:

> Monsieur Moitte attracted notice from the beginning in his profession by numerous drawings that clearly reveal his style and also propagated the taste for antiquity for which he was an advocate. His manner generally expressed and inspired greatness; and the style of his design contrasted with that of the generation preceding him, announcing a change of taste.[9]

K.C.

1 Joachim Le Breton, *Notice historique sur la vie et les ouvrages de Moitte* (1812): 30; and Michael Snodin and Malcolm Baker, "Beckford's Silver [Parts 1–2]," *Burlington Magazine* 122 (1980): 734–48, 820–34: esp. 739, n. 27.

2 Richard James Campbell, *Jean-Guillaume Moitte: The Sculpture and Graphic Art 1785–1799*, (Ann Arbor, 1983): 189–90. For sketches from Moitte's early years in Rome, see Gisela Gramaccini, *Jean-Guillaume Moitte (1746–1810) Leben und Werk* (Berlin, 1993): vol. 2, 158–84. In addition to studying Roman antiquities, Moitte must have been influenced by popular publications and printed images related to these ancient cultures such as the Comte de Caylus's *Receuil d'antiquites...* (1756–67), Jean-Charles Delafosse's *Nouvelle Iconologie historique* (1786), and Joseph-Marie Vien's *Suite des vases* (1760); see Derek E. Ostergard, ed., *William Beckford, 1760–1844: An Eye for the Magnificent* (New Haven/London, 2001): 418, cat. 50.

3 Frégnac 1966: 270, fig. 2.

4 See Myers 1991: 121–23.

5 Myers 1991: 122. See Christie's New York, Feb. 26, 1986: lot 184, for an illustration of Odiot stamp no. 294. See also Campbell (note 2): 186–87, 301, n. 30.

6 Myers 1991: 1–2.

7 Auguste could have had copies of Moitte's drawings made for his artisans to use or kept the originals to show to clients, which may explain the absence of signatures. See Myers 1991: 124.

8 Campbell (note 2): 197. Campbell posited (123) that Moitte's drawings probably date to the late 1780s, because actual silver objects made for Napoleon were based on designs that Moitte executed between 1784 and 1789.

9 Antoine-Chrysostome Quatremère de Quincy, "Institut de France. Funerailles de M. Moitte, le 3 mai 1810, Discours de M. Quatremère," *Gazette nationale ou Le Moniteur universel* 125 (May 4, 1810): 495–96, quoted in Campbell (note 2): 178.

64. DESIGN FOR A WINE COOLER

Workshop of Henri Auguste (1759–1816)
Late 1780s
Pen, black ink, and gray wash
17 3/8 × 16 1/8 in. (441 × 410 mm)
Inscribed [lower right, in pencil] *231*; [lower right] Odiot stamp no. 231
Rogers Fund and The Elisha Whittelsey Collection,
The Elisha Whittelsey Fund, 1978
1978.521.5

Ten drawings (see also cat. nos. 65–66) purchased by the Metropolitan Museum from the Maison Odiot archives are a valuable tool for the study and appreciation of the work of Henri Auguste and for the understanding of early Empire silver.[1] The presence of Odiot's stamp on these drawings clearly establishes their provenance from the Auguste workshop and dates them to the late 1780s. As we shall see, this drawing exhibits several features found on a drawing by the sculptor Jean-Guillaume Moitte featured here (cat. no. 63), which also helps pinpoint the date of this wine-cooler design.

The form and decoration of the wine cooler are consistent with most of the silver produced by the Auguste workshop. The stern Medusa mask; the long, tensile curve of the handles; the flowing frieze encircling the rim; and the flared sculptural palmettes and deeply carved gadrooning at the base all enliven the urn's austere form. Two different handle designs offer clients imaginative alternatives; one ends in a classical rosette, whereas the other features a bearded male mask. This and the two other drawings attributed to Auguste featured here are considerably larger than a signed Auguste drawing (cat. no. 67) and Moitte's silver designs, cited above. The bigger drawings probably represent full-scale images, meant to facilitate production.

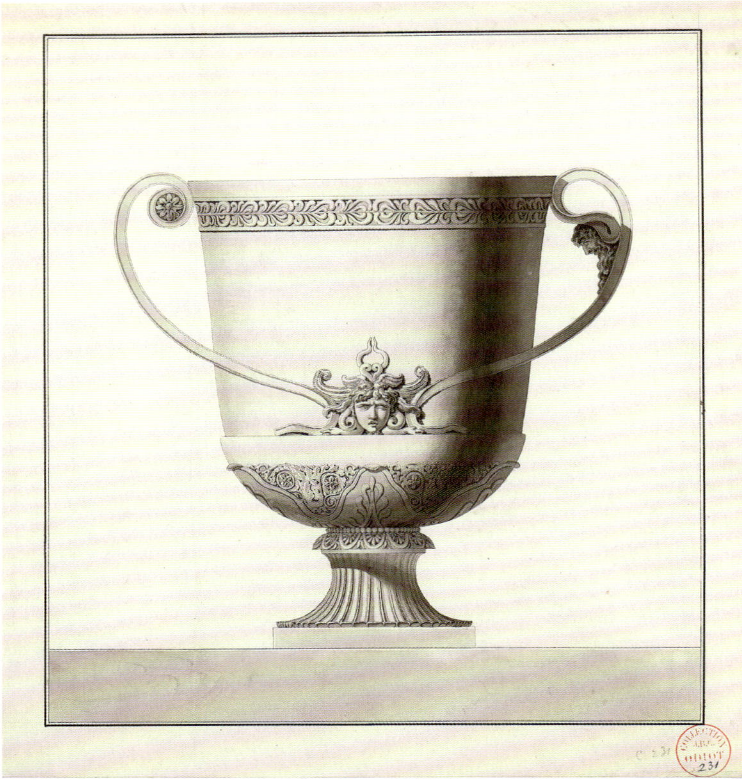

Cat. no. 64

The wide-bodied urn shape is based on the Greek volute krater and can be compared to two similar wine coolers on the drawing by Moitte. The basic shape, proportions, and the distribution of the ornament are very close.[2]

There are additional similarities between this design and the two wine coolers in Moitte's drawing. Their decoration is restricted mainly to the handles and the upper and lower parts of the object. Like the wine cooler in the present sheet, the cooler in the top row employs a frieze around the mouth; that in the bottom row is decorated with a band of swags, rosettes, and masks. As these drawings demonstrate, by the early Empire period, popular classical motifs were applied in various combinations to similar types of objects to produce slight variations that obviated the need to redesign each vessel completely. K.C.

[1] See Phillipe de Montebello, "Foreword," *Notable Acquisitions, 1975–1979, MMA Bulletin* (1979): 61–62; and Myers 1991: 7–8.

[2] A similar head features on both handles of a heavily decorated wine cooler bearing the engraved signature of Auguste and a date of 1809; see Frégnac 1966: 272–73, figs. 1–2.

65. DESIGN FOR A TUREEN

Workshop of Henri Auguste (1759–1816)
About 1800
Pen, black ink, and gray wash
$17\frac{13}{16} \times 23\frac{1}{16}$ in. (452 × 585 mm)
Inscribed on recto: [lower right] Odiot stamp no. 138. Inscribed on verso: [on lining at upper right, in brown ink] *Collection de desseins / par Auguste / ancien orfèvre de / L'empereur Napoléon*; [below and partly underneath, in lightly effaced pencil] *Soupière pour M. Damidoff*
Rogers Fund and The Elisha Whittelsey Collection, The Elisha Whittelsey Fund, 1978
1978.521.1

Henri Auguste is credited for transforming neoclassical silver into what would be called the Empire style. Abandoning the heavily ornamented manner of his father, master goldsmith Robert-Joseph Auguste, he lightened and refined his designs and decorative vocabulary. The basic shape and proportions of this tureen, although somewhat heavier and more squat than others by the younger Auguste, derives from the ancient Greek skyphos and kylix.[1] Auguste fitted this bowl shape with a large stand elevated on winged lion's paws and a tapering lid terminating in a half-plant/half-female caryatid. Typical of his work is the balance he achieved between unembellished surfaces and tight, linear bands of ornament. The tureen includes some of Auguste's favorite neoclassical motifs: Medusa-like masks combined with snake handles, a base sheathed with acanthus leaves, and the above-mentioned winged lion feet. Especially inventive are the frieze's intertwined serpents, which create round openings for rosettes or medallions, and the lid's fantastic caryatid finial. The clean profile of the stylized female heads beneath the handles, their lips slightly parted and their expressions blank, almost other-worldly, enhances their enigmatic nature.

This drawing closely resembles a gilt-silver round tureen (*pot á oille*) by Auguste that is in the Musée National du Château, Fontainebleau (see Ottomeyer essay, Fig. 13). It formed part of the famous service presented to Napoleon by the city of Paris at his coronation in 1804. But like many of Auguste's designs and objects, the tureen exhibits modifications that occurred in transforming the idea into reality. Instead of the drawing's anthropomorphic finial, we find an allegorical seated woman; the animal feet supporting the stand now include busts of lion sphinxes. The tureen retains the low-bodied, classical shape and pedestal depicted in the drawing; the snake handles, although curved differently, emerge from somber, braided masks, as in the drawing. Both design and object feature

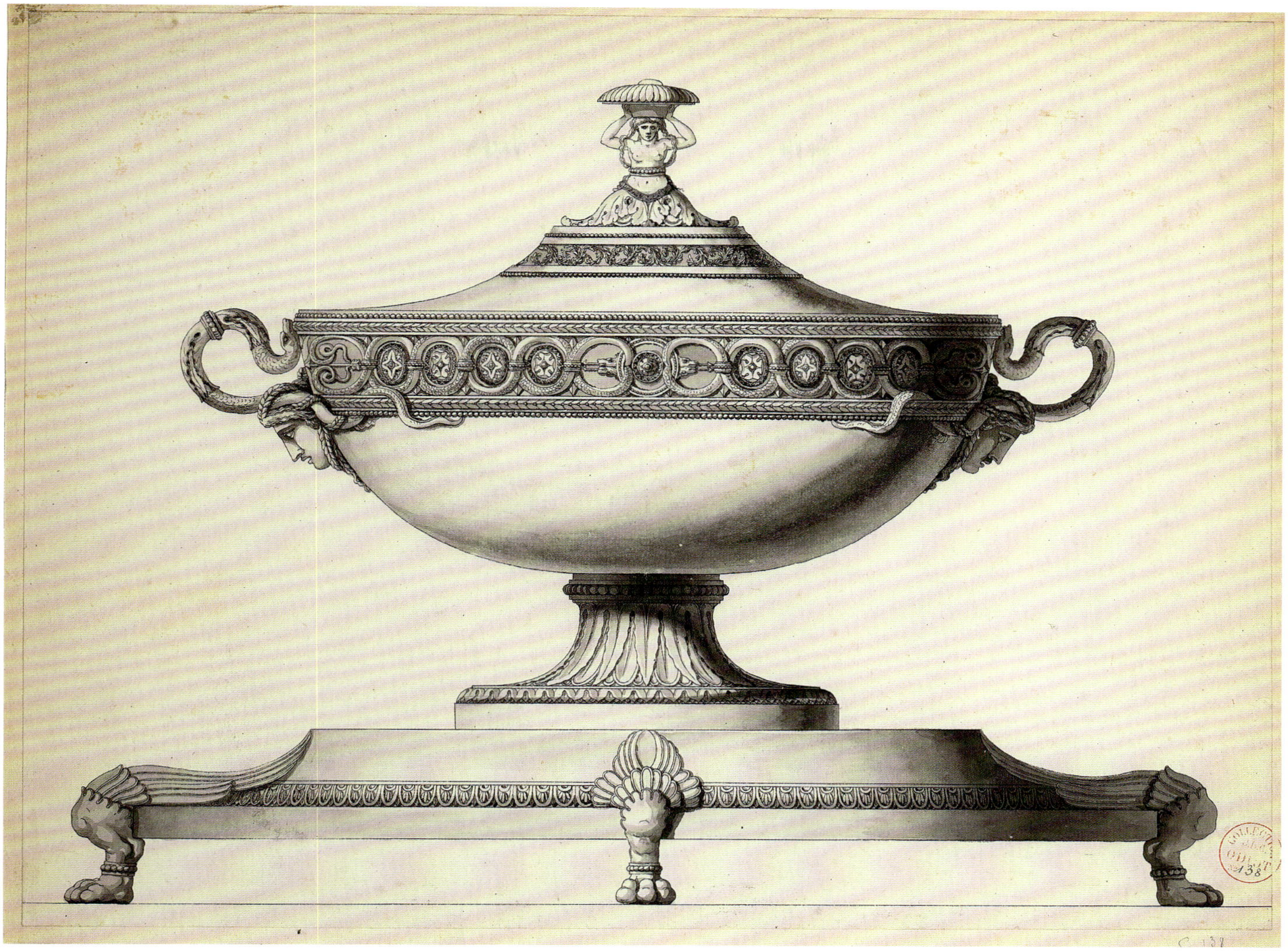

Cat. no. 65

running bands of decoration on the stand and near the top of the vessel. The circular forms of the uppermost friezes echo one another, and the bands around the stands of both the drawn and actual tureen contain scalloped leaf forms.

The inscription on the verso of the drawing probably refers to Count Nicholas Demidoff, who lived in Paris around 1817.[2] Demidoff's signature of approval appears on another signed drawing by Auguste that was once part of the Odiot collection.[3] Because of significant stylistic similarities to the 1804 Fontainbleau tureen, one can probably date this drawing to about 1800. As Auguste's designs continued to be referred to for production, this one may have been utilized for a 119-piece service, executed by the Odiot firm, that Demidoff purchased in 1817.[4] K.C.

[1] See Myers 1991: 6–7.

[2] James Draper and Clare Le Corbeiller, *The Arts under Napoleon* (New York, 1978), cats. 156–59. A bill of sale from Dec. 5, 1817, records the purchase by "M. de Demidoff" from the Odiot firm.

[3] Sotheby Parke-Bernet Monaco, sale, Nov. 26, 1979: lot 614, illus. in Myers 1991: 7.

[4] Count Demidoff probably gave the set to a Madame de la Chapelle, who later had her coat of arms engraved on it. See Draper and Le Corbeiller (note 2).

66. DESIGN FOR A TABLE FOUNTAIN

Workshop of Henri Auguste (1759–1816)
About 1800
Pen and black ink, with brush and gray wash
10 ¼ × 7 ⅛ in. (260 × 181 mm)
Inscribed: [bottom right, in pencil] *C 47*, over Odiot stamp no. 47
The Elisha Whittelsey Collection, The Elisha Whittelsey Fund, 1978.
1978.521.10

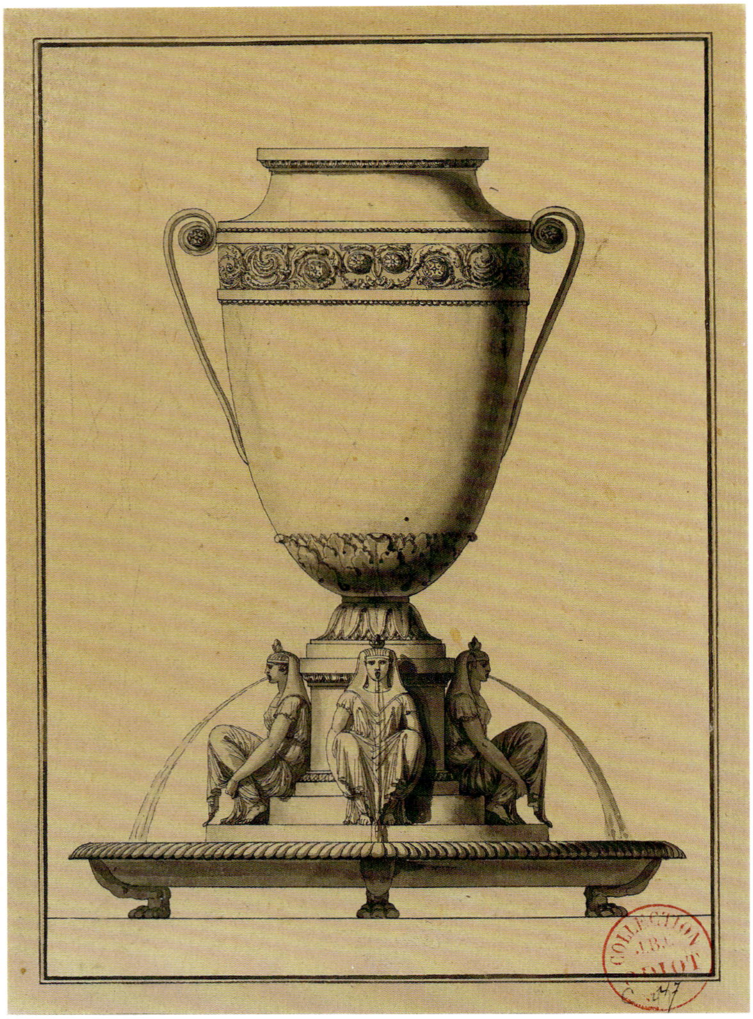

Cat. no. 66

After Napoleon's North African campaign in 1798, France, followed by the rest of Europe, developed a craze for Egypt and its ancient past. Publications by Charles Percier and Pierre-François-Léonard Fontaine, as well as Baron Dominique Vivant Denon, popularized the use of Egyptian motifs in interiors and on a variety of objects.[1] The proliferation of Egyptian decorative elements and themes carried over to theater and music; for example, *Les Mystères d'Isis*, an adaptation of Wolfgang Amadeus Mozart's *Magic Flute* was performed at the Paris Opera in 1801.[2]

Thus, late neoclassical and early Empire designs, such as this idea for a table fountain, often feature Egyptian motifs. Below the monumental vase, its shape based on the classical amphora, sit three women, clad in Grecian robes and wearing Egyptian headdresses. From the mouths of these figures, who are seen frontally and in profile, issue streams of either wine or water. The strict symmetry of the design suggests the existence of a fourth figure, seated beyond our view at the back of the vase. Despite the Egyptian touches, the table fountain, with its monumental vessel, remains a tribute to neoclassicism.

As royal gold- and silversmith, Henri Auguste paid attention to Napoleon's fondness for things Egyptian. Always sensitive to the propagandistic power of symbols and imagery, the French leader seized upon one Egyptian motif in particular, the goddess Isis.[3] In 1806 a seated Isis was installed, presumably with Napoleon's approval, as part of a large relief in the east wing of the Cour Carrée in the Louvre. The sculpture was executed by Jean-Guillaume Moitte, whose work is closely linked with Auguste's (see cat. no. 63). The Auguste drawing for a table fountain includes a female figure with the same finial on her headdress as on the Louvre's Isis.[4] Similar drapery in other sculptures by Moitte supports the argument that this drawing may reflect an original study by the sculptor that related directly to the Louvre relief. Stylistically, the drawing displays Moitte's sculptural sense of weight and monumentality, but because of the difficulty of distinguishing his drawings from those of Auguste, the attribution is best left with a more general designation to the Auguste workshop. The drawing can be dated to around 1800, when, in the wake of Napoleon's appointment as First Consul, both Auguste and Moitte were contributing actively to the increased interest in Egyptian art and mythology.
K.C.

1 Percier and Fontaine 1801–12 (1812, all volumes in one); Dominique Vivant Denon, *Voyage dans la Basse et la Haute Égypte* (Paris, 1802).
2 See Jean-Marcel Humbert, *Egyptomania* (Ottawa, 1994): 252.
3 Ibid.: 255.
4 See ibid. For a plate in illustrating Moitte's relief, see Frédéric de Clarac, *Musée de sculpture antique et moderne* (Paris, 1841–53), vol. 1, pl. 32.

67. DESIGN FOR A TUREEN WITH SWAN HANDLES

Henri Auguste (1759–1816)
1800–09
Pen, brown ink, and gray-brown wash
7 13/16 × 5 7/8 in. (198 × 149 mm)
Signed: [below design at right, in light brown ink] *H. Auguste*
Inscription: [upper left, in pencil] *12*
The Elisha Whittelsey Collection, The Elisha Whitteley Fund, 1978
1978.638.2

Although this drawing bears Henri Auguste's signature, it does not exhibit the Maison Odiot stamp, which generally establishes the authenticity of a drawing by this artist (see cat. no. 63).[1] Historical and stylistic evidence do, however, support an attribution to the silversmith himself.

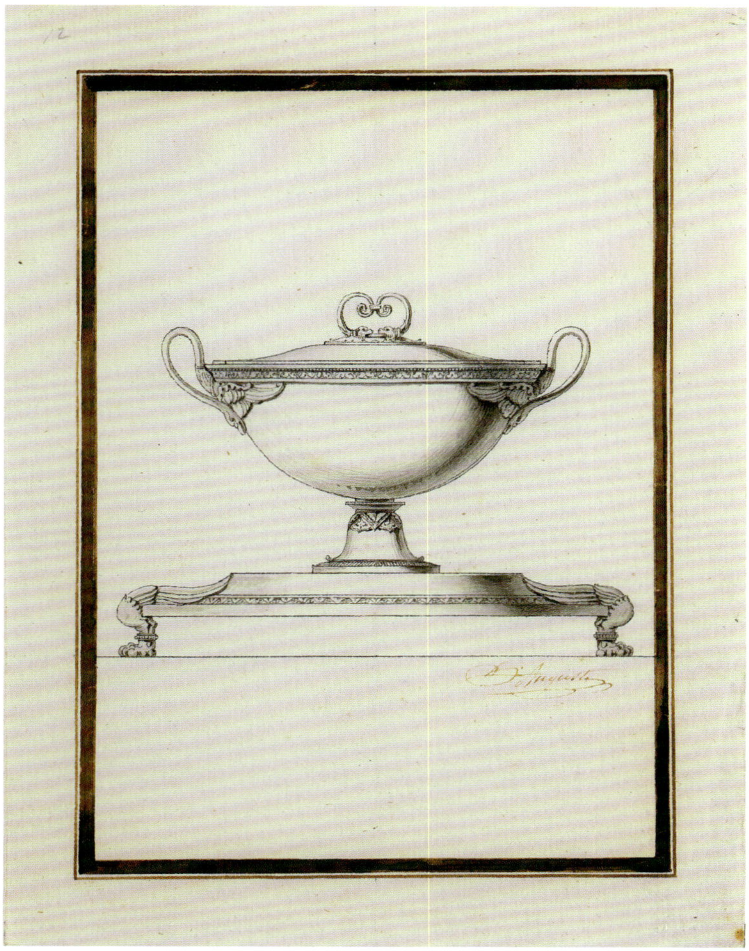

Cat. no. 67

The tureen's smooth form and restrained ornamentation convey Auguste's uncluttered aesthetic and his characteristic delicate drawing style. The graceful swan handles can be associated with Napoleon and Josephine's emblem, indicating a date of the first decade of the nineteenth century, when this motif was particularly popular.[2] Auguste, however, used swans not only as a propagandistic emblem on commissions made for the emperor and empress but also as purely decorative elements.

The model for the footed-bowl shape of this tureen is the Greek kylix; its handles more closely resemble those of a *kantharos*. This drawing is related to another, larger one by the Auguste workshop, *Design for a Tureen* (cat. no. 65), in that they display similar ornamental bands and their stands both feature lion-paw legs. The difference, however, in the amount of decoration in the designs is striking. On the tureen in the present drawing, the prominent ornamental features are restricted to the swan handles and the opposed C-scroll handle on the cover. That depicted in the larger drawing employs many more decorative motifs, including snakes, masks, and rosettes. In addition, the proportions of the swan-handled tureen are more elongated, because of the slender bell-shaped foot and tall, thin handles. These differences suggest the evolution of the Empire style toward a lighter, more delicate sensibility. This drawing can be assigned to the end of Auguste's artistic activity, between 1800 and 1809. K.C.

1 The issue is further complicated because Auguste employed, in addition to Jean-Guillaume Moitte, other draftsmen, such as the architect Charles de Wailly and the sculptor Félix Lecomte. See Myers 1991: 1–2.
2 Myers 1991: 1–2, n. 8.

The Vase as Ornament in Furniture and Furnishings

Stefanie Walker

Unlike ceramics or silver, furniture largely lacked ancient antecedents and was a less obvious medium to incorporate the vase motif. Consequently, the vase was more readily adapted as ornament rather than as form, although a number of furniture types with specific purposes were based on vase shapes.

In the seventeenth century, elaborate floral marquetry had become a hallmark of luxurious case furniture, and vases appeared as containers for enormous bouquets that filled large panels on cabinet doors or table tops. Wood-inlay decoration persisted throughout the eighteenth century, but the types and appearance of motifs were updated to reflect the latest wave of classicism. In comparison to the floral elements, vases became more prominent, often assuming the fashionable "spindle" shape with a long, tapering neck or lid that corresponded to similar forms in the foot. They were frequently accompanied by ornaments drawn from the whole repertory of neoclassical motifs, such as Bacchic staffs (thyrsoi), laurel garlands, decorative trophies, and acanthus fronds. Vase designs on furniture usually derived from the multitude of prints that circulated in all the trades in England and continental Europe and helped to spread this form so widely (see cat. nos. 71–72).

Both French and English designers and furniture makers sought to integrate vases into their interiors as the most readily recognizable symbols of the admired classical past. Architects Robert Adam and James "Athenian" Stuart invented a variety of utilitarian vase forms meant to complement the mostly decorative creations of Josiah Wedgwood, Matthew Boulton, and their followers. Between 1757 and 1762, Adam and Stuart's designs for a sideboard at Kedleston Park, Derbyshire, included a number of early vase-shaped urns, knife boxes, and perfume burners, exactly the types featured here. The two grand mahogany urns (cat. no. 69) on pedestals from Delapre Abbey, Northamptonshire, dispensed water and wine in the dining room, while their classical form and ornament enhanced their stateliness and appeal. An English knife box (cat. no. 68) exhibits the same crisp outlines and moldings of turned silver objects. In 1773 Jean-Henri Éberts first published a design for a large vessel, supported by a classicizing tripod, that could function in a variety of ways: for example, as a plant stand, brazier, and washbasin. Eberts called it an *athénienne*, probably inspired by *The Virtuous Athenian Maiden* (Musée du Louvre, Paris), a 1763 painting by Joseph-Marie Vien, who in turn may have adapted it from published designs by Stuart. This versatile and popular type of furniture is exemplified here by a pair of small French models for perfume burners (cat. no. 70). Ranging from the miniature to the monumental, these works represent the vase form's astonishing adaptability.

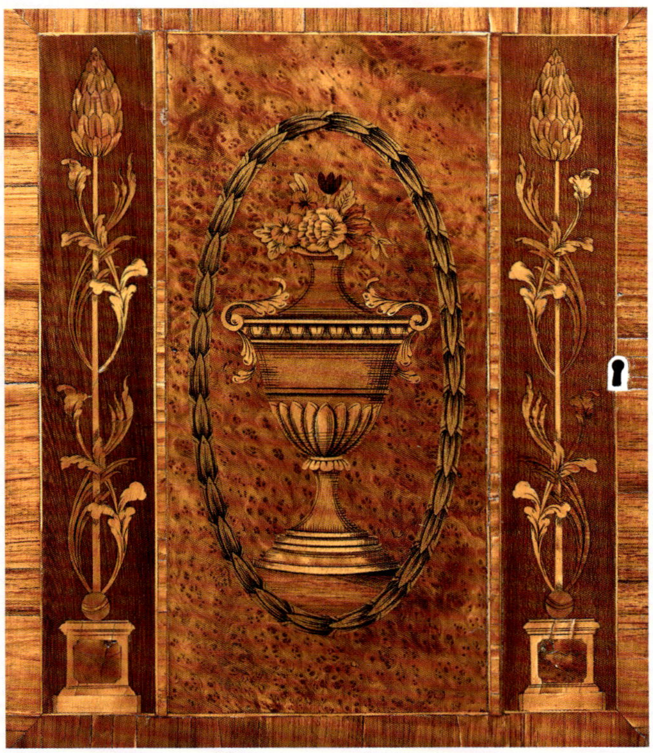

Detail of cat. no. 72

68. KNIFE BOX

English
About 1770–80
Mahogany inlaid with boxwood
22 ¾ × 13 ¼ in. (57.8 × 33.7 cm)
Rogers Fund, 1911
11.117.8

An elegant integration of form and function, this vase-shaped knife box was suitable for display and provided storage for dining accouterments. The container's basic shape derives from a krater, the vertical fluting recalls classical columns, the upturned handles appear on antique vases, and the guilloche and floral marquetry resemble Greek vase decorations. The same simplified classical vocabulary can be seen in the border designs developed by Pierre d'Hancarville for his publication of Sir William Hamilton's first collection of antiquities (see cat. no. 1). In addition to classical sources, the designer may also have been inspired by the works of Stefano della Bella, who depicted a vase with similarly curved handles in plate 6 of his *Raccolta di vasi diversi...* (1646).[1]

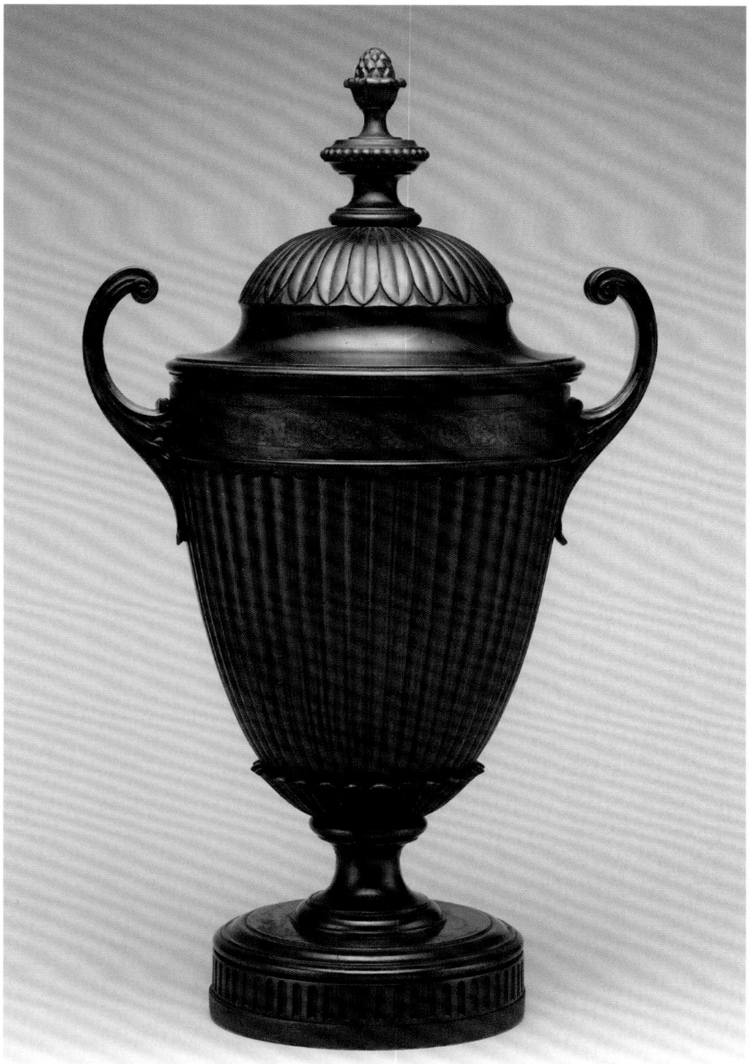

Cat. no. 68

Most likely one of a pair, this knife box resembles the pedestal urns from Delapre Abbey (see cat. no. 69), intended to adorn a dining-room sideboard. By the mid-eighteenth century, matching cutlery sets were made in order to provide each diner with knife, fork, and spoon. New specialist knife designs, utilizing better quality, low-priced steel alloys, were introduced. Nevertheless, table knives remained expensive, and, consequently, they were stored in special boxes, objects that, as the century progressed, became stylish decorative pieces in their own right, displayed on sideboards, along with other high-priced plate.[2]

The third edition of George Hepplewhite's *Cabinet-Makers and Upholsterer's Guide* (1794) includes four knife cases shaped as vases, but without handles. By this time, the form must have been commonly used for knife cases, because his widow, Alice Hepplewhite, claimed to eschew "mere novelty" and present "such articles only as are of general use and service."[3] Thomas Sheraton also illustrated a sideboard outfitted with vase knife boxes. These are more bulbous in nature, lack handles, and rest on square plinths.[4] Although more rare than slant-lidded rectilinear knife boxes, other vase-shaped examples survive.[5] However, the inclusion of handles, the round plinth, and the suggestive relationship to Greek models appear to make this knife box unique. J.L.

1 This plate is illustrated in Clifford 1978: pl. 76a.
2 Sara Paston-Williams, *The Art of Dining: A History of Cooking and Eating* (London, 1993): 16, 250; and Peter Brown, ed., *British Cutlery, An Illustrated History of Design, Evolution and Use* (London, 2001): 101.
3 Hepplewhite 1969: vi, pl. 39.
4 Thomas Sheraton, "Appendix to the Cabinet-Maker and Upholsterer's Drawing Book," in *The Cabinet-Maker and Upholsterer's Drawing Book* (1802; reprint, with a new intro. by Joseph Aronson, New York, 1972): 41, pl. 21.
5 See for example Sotheby's London, *Collection of Baroness Bentinck-Thyssen*, sale cat., July 18–19, 1996: lot 154, for a knife box described as "a mahogany and boxwood strung urn knifebox...circa 1780, the domed lid with an urn finial enclosing a fitted interior on a trumpet base and square plinth, 70 cm high." The entry does not mention handles; unfortunately, the knife box is not illustrated. Two others, one of mahogany and another of satinwood, also lack handles; see Sotheby's London, *Fine English Furniture*, sale cat., June 4, 1999: lot 230; and Christie's London, *Works of Art from Country Houses*, sale cat., Jan. 24, 2001: lot 421.

69. SIDEBOARD PEDESTALS WITH URNS

English
About 1775
Mahogany with mahogany veneer
Each 74 × 22 ½ × 23 ½ in. (188 × 57.1 × 59.7 cm)
Gift of Irwin Untermyer, 1964
64.101.894-.895

The 1760s witnessed the final transformation of the sideboard from a single, stand-alone object into a suite, including a long table for the display of plate and knife boxes, two flanking pedestals with urns, and a separate wine cistern or cooler.[1] Each of the present pedestals, part of a suite from Delapre Abbey in Northamptonshire, is fitted with a storage cupboard and a plate warmer. The urn with a brass spigot was filled with water for the cleaning of silverware and rinsing of glasses between courses, while the other contains a wine cistern. These pedestals and the matching breakfront sideboard (also in the Metropolitan Museum's collection) would have functioned as the general beverage and service station.[2] Dining rooms, like other public spaces for entertaining, were usually located far from the kitchen, and wine bottles were not placed on the table until the beginning of the nineteenth century.[3]

The essential aspects of the suite derived from the work of Robert Adam. For the dining room of Kedleston Park, Derbyshire, Adam designed in 1762 a sideboard decorated with urns, some fitted out for the service of beverages, a tripod perfume burner by James Stuart, and six or eight knife boxes. Low urns sat on separate plinths possibly for use as wine coolers.[4] In 1767 at Osterly Park, Hounslow, and again at Kenwood, Hampstead, in 1768–69, Adam repeated the arrangement of a sideboard flanked by pedestals supporting urns with a modified amphora shape (it would become the standard form). Adam followed the same basic design in numerous commissions he carried out through the 1770s and 1780s.[5]

Like the knife box and armchairs featured here (cat. nos. 68, 71), the sideboard pedestals appear in six variations in George Hepplewhite's *Cabinet-Makers and Upholsterer's Guide* (1788). Hepplewhite's designs include pedestals arranged for storage or for warming plates, as is the case with this pair; all feature spigots for dispensing water. However, the pedestals from Delapre Abbey predate the Hepplewhite examples by a decade; the latter lack the vigorous form and robust decoration of the former, reflecting perhaps a growing preference for the lighter touch of Adam's later style.[6]

Adam popularized the use of motifs derived from archeological excavations in the design of every conceivable object. This suite

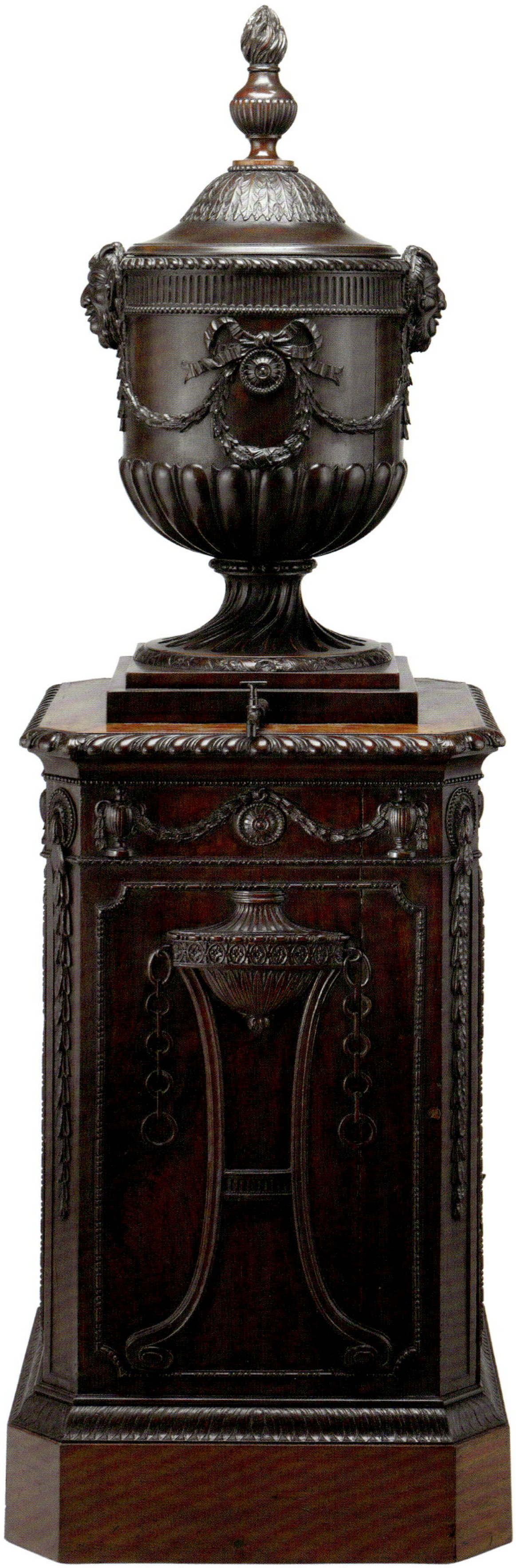

Cat. no. 69

provides a veritable dictionary of classical ornament—vases, swags, and round paterae; pendant husk ornament at the corners and deep gadrooning—all realized in energetic, high-quality carving. The vase covers, further embellished with acanthus leaves, are surmounted by turned and fluted flame finials, and the urns are carved with fluting, egg-and-dart molding, pearl ornament, festoons suspended from bows, and two satyr masks. Such masks appear as decoration on ceramic vases, as well as vessels depicted in prints and in woodcarvings (see cat. nos. 34, 39, 74).

The sideboard urn replicates the form of a funerary vessel on an altar.[7] Atop the container, an eternal flame rises from the finial; the tripod perfume burner on the door further reinforces the allusion to ancient memorials. The dining room was considered an arena for men and a display space for entertaining.[8] As such, it became an especially appropriate site for the display of classical artifacts that could be associated with male virtues and concerns. Tables were often decorated with porcelain and sugar figurines, arranged in miniature landscapes complete with temples and urns; classical sculpture adorned the walls. Publisher Pierre d'Harcanville may have put his finger on the motive for arranging a dining-room sideboard with flanking urns when he stated, "It was upon these buffets that the ancients used to place their richest vases."[9] J.L.

1 Sara Paston-Williams, *The Art of Dining: A History of Cooking and Eating* (London, 1993): 246.
2 For the sideboard (64.101.893), see Yvonne Hackenbroch, *English Furniture, with Some Furniture of Other Countries in the Untermyer Collection* (Cambridge, Mass., 1958): 16, figs. 50–52.
3 Paston-Williams (note 1): 224, 251–52.
4 Clifford Musgrave, *Adam and Hepplewhite; and Other Neo-Classical Furniture* (London, 1966): 181, pl. 2.
5 Eileen Harris, *The Furniture of Robert Adam* (London, 1963): 48–57.
6 Hepplewhite 1969: pls. 35–36.
7 See for example a 1777 design for a funerary monument by James Craig illustrated in Timothy Clifford, *Designs of Desire: Architectural and Ornament Prints and Drawings, 1500–1850* (Edinburgh, 1999): cat. 135.
8 In 1685 Sir George Courthoppe willed his wife the furniture of all of the chambers in their house except the dining room, "which is no way suitable for her." See Christopher Hartop, *The Huguenot Legacy: English Silver 1680–1760, From the Alan and Simone Hartman Collection* (London, 1996): 33. By 1788 men and women were still separated from one another at table—with women at one end and men at the other. But around this time, a "new promiscuous mode of seating," with "a gentleman and lady sitting alternatively round the table" was being introduced; see Rev. John Trusler, *Honours of the Table* (1788), quoted in Paston-Williams (note 1): 249.
9 Quoted in Michael Vickers, "Value and Simplicity: Eighteenth Century Taste and the Study of Greek Vases," *Past and Present* 116 (Aug. 1987): 112. See also Robert Wemyss Symonds, in *Antique Collector* 13 (Jan.–Feb. 1942): 13-17.

70. A PAIR OF PERFUME-BURNER MODELS

French
About 1775–1800
Wood, painted and gilded
10 ½ in. (26.7 cm)
Gift of J. Pierpont Morgan, 1906
07.225.385a–b

While perfume burners, or cassolettes, were usually made of heat-resistant materials such as bronze, porphyry, agate, earthenware, or porcelain, this pair is composed solely of painted and gilded wood, suggesting that these pieces may have been models to show commissioning or prospective clients.[1] The red and orange on the vases and green and reddish brown paint—on the tops and bottoms of the pedestals—was probably intended to imitate colored stones, such as porphyry or marble, which would have

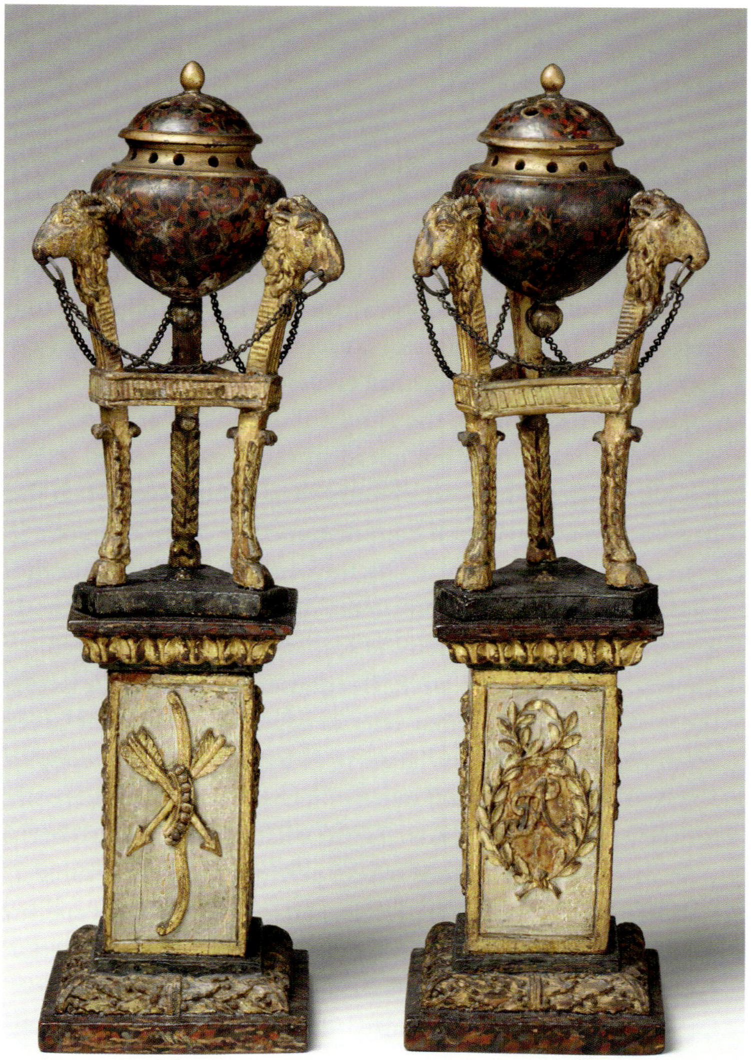

Cat. no. 70

The Vase as a Design Element in Interiors

Stefanie Walker

As many vase designers were primarily active as architects, it is not surprising that the form found its way into their proposals for interior decoration. Featured here is such work by three of the most prominent French architects of the later eighteenth-century, Jean-Charles Delafosse, Claude-Nicolas Ledoux, and François-Joseph Bélanger. In a large drawing for a gallery (cat. no. 73), Delafosse found opportunities to place vases in a number of locations. Six overdoors designed by or made under the supervision of Ledoux and Bélanger (cat. nos. 74–75) illustrate another popular site for employing the vase form as a central organizational motif. Another designer, Michelangelo Pergolesi, produced elevations for interiors (see cat. no. 81) that reflect ideas of architect Robert Adam, the most important English promoter of vases as decorative elements.

As wall decoration, the vase appeared in a number of sites: in vertical paneling, in horizontal overdoors, and in the running design of friezes. The form could be gilded, painted, or treated monochromatically, or could be sculpted in wood or stucco. Vases also frequently figure in fabrics used to cover walls and upholster furniture, unifying the appearance of a room.

The motif might be a simple ornamental feature, one among many. More often, however, it serves to anchor or punctuate an ornamental scheme, such as in the paneling designs of Gilles-Paul Cauvet (see cat. no. 79) and Jules-François Boucher (see cat. no. 80). Its use is particularly refined in the above-mentioned overdoors (cat. no. 74), where repetition of the form succeeds in pulling together the various attributes that represent each of the Four Seasons. A silk fabric designed by Jean-Demonsthène Dugourc (cat. no. 78) and Pergolesi's interior demonstrate how central the vase form was to the mature neoclassical style, or the *goût étrusque*, in vogue in the 1780s. In Friedrich August Leo's elevation of a garden building (cat. no. 82), the quotation of classical vases is more specific and serves to enhance the evocation of the almost elegiac aura of the "purer" eras of the distant past.

Two-dimensional vases, in wall paintings or on fabrics, complemented free-standing ones, in ceramic, metal, stone, or wood. Neoclassical interiors in fact resonated with the pervasive and insistent presence of the motif. Beyond mere ornament, vases in interiors symbolized the occupants' extensive knowledge of antiquity and, thus, their erudition and good taste.

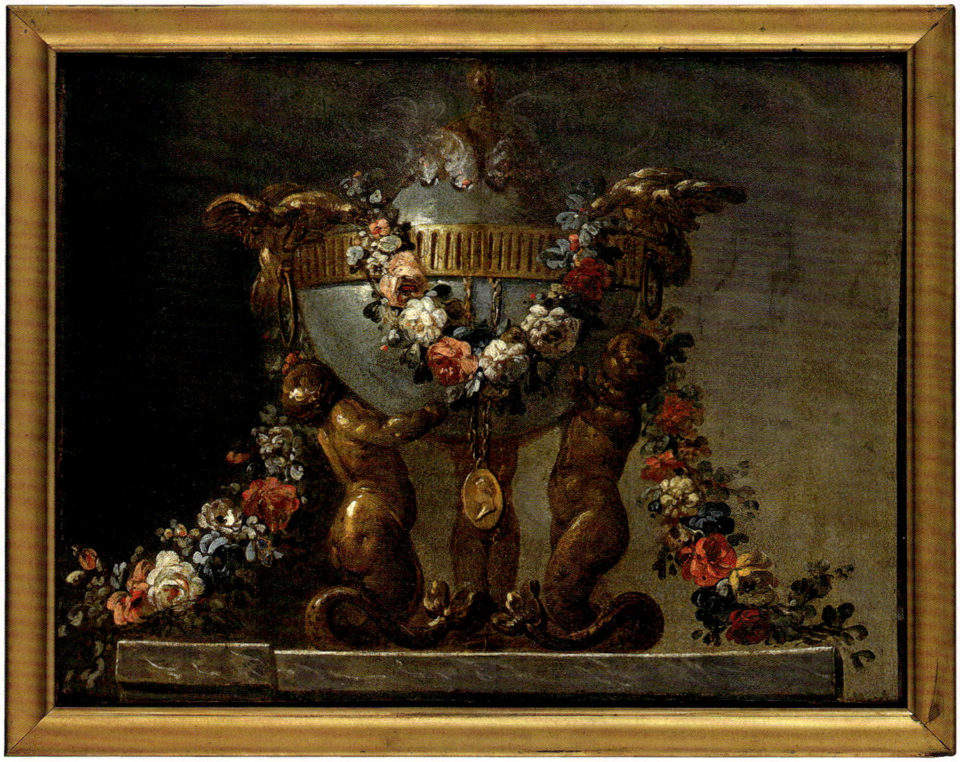

Cat. no. 85 (07.225.457a)

73. ELEVATION DRAWING OF A LARGE GALLERY

Jean-Charles Delafosse (1734–1789)
1769
Pen and black ink, with gray and brown washes
14 3/4 × 32 7/16 in. (375 × 824 mm)
Inscribed: [lower left] *J. C. Delafosse* [illegible] *faite. 1769*; [lower middle] *Grand Galerie de L'ordre Corinthien avec Sallon aux Extremités/Echelle de 12 pieds*
Purchase, Joseph Pulitzer Bequest, 1973
1973.638

In his many engravings that feature vases, Jean-Charles Delafosse frequently depicted them inventively and provocatively. Exaggerating their scale and rendering them in unfamiliar, hypothetical contexts, the artist challenged the imagination while providing innovative designs for the use of artists and craftsmen (see cat. nos. 9–10, 87). By contrast, in this 1769 drawing of a grand gallery and adjoining salon, Delafosse illustrated some of his favorite vase designs as they have might have appeared in an actual interior, liberally using them as a recurring motif in the elaborate woodwork, frieze, and ceiling cove, and displaying them like sculpture on Roman-inspired pedestals. One of the few surviving drawings of interiors by Delafosse, this example is exceptional for its size and completeness, providing a rare glimpse of his grand and sumptuous vision of early neoclassical room decoration.[1]

Although trained as a sculptor and best known as a prolific designer of engraved ornament, Delafosse considered himself an architect, using that title in publications such as his *Livres de trophées* (see cat. no. 87) and the influential *Nouvelle Iconologie historique* (1768).[2] Sadly, except for two Paris residences, little is known of the buildings on which he worked. Thus, drawings like this one—with functional details, such as ceiling profiles and a scale—offer clues to his work as a practicing architect, in contrast to his more speculative designs for fantastic tombs, monuments, and other structures.[3]

Hinting at his ambitiousness, Delafosse's drawing depicts an extremely rich gallery in which paintings, sculpture, and furnishings—including vases—are fully integrated with the architecture. Conceived like a neoclassical façade, the elevation features a portico-like projecting colonnade of soaring Corinthian columns, with double-columned piers defining a smaller salon to the right. Painted panels and relief ornament embrace a wide range of iconographical themes similar to those in the architect's encyclopedic *Nouvelle Iconologie*, published the year before he executed this drawing. Maidens engage in artistic and scientific pursuits on large panels and in friezelike compositions in the ceiling's cove, their activities echoed by scenes of putti below the main panels. On free-standing pedestals, more putti, pursuing

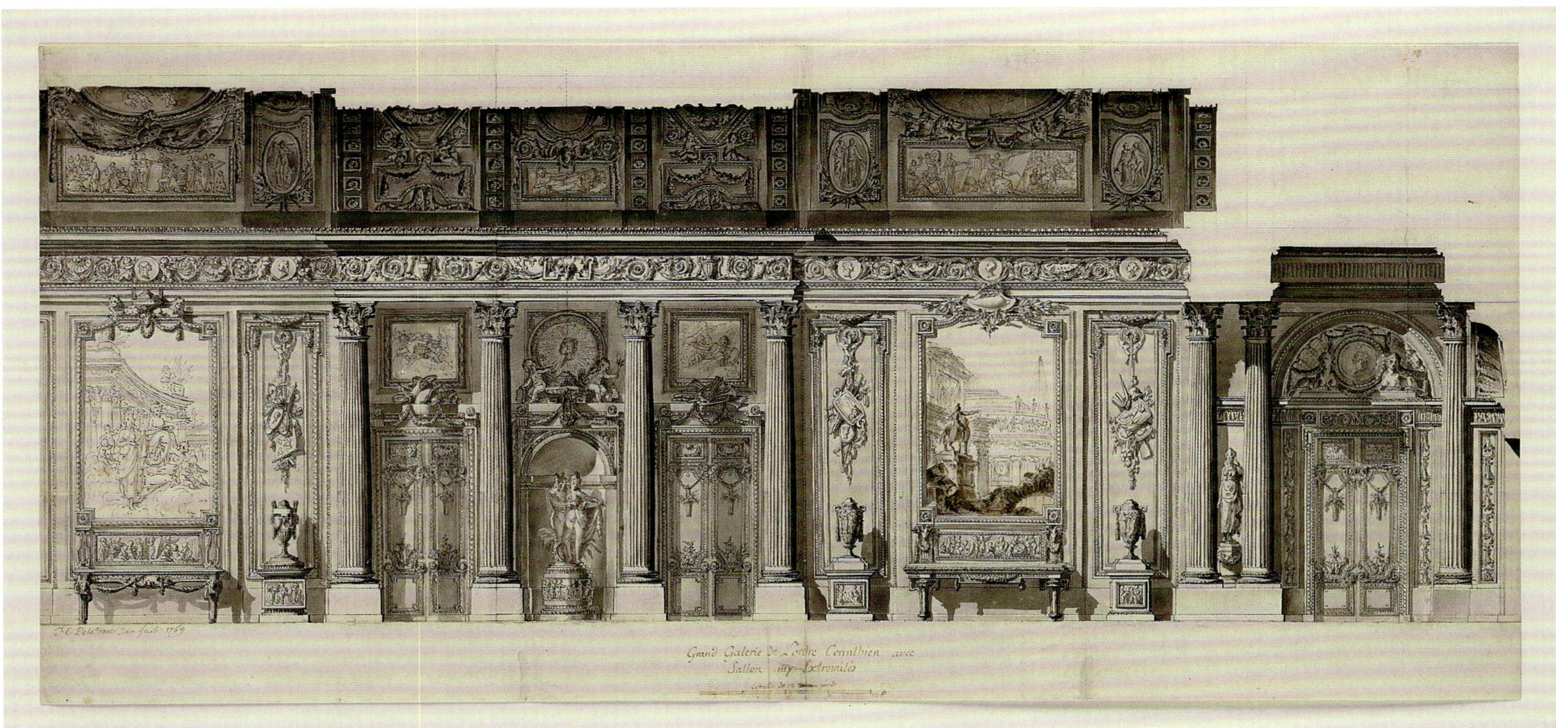

Cat. no. 73

occupations like threshing and wine making, represent the seasons. Pendant trophies incorporate symbolism familiar from Delafosse's allegorical representations of nations. An image of Athena, goddess of wisdom and patroness of the arts, surmounts the salon door.

Delafosse routinely executed designs related to student assignments at the Académie de Saint-Luc, Paris, where he taught. Given this drawing's generic title, it seems plausible that this tour de force may have been used in this way.[4] Intriguing physical evidence, however, indicates that the drawing's history is complicated. Close examination reveals several changes made after Delafosse finished it, possibly to adapt it for a specific purpose. For instance, the entire central bay of the colonnade containing the gallery's main sculptural group was cut away and a new drawing of the sculpture adhered from the back. The partially nude female statue at the far right of the drawing is another addition, applied to the paper's surface in the narrow space between columns. The reasons for these changes are unclear, but they may have been made to adapt the design to a patrons's collection.[5] H.J.M.

1 For this drawing, see *Notable Acquisitions, 1965–1975 MMA Bulletin* 20 (1975): 204; and Myers 1991: cat. 30. For other examples of Delafosse's drawings, see Armand Guérinet's five-volume series of the artist's work, *J. C. Delafosse, Le Style Louis XVI: Oeuvre de Delafosse* (Paris, n.d.). I would like to thank Perrin Stein, Associate Curator of Drawings and Prints at the Metropolitan Museum, for her generous help in examining this drawing, and credit her many useful observations about its design and physical condition.

2 See Eriksen 1974: 170. For example, Delafosse signed as "Jean Charles Delafosse architecte. Décorateur et Professeur en Desseins," in his *Nouvelle Iconologie historique* (Delafosse 1768).

3 The adjacent hôtels were built for a Madame Delbarre during the 1770s and 1780s. Located at numbers 58 and 60, rue du Faubourg-Poissonière, they have long been known as the hôtels Titon and Goix. Delafosse also apparently designed houses in Pantin and in the rue Sainte-Apolline, and two groups of barracks in Paris; see Michel Gallet, "Jean-Charles Delafosse, Architecte," *Gazette des beaux-arts* 61 (1963): 160; and Eriksen 1974: 170.

4 As suggested by Christopher Riopelle and cited in Myers 1991: 55.

5 The drawing, on five joined pieces of paper, has a title with the artist's signature and a scale. They are inscribed on two long strips of paper affixed to the lower edge, probably after part of the image was lost. Clearly cut at its left edge, the design—meant to be balanced but not to repeat—is missing elements, such as a vase and plinth illustrating the fourth season. The gallery's center line, marked by the artist, also shows that part of the concept is missing. Delafosse's signature, however, is aligned with the severed left edge, suggesting that it was added after the drawing was reduced.

74. OVERDOOR PANELS DEPICTING THE FOUR SEASONS

Possibly made under the direction of François-Joseph Bélanger (1744–1818)
About 1777–81
Wood, carved and gilt
Each 24 ½ × 52 ¾ in. (62.2 × 134 cm)
Gift of J. Pierpont Morgan, 1906
07.225.127a–d

Because the quality of carving and design in each of these four overdoor panels is unusually high, they traditionally have been thought to originate from the famed Château de Bagatelle in the Bois de Boulogne, Paris.[1] Made of oak, they are carved in minutely detailed low and high relief, with their backgrounds painted light gray and ornament originally gilt to suggest ormolu.[2] The sophisticated design of each panel centers around a footed vase decorated with twisted fluting, angular *goût grec*-style handles, and a naturalistic floral garland. Except for the pairs of masks at the shoulders, the vases are identical in all four overdoors, establishing a visual coherence within the group while simultaneously allowing for variation in the details of each panel.

The differences between the overdoors are found in the background vegetation, in the arrangement of objects at the bases of the vases, in the flora of the swags, and in the pairs of profile masks below the handles. The changing elements in the panels correspond to each of the four seasons. Spring is characterized by lush, blooming roses and lilacs, attributes of love (flaming torch, rose wreath, quiver, and arrows), and the heads of Flora or Venus at the shoulders of the vase. A plentiful harvest is indicated in the panel depicting summer, with abundant shafts of wheat mixed with poppies and cornflowers, tools of the harvester at the base of the vase, and the heads of Ceres (Roman goddess of the harvest and of fertility). Autumn is represented by attributes of the cult of Bacchus, including vines and grapes, a thyrsus, and cymbals. The goat heads at the base of the vase's handles are associated with the pleasure-loving god. In the fourth panel, ivy and oak leaves, acorns, and the bundled head of an old man on each side of the vase signify winter, while the masks and instruments at the base refer to Carnival.

It is not possible to definitively trace the design of these overdoors; however, there are antecedents that seem to have influenced it. The designers most likely derived the vase motif from works by Jean-Charles Delafosse and Jacques-François-Joseph Saly, translating their engraved and etched designs into carved, almost three-dimensional interior decoration. Delafosse's publications, such as the important *Nouvelle Iconologie historique* (1768), may have inspired the angularity of the handles and overall design; the densely packed, deep striations in the body of

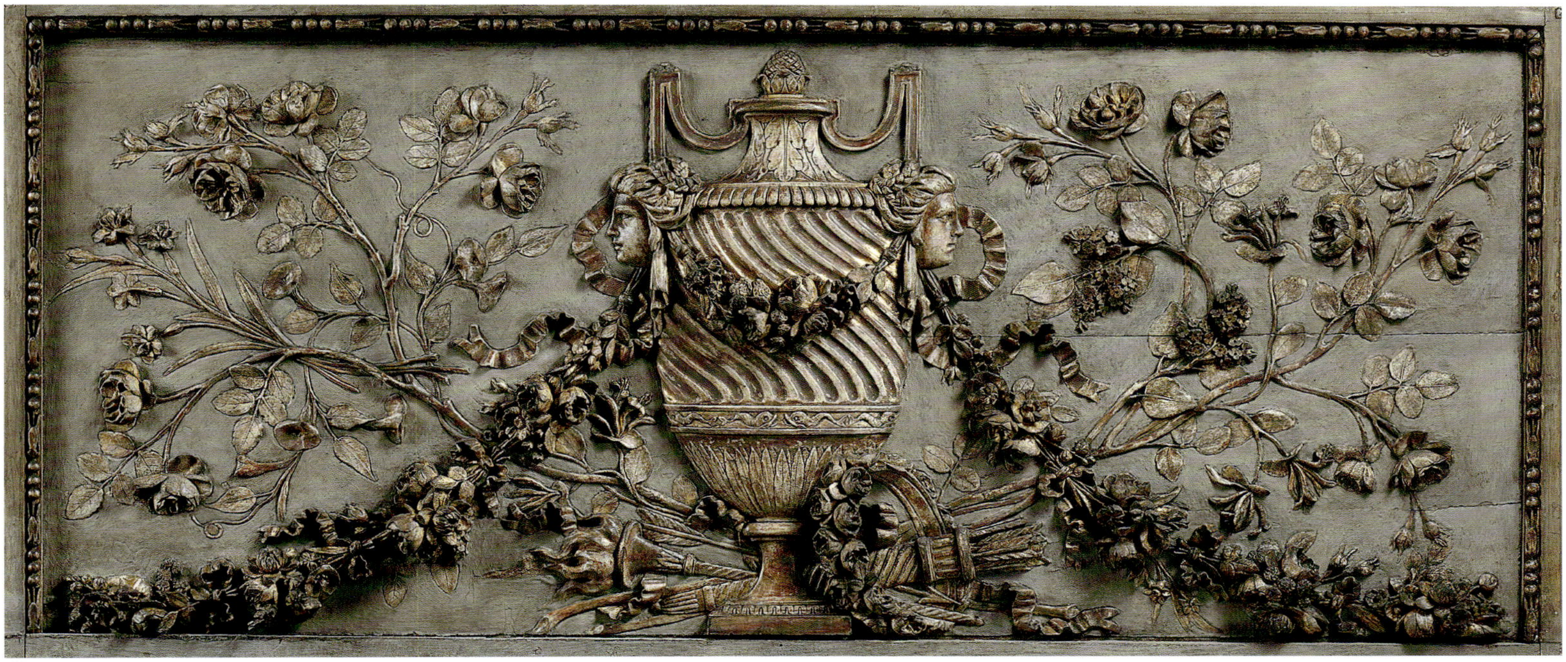

Cat. no. 74 (spring)

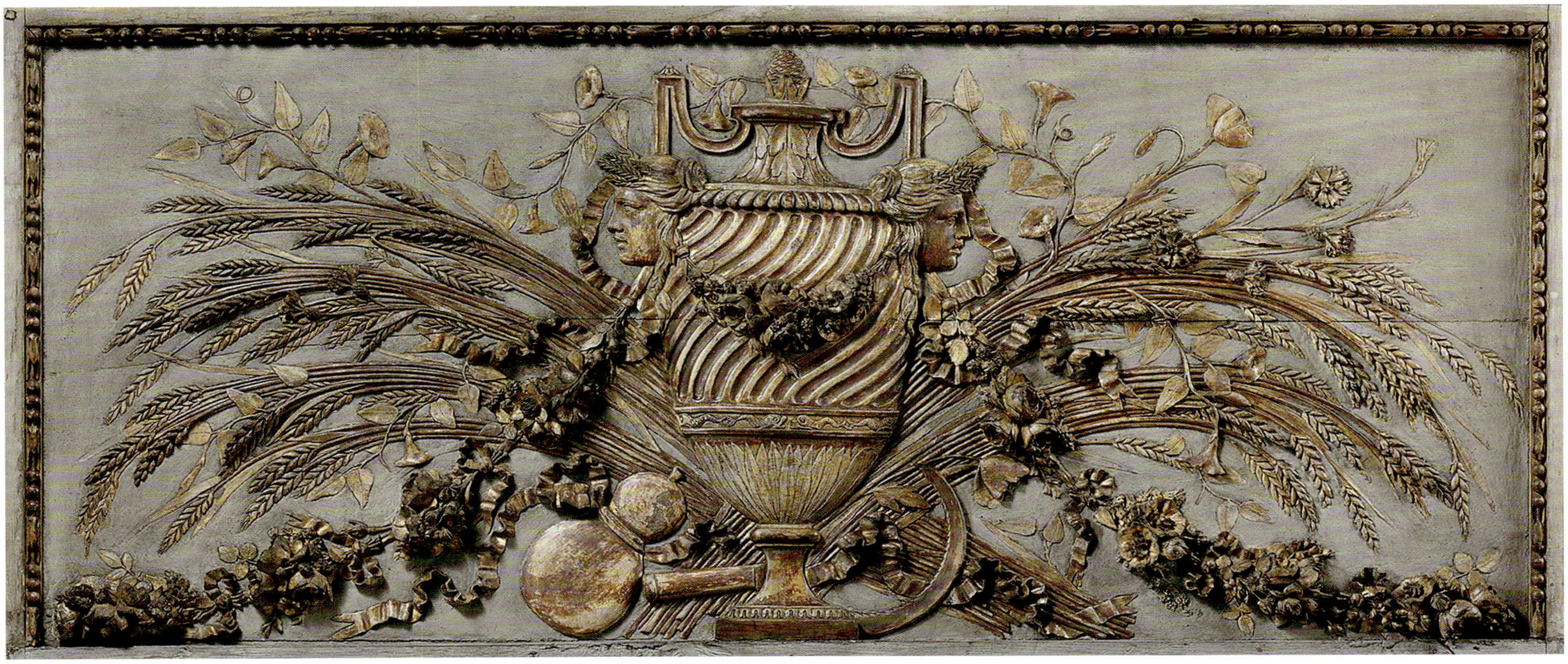

Cat. no. 74 (summer)

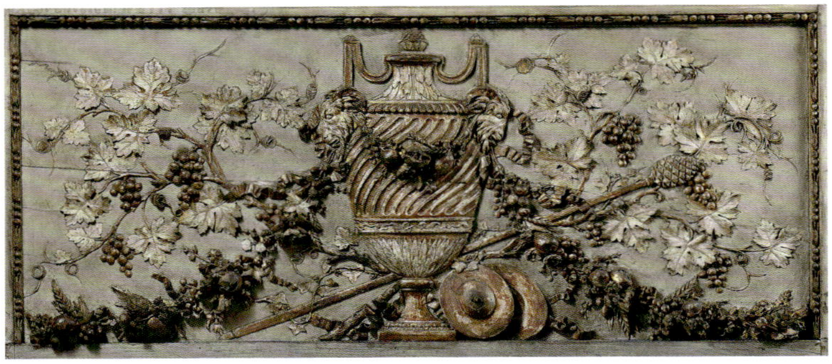

Cat. no. 74 (autumn)

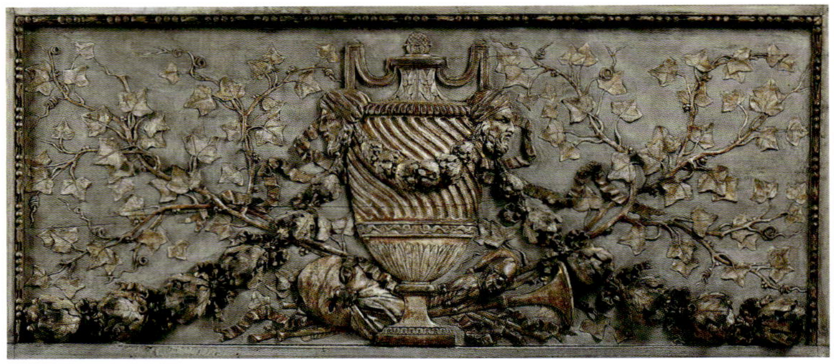

Cat. no. 74 (winter)

the vase; and the lancet leaves along its base. Plate 21 of Saly's *Vasa a se inventa atque studii causa delin. et incisa* (1746) presents a vase design that features lancet leaves at the foot of the vessel and deep, curved, diagonal channels, which are softer and wider than those in Delafosse's work and these overdoors. In this etching, Saly also included at the shoulders of the vase two heads joined by a floral garland.[3]

The Château de Bagatelle was the folly of the Comte d'Artois, youngest brother of Louis XVI and the future Charles X. In 1777 d'Artois won a bet with his sister-in-law Marie-Antoinette, that he could completely reconstruct this château during the nine-week period in which the royal family visited Fontainebleau. Architect François-Joseph Bélanger drew up plans in forty-eight hours, and some nine hundred workers completed the main construction of the pavilion, its Court of Honor, the Building of Pages, various outbuildings, and the gardens in sixty-four days, while taking another few years to finish the interior decoration and modify the landscaping.[4]

Although Bélanger was responsible for many of the designs for Bagatelle, he hired others to assist with the ornamentation of the château. His brother-in-law, Jean-Démosthène Dugourc, worked with him to coordinate the decorative schemes for each of the rooms. Painters Jean-Marie Dusseaux and Hubert Robert were responsible for the ornamental paintings, and several sculptors contributed to the project, including Daniel Aubert, Nicholas-François-Daniel Lhuillier, and Philippe-Laurent Roland.[5]

Current evidence does not allow a secure attribution of the panels' composition to Bélanger or their origin to Bagatelle. The extremely high quality of design and execution of the carving, as well as the iconography of the panels, does connect them to other decorative elements at the château. Attributes of the cult of Bacchus, such as thyrsi, goat heads, and grape leaves, feature in the decoration of several chimneys on the second floor of the main pavilion, while comparable motifs and heads of jesters (similar to the one in the winter panel) appear in several of the overdoors in the billiard room.[6] Such ornaments indicate d'Artois's use of Bagatelle for both public and private entertainments and match the overall decorative scheme of the structure, which was intended to reflect the sensual and theatrical personality of the pavilion's owner. M.H.

1 *Collections Georges Hoentschel; notices de André Pératé and Gaston Brière* (Paris, 1908): vol. 3, 31–22, pls. 118–19. The connection of the panels with Bagatelle probably stems from information in the shipping inventory of the Georges Hoentschel Collection (Metropolitan Museum Business Archives), of which they were part, before they arrived at the Metropolitan Museum. Hoentschel may have acquired the overdoors during renovations to the château (1851–60 and 1872) by its later owners, Lord Richard Seymour, the fourth marquess of Hertford, and his adopted son, Sir Richard Wallace; see J. Vacquier, *Les Anciens Châteaux de France. L'Île-de-France. Bagatelle…*, 3rd ser. (Paris, 1920): 2. These four overdoors were illustrated in E. R., "The Hoentschel Collection," *MMA Bulletin* 2 (June 1907): 96.

2 Lester Margon, *Masterpieces of European Furniture, 1300–1840* (New York, 1968): 194.

3 For examples of Delafosse's designs, see Oechslin 1982: figs. 196b–c, 203. Saly's print in the Victoria and Albert Museum, London, is illustrated in Clifford 1978: pl. 81a. Another Saly design, from the same 1746 collection of vase engravings, depicts goat heads at the shoulder of the vessel, also joined by a garland of flowers; see Eriksen 1974: fig. 293. I am grateful to Heather Jane McCormick for suggesting Saly's influence.

4 For more on Bagatelle, see Henri-Gaston Duchesne, *Histoire du Bois de Boulogne. Le Château de Bagatelle…* (Paris, 1909); Jean Stern, *À l'Ombre de Sophie Arnould: François-Joseph Bélanger, Architecte des menus-plaisirs: Premier architect de Comte d'Artois*, 2 vols. (Paris, 1930); C. Arnauld et al., *La Folie d'Artois* (Paris, 1988).

5 If these particular panels did originate at Bagatelle, it is unclear who designed and executed them, but Roland created a model for the overdoor of the entrance portal that Lhuillier carved; it is in the Metropolitan Museum (07.255.18). Roland also designed for the billiard room a set of overdoors, whose execution is attributed to Aubert; see Jean-Jacques Gautier, "L'Art des sculpteurs de Bagatelle," *Estampille: L'Objet d'art* 252 (Nov. 1991): 64–82, fig. 1; and James David Draper, "Philippe-Laurent Roland in The Metropolitan Museum of Art," *MMA Journal* 27 (1992): 129–47, fig. 6. According to both Gautier and Draper, Aubert asked for a total of 20,652 *livres* for his sculptural contribution to Bagatelle, suggesting that he was probably responsible for the majority of such work on the project.

6 For images of the chimneys, see *Le Château de Bagatelle: Étude historique et descriptive…* (Paris, [1909]), pls. 61–65. The overdoors, which are attributed to Aubert, are illustrated in Gautier (note 5): figs. 5–6, 8–9.

75. OVERDOOR PANELS FROM THE CHÂTEAU DE MONTFERMEIL

Design attributed to Claude-Nicolas Ledoux (1736–1806)
About 1762–65
Plaster, carved and painted
25 ½ × 50 ¼ in. (64.8 × 127.6 cm)
Gift of Mrs. Benjamin Moore, in memory of her mother, the Hon. Mrs. Alfred Anson, 1956
56.202.1–.2

Although at first glance they appear to be identical, these two overdoor panels differ slightly in the foliage that bursts forth from their central element, a bowl-shaped vase. Unlike overdoors made of more expensive carved and gilt wood (see cat. no. 74), these panels, which were intended for the Château de Montfermeil, in the Seine-et-Oise region, were carved and painted in plaster.[1] A number of classicizing eighteenth-century architects, including Robert Adam, François-Joseph Bélanger, and Claude-Nicolas Ledoux, substituted stucco for wood in the decoration of their interiors, praising the material's low price, ability to imitate marble when polished, and adaptability to multiple spaces and designs.[2] Stucco could easily be carved in high and low relief, and indeed these overdoor panels present a distinct sculptural quality, heightened by the undercutting of the carving. Such workmanship displayed the design to best effect when the panels were placed in their intended positions over the eight-foot high doors at Montfermeil.[3]

The rural estate of Montfermeil was acquired in 1735 by Jean-Hyacinthe Hocquart, a member of the powerful Ferme Générale, an extremely wealthy and tight-knit organization that profited considerably from collecting sales taxes for the state. Between 1762 and 1765, Hocquart's son, Jean-Hyacinthe-Louis-Émmanuel, hired Ledoux to modernize the seventeenth-century château.[4] Ledoux's renovations were relatively modest and consisted mainly of the addition of terraces, Doric pilasters and pediments to the entry and garden façades, and two wings on each side of the original structure. These overdoors hung in the lower level of one of the new wings.[5]

As the project's architect, Ledoux probably did not execute the overdoors himself. He may have provided the design and then contracted stucco sculptors such as Joseph Métevier or Jean-Baptiste Feuillet, to ornament Montfermeil's interiors. Ledoux was certainly attentive to detail, and presumably these panels, among the few decorative architectural elements in the lower level of the structure, would have had to meet with his approval and reflect his style and vision.[6]

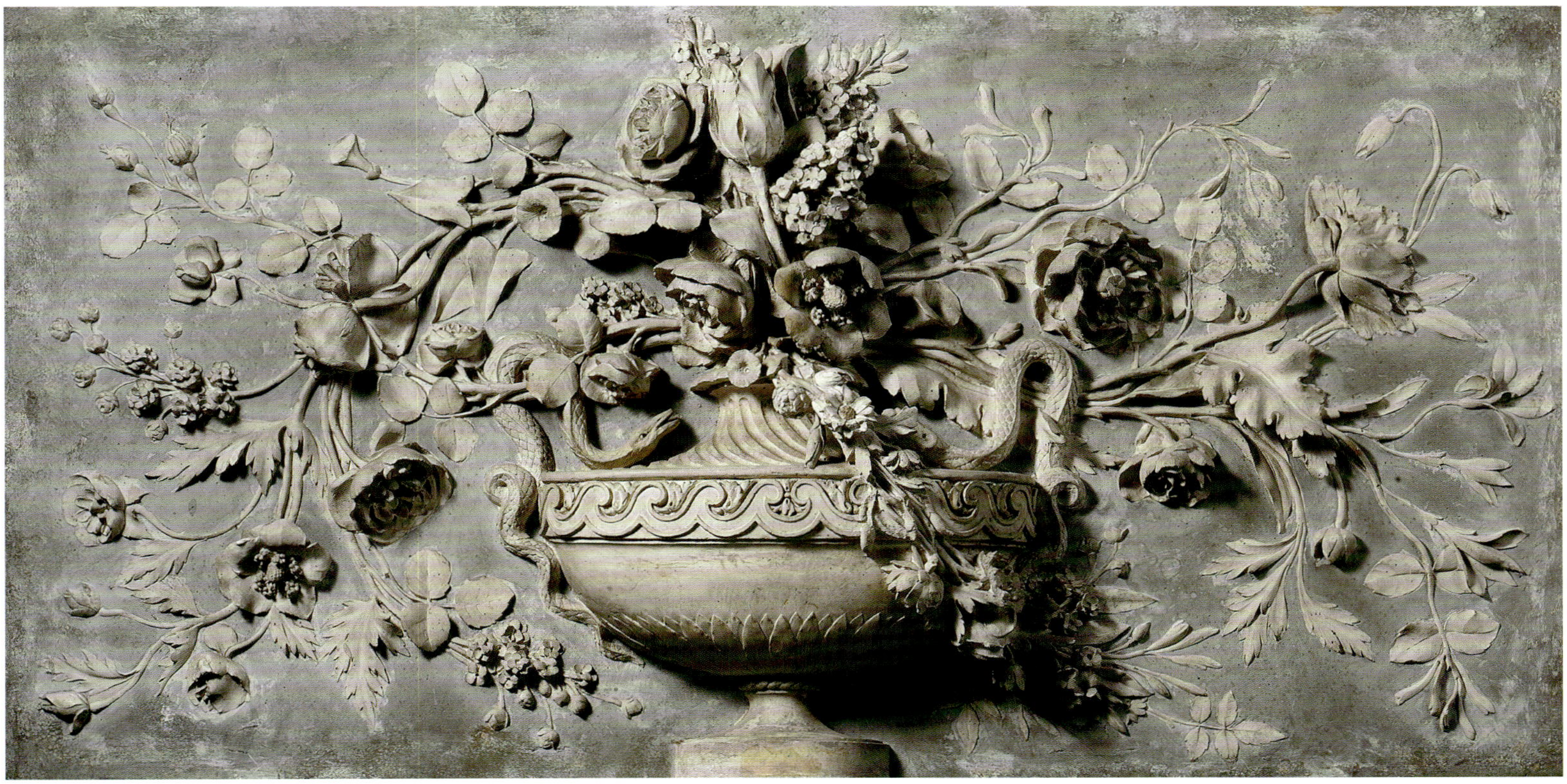

Cat. no. 75

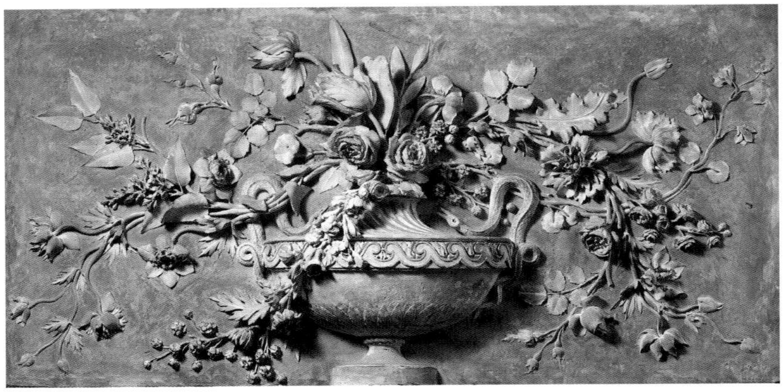

Cat. no. 75

Montfermeil, one of Ledoux's early commissions, represents a transitional stage in his evolution toward the functional classicism for which he would become known. Architectural ornament, such as the fluted pilasters and arched niches in the château's Large Hall and the abundant vegetal curves and guilloche band on the overdoors, seems to have been borrowed from Pierre Contant d'Ivry's transitional architectural designs for the Palais Royal, Paris (1755–60), which were engraved by one of Ledoux's mentors, Jacques-François Blondel.[7] Despite this reflection of the past, Ledoux revealed the roots of his incipient style in Montfermeil; this can be seen in the present overdoors' almost perfect symmetry and pared-down, neoclassical ornament. Utilizing a vase as the central motif, he created majestic, sculptural forms that, when realized in high and low relief, achieved an interplay of contrasting volumes in space.[8]

The overdoors also illustrate Ledoux's interest in nature, which stemmed from his admiration for the philosophy of Jean-Jacques Rousseau, specifically his call for a society grounded in the natural world. In many of his domestic commissions outside of Paris, Ledoux attempted to take the purifying effects of a pastoral life indoors through his decorative choices.[9] At Montfermeil numerous glass doors and windows that are almost floor-length look out onto the gardens, while architectural accents with vegetal or floral themes, marble floors, and a marble fountain in one of the salons suggest the outdoors. In the overdoors, the gray vase, abundant and realistic flowers, and serpent handles (symbolizing earth's fertility), all set against a pale blue-green background, may have been intended to evoke images of garden statuary. Such allusions express Ledoux's perception of the architect as a thinker and reformer. This approach to early neoclassicism contrasts with the overdoors of the Four Seasons (cat. no. 74). While the latter's *boiseries* are playful, erotic, and intended to amuse, Ledoux's stucco overdoors reflect a philosophy as well as an aesthetic. M.H.

1 The provenance of Montfermeil is based on correspondence and photos in the curatorial files of the Metropolitan Museum's Department of European Sculpture and Decorative Arts.

2 Geoffrey Beard, *Stucco and Decorative Plasterwork in Europe* ([London], 1983): 187–88; and Whitehead 1992: 102. Whitehead (101) quoted a September 1775 letter from Horace Walpole describing a stucco interior: "Mad. de Mirepoix's house, where I supped last night is charming.... The *salle à manger* is all of stucco, highly polished, representing white marble with panels of vert antique."

3 For the door height, see C. Hamilton Preston, "Details of the Salons, Château de Montfermeil, Paris: Two Simple Interiors in the Style of Louis XIV," *Architectural Forum* 41 (July 1924): 36–41.

4 Michel Gallet, *Claude-Nicolas Ledoux: 1736–1806* (Paris, 1980): 31; and Anthony Vidler, *Claude-Nicolas Ledoux: Architecture and Social Reform at the End of the Ancien Régime* (Cambridge, Mass., 1990): xi–xii, 24. Unfortunately, very little is known about Montfermeil; the dates assigned to its renovation vary between 1762 and 1765. For example, Michel Gallet, "Claude Nicolas Ledoux," in *Macmillan Encyclopedia of Architects*, ed. Alfred K. Placzek (New York, 1982): 653, dated Montfermeil to circa 1762–64, while Vidler (24) proposed 1765.

5 Gallet, *Claude-Nicolas Ledoux* (note 4): 31; and Vidler (note 4): 40. According to Gallet and Philippe Lefrançois and Patrice Boussel, *Petits Musées et Joyaux de l'Île-de-France au Nord de la Seine et de la Marne* (Paris, [1951–52]): 64–65, the château fell into disrepair and was demolished between 1926 and 1928. For engravings and a photograph of the structure, see Lucien Noël, *Montfermeil et sa région: Fragments historiques* (Aurillac, 1931–35), vol. 1, frontis.; vol. 2, frontis., 30, 36, 53; vol. 3, title page, page in-between 6 and 7, 94. The room provenance is based on comparing photographs in the object files in the Metropolitan Museum's Department of European Sculpture and Decorative Arts to elevations in Preston (note 3); and idem, "Details of the Large Hall, Château de Montfermeil, Paris: Classic Influence in Louis XIV Interiors," *Architectural Forum* 41 (Aug. 1924): 81–86.

6 Beard (note 2): 187–88. Ledoux hired Métevier and Feuillet to do stucco work on many of his projects, including the Hôtel Tabary, Paris, in 1773; however, it is uncertain whether they participated in the construction of Montfermeil. See Preston (notes 3, 5). Wolfgang Herrmann, "The Problem of Chronology in Claude-Nicolas Ledoux's Engraved Work." *Art Bulletin* 42 (Sept. 1960): 196, discussed Ledoux's attention to detail in the Château de Bénouville, for which a local architect named Gilet was required to submit even minor alterations to him for approval, even after the main construction of the building was completed.

7 Two designs by Contant d'Ivry for the Palais Royal are illustrated in Wend von Kalnein, *Architecture in France in the Eighteenth Century* (New Haven/London, 1995): figs. 226–27. Ledoux's contributions to Montfermeil are clearly neoclassical in inspiration, but his designs may have been influenced by aspects of Contant d'Ivry's work.

8 For information on Ledoux's mature style, see for example Gallet, "Claude Nicolas Ledoux" (note 4): 648–54.

9 Ibid.: 652; Vidler (note 4): ix–xv.

76. FABRIC FOR A CHAIR BACK

French, unknown designer and maker
1780–90
Silk: plain compound satin; ground pale blue; design white
20 ¾ × 19 ½ in. (52.7 × 49.5 cm)
Rogers Fund, 1940
41.134.11b

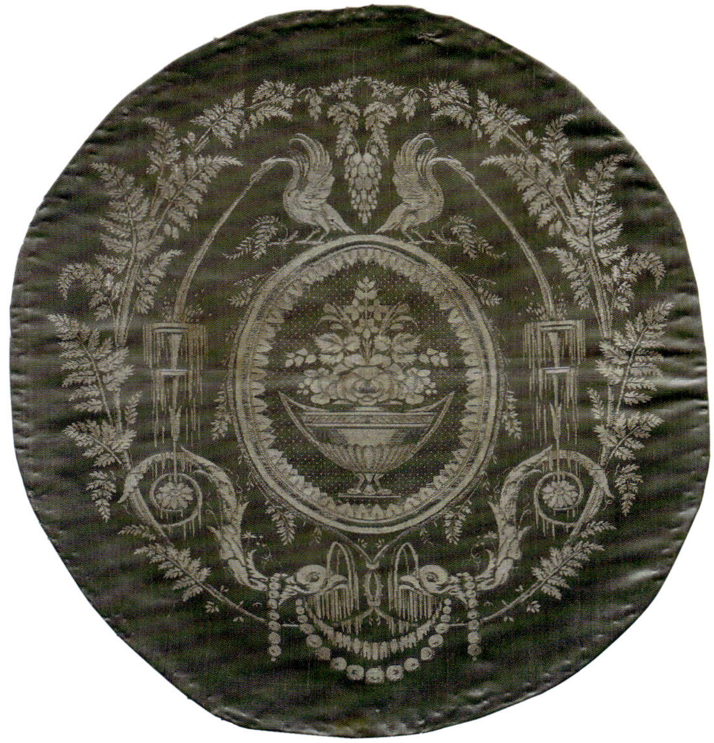

Cat. no. 76

77. TEXTILE PANEL

French, unknown designer and maker
Eighteenth century
Silk: fancy satin; ground crimson; design white
20 ½ × 20 in. (52.1 × 50.8 cm)
Gift of the United Piece Dye Works, 1936
36.90.1427

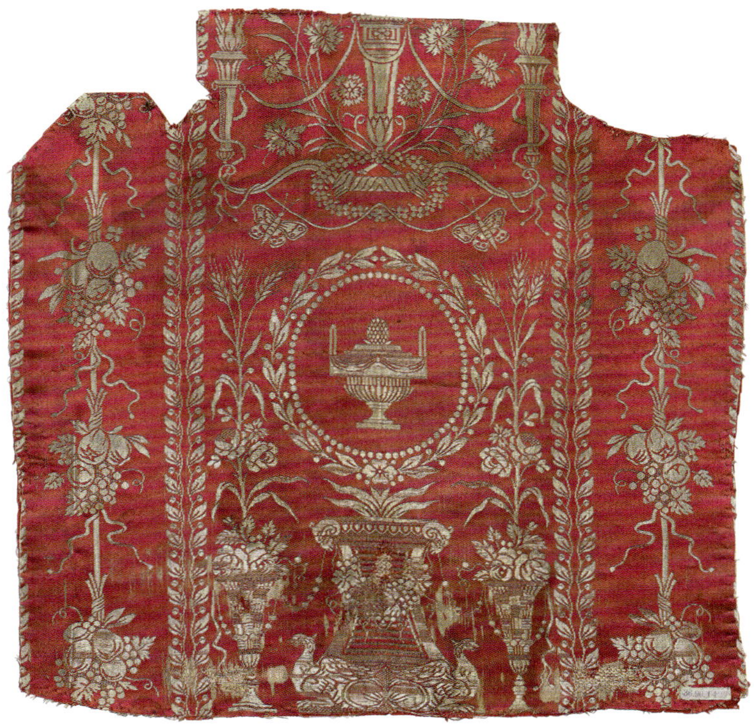

Cat. no. 77

78. TEXTILE PANEL WITH VASE MOTIF

Designed by Jean-Démosthène Dugourc (1749–1825); manufactured by Camille Pernon (1753–1808)
About 1780–1800
Silk: plain compound satin; ground pink (much faded); design silvery white
96 × 22 in. (243.8 × 55.9 cm)
Purchase, Joseph Pulitzer Bequest, 1933
33.140

Surviving designs for silk wall panels from the late eighteenth century often include classical figures or mythological animals. Featuring the latter, as well as a complex network of garlands and ribbons, this design consists of elements arranged in strict symmetry around a central, vertical axis. It is divided into three registers, each framed by blossom chains. The lower section consists of a tapering base decorated with two snakes entwined around a thyrsus (a fanciful interpretation of the caduceus) and supporting a footed bowl of fruit. The middle, and largest, register centers around a canopy over a burning altar and a neoclassical representation of a krater vase in a roundel decorated by a border, the two motifs separated by a lyre (Fig. 1). The top portion of the textile depicts a floral bouquet within a circle framed by an octagon, on either side of which stand two ewer-shaped vases on caduceus staffs. While the classical motifs and overall design derive from Renaissance grotesques, particularly Raphael's pilaster decorations in the Vatican Loggie, the flowers and branches hanging from ribbons or issuing from other elements are related to the more naturalistic flavor of mid-eighteenth-century rococo concepts.[1] This subtle blend of styles, often called the *goût étrusque*, results in a light, delicate neoclassicism, in contrast to the heavier Vitruvian scrolls, Greek-key patterns, and other relatively ponderous motifs typical of the preceding phase, known as the *goût grec*.[2]

The panel's palette reflects a preference for bright colors over the sedate earth tones of the classical prototypes from which such designs in part derived.[3] The pink ground, now faded, was originally more vivid, providing a pronounced contrast to the silvery white figures.[4] The subtlety of the fabric's weave, which gives dimension to even the smallest motifs, highlights the successful collaboration between designer, Jean-Démosthène Dugourc, and manufacturer, Camille Pernon. The latter's initials, "MPC," can be seen in the lower register of the panel, incorporated into small medallions held by swans.

Dugourc, who claimed in his autobiography to have invented the *goût étrusque*,[5] was appointed Designer of Royal Furnishings in 1784.[6] The character of many of his furniture and textile designs anticipated the Empire style by nearly two decades.[7] He worked extensively for the Lyon-based Maison Pernon & Cie.,

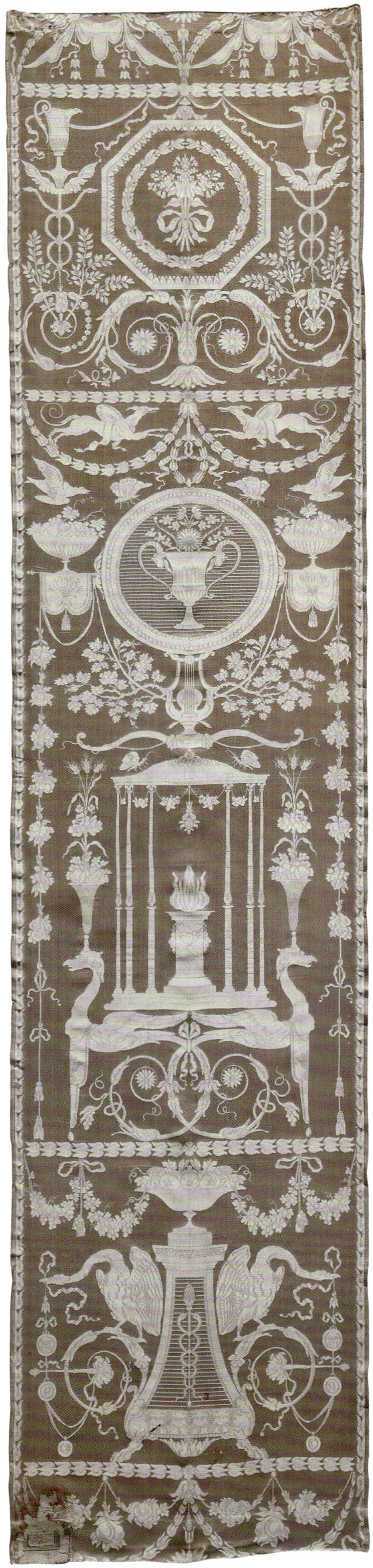

Cat. no. 78

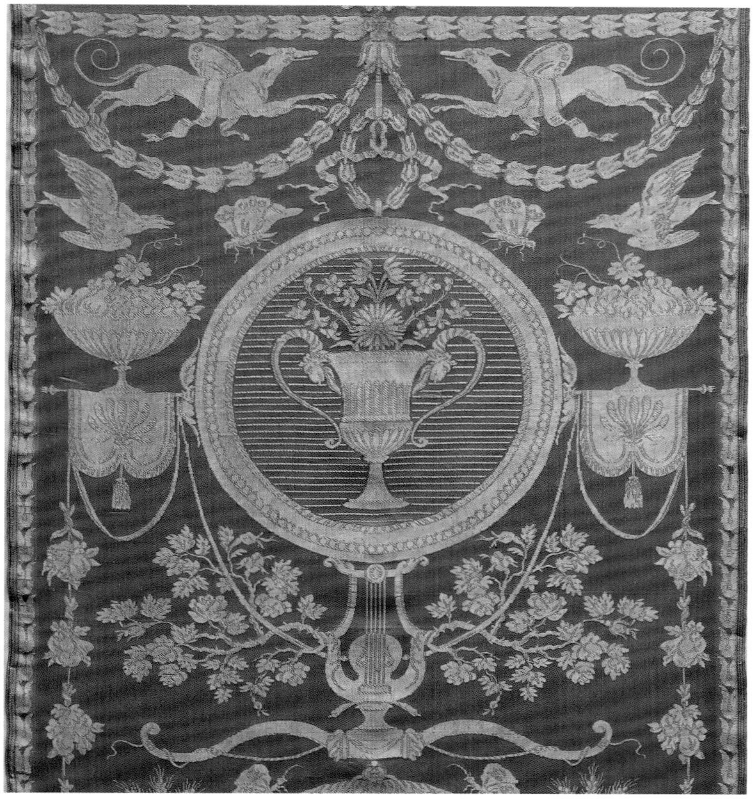

Fig. 1: Detail of cat. no. 78

founded in 1680.[8] Camille Pernon revitalized his family's textile firm by engaging the leading designers of the day and introducing technical innovations to produce intricately woven fabrics for both aristocratic and commercial clients.[9] In addition to royal commissions, Duguorc and Pernon produced classicizing designs like this wall panel for upper-class patrons in the late eighteenth century, and their work continued to be fashionable in the decades to come.[10] K.M.

1 Whitehead 1992: 70.
2 Whitehead 1992: 67.
3 Alain Gruber, *Grotesques: Un Style Ornamental dans les arts textiles du XVI–XIX siècle* (Riggisberg, 1985): 40.
4 A piece of the same fabric design, with a yellow ground, is owned by Cora Ginsberg, Inc., New York.
5 Léon de Groër, *Decorative Arts in Europe 1790–1850*, trans. Aileen Dawson (New York, 1986): 310.
6 Chantal Gastinel-Coural, *The Dictionary of Art* 9 (New York, 1996): 378–79.
7 F. J. B. Watson, *Louis XVI Furniture* (London, 1973): 83.
8 Gastinel-Coural (note 6).
9 Pilar Benito, "Camille Pernon y El "Tocador" de la Reina Maria Louisa en el Palacio Real de Madrid," *Reales Sitios* 30, 116 (1993): 17–24.
10 Gruber (note 3).

79. DESIGN FOR AN INTERIOR PANEL WITH A VASE SUPPORTED BY NEREIDS

Designed by Gilles-Paul Cauvet (1731–1788); etched by Simon-Charles Miger (1736–1820)
1777
Etching, from *Recueil d'ornemens á l'usage des jeunes artistes qui se destinent á la decoration des bâtiments*
Signed at lower left: *G.P.Cauvet inv. del.*; at lower right: *S.C.Miger Sculp.*
Plate: 20 3/16 × 9 5/16 in. (513 × 238 mm)
Rogers Fund, 1920
20.60.1 (5)

By 1777 the classical vase had become an iconic form in the arabesque panel used to decorate interiors.[1] Cauvet's design features a vase with a wide flared neck atop an elongated body. Its gadrooned shoulder is set off by a broad band of foliate scrolls with a central leafy mask balanced by a guilloche band and radiating flattened leaves on the lower portion above the stem. The foot is hidden by the heads of a pair of mermaids, who hoist the colossal vase above them while seated on festoon loops. Their supple torsos terminate in fecund foliate scrolls that taper into tendrils and wind their way up narrow poles, supporting a triangle and pair of castanets on either side. Garlands of roses hang from the mouths of a pair of dolphins, twisted like handles on the rim of the vase's neck. They hang down as pendants, accentuating the floral festoon and pendant on the central portion of the vase. Issuing from the mouth of the vase and filling the upper third of the design is an acanthus plume and another set of foliate scrolls from which dangle a tall basket of flowers, a pair of horns, and bunches of fruit.

While the vase is the central motif, situated just below midpoint, the abundance and rendering of the other motifs nearly obfuscate it. The quality of the etching is particularly high, for example, in the modeling of the nude females using delicate curved and dotted lines. As the added signature of Simon-Charles Miger indicates, Cauvet's design was etched by another artist, a common practice in the eighteenth century.

A proponent of the "arabesque" style, the sculptor, architect, and ornament designer Cauvet drew from sixteenth-, seventeenth-, and eighteenth-century sources in this elaborate composition.[2] In the 1770s, he joined others in shifting away from the severe, geometric *goût grec*. Instead, they combined motifs from European design of the previous three centuries with elements of classical decoration revealed by recent archeological discoveries in Italy. The resulting designs are often symmetrical, and offer a renewed sense of grace and lightness.[3] Bound together by scrolling tendrils derived from Renaissance grotesques, Cauvet's elegant neoclassical motifs here reflect this type of fusion (Fig. 1).

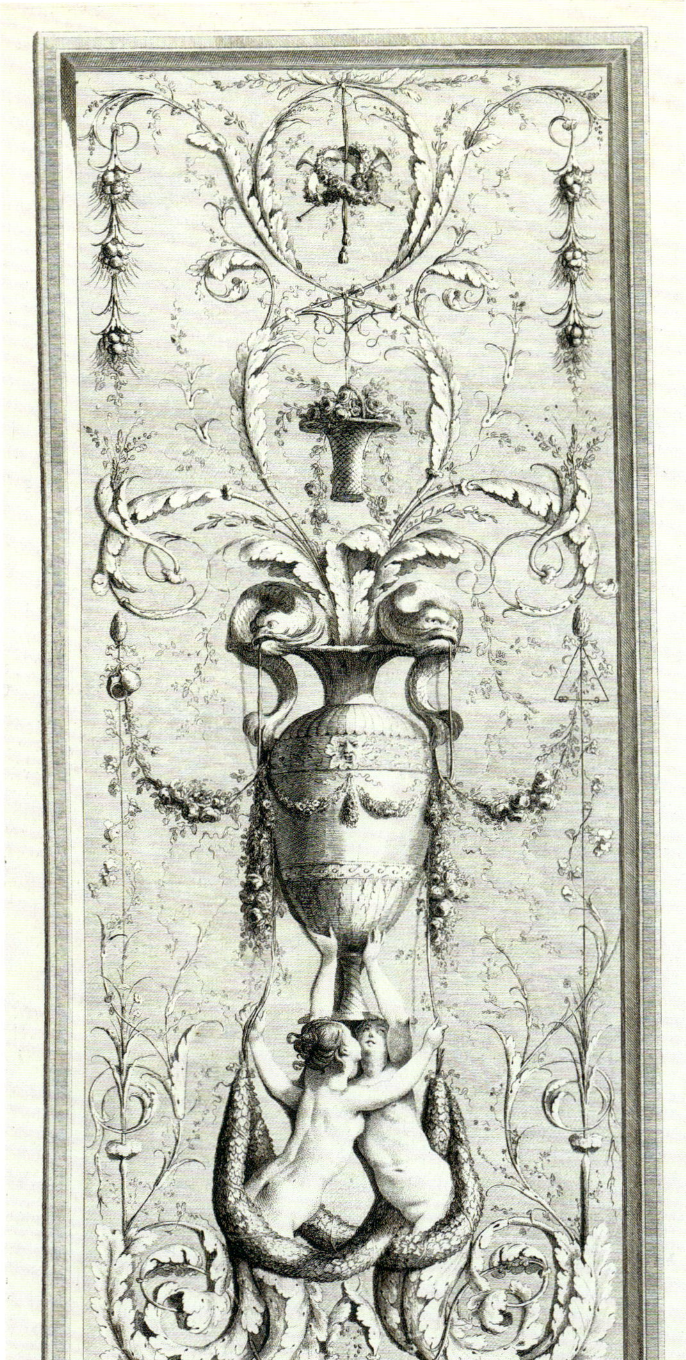

Cat. no. 79

Cauvet's oblong design could have been adapted equally to a door panel or a full-scale wall treatment, as the delineation of the frame's inset molding suggests. In fact, the preceding plate of *Receuil d'ornemens* illustrates how such panels would have been paired to

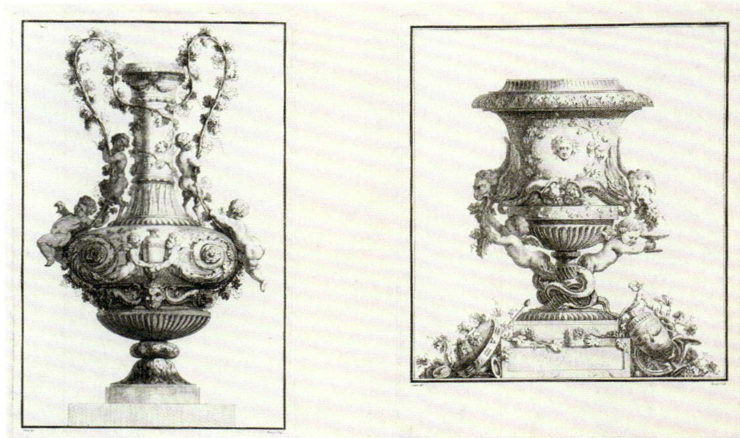

Fig. 1: *Two Designs for Vases*, 1777. Designed by Gilles-Paul Cauvet, etched by Hemery. Etching, from *Recueil d'ornemens*; plates: 10 $\frac{1}{4}$ × 6 $\frac{1}{2}$ in. (261 × 165 mm) and 9 $\frac{15}{16}$ × 7 $\frac{15}{16}$ in. (252 × 202 mm). Rogers Fund (1920), 20.60.1 (57).

fit into a doorway. The amount of detail suggests that the artist intended the panel to be painted, but it could as well have been carved in wood or modeled in stucco, techniques that were gaining favor in the two decades preceding this volume's publication.[4]

Similarly ornate patterns for all sizes of panels and friezes, copiously realized details of architectural ornament, and a series of individual fanciful vases fill Cauvet's substantial volume. While he relied on the vase as a major element in the majority of his designs, other motifs seen here demonstrate his tremendous range and understanding of classical ornament. As director of the Académie de Saint-Luc, the guild of decorative painters and sculptors in Paris, a position he assumed in 1766, Cauvet, by publishing ornamental designs for interiors, furniture, and other works of art, demonstrated his interest in inspiring his students as much as promoting himself.[5] C.H.

1 For example see a late-eighteenth-century drawing by Étienne de Lavallée (Lavallée-Poussin) illustrated in Myers 1991: cat. 58; and a 1731 design for Cardinal Albani in Alain Gruber et al., *L'Art décoratif en Europe: Classique et baroque* (Paris, 1992): 212.

2 See for example a sixteenth-century engraving by the Master of the Die, after Perino del Vaga, illustrated in Timothy Clifford, *Designs of Desire: Architectural and Ornamental Prints and Drawings, 1500–1800* (Edinburgh, 1999): cat. 11a; and a circa 1660 panel from the Château de Vaux-le-Vicomte, in Gruber et al. (note 1): 180. Cauvet's early projects included sculptural ornament for the interiors of various Parisian residences, including the Hôtels de Kinsky, de Mailly-Nesle, and de Noailles.

3 For an elaboration of this shift, see Myers 1991: 35.

4 Myers 1991: 34.

5 Myers 1991: 34. In 1774 Cauvet was named sculptor to Louis XVI's brother the Comte de Provence (later Louis XVIII), to whom he dedicated his *Receuil d'ornemens*....

80. DESIGN FOR AN ARABESQUE PANEL WITH A VASE ON A CLASSICAL BASE

Jules-François Boucher (Boucher the Younger or Juste-Nathan Boucher) (1736–1782)
After 1780
Etching, from *VIIIe Cahier d'Arabesques*
Plate: 8 ½ × 6 in. (215 × 153 mm); image: 6 ½ × 4 5/16 in. (165 × 110 mm)
Signed: [lower left] *F. Bo Fecit*
The Harris Brisbane Dick Fund, 1941
43.5 (4)

In this symmetrical design, presumably for a wall panel, Jules-François Boucher featured the classical vase as the main component. The amphora-shaped vessel exhibits a familiar classicizing vocabulary: high looping handles that terminate in satyr heads on the foliate band just below the shoulder, and gadrooning on the bowl and rim that complements the fluted neck and foot. Cresting with a bouquet of flowers, the vase sits on an ornate, gueridon-like pedestal attached to a tapering base, decorated with a female figure in classical dress. The precarious composition is balanced by the curling armature of an acanthus cup, blossom chains, scrolling vines, and a single central festoon. The three main parts—vase, armature, and base—are encased by a rectilinear frame of laurel leaves and corner rosettes. The massing of elements in the center of a panel is a hallmark of Boucher's style.

Boucher's small-format etchings filled eight *Cahiers des arabesques*, issued in sets of five or six. Most of his designs in this series were intended to be realized in paint, plaster relief, or wallpaper. Two suites, both entitled *Nouvelle Livre des vases*, followed these published works. The designer drew from a stock of classical elements in all his designs, mixing and matching motifs to create new compositions. For example, the frontispiece and first etching in the fourth *Cahier* are similarly conceived with related details. Unlike the whimsical and lively treatment of a similar visual vocabulary in sixteenth-century prototypes, or even in the art of such neoclassical designers as Robert Adam, Boucher's style seems self-conscious and stiff.[1]

Boucher signed this plate and the others in shorthand, "F. Bo," but he also used other monikers: Boucher fils, Boucher fil, and François Boucher.[2] In this way, he identified himself as the son of the preeminent rococo painter of the same name, whose deft and sentimental depictions of shepherdesses and female nudes embody the lighthearted spirit and sensuousness of the mid-eighteenth century. The stylistic division between father—the epitome of the rococo—and son—*ornamentaliste* working in the succeeding, neoclassical style—could not be more distinct. C.H.

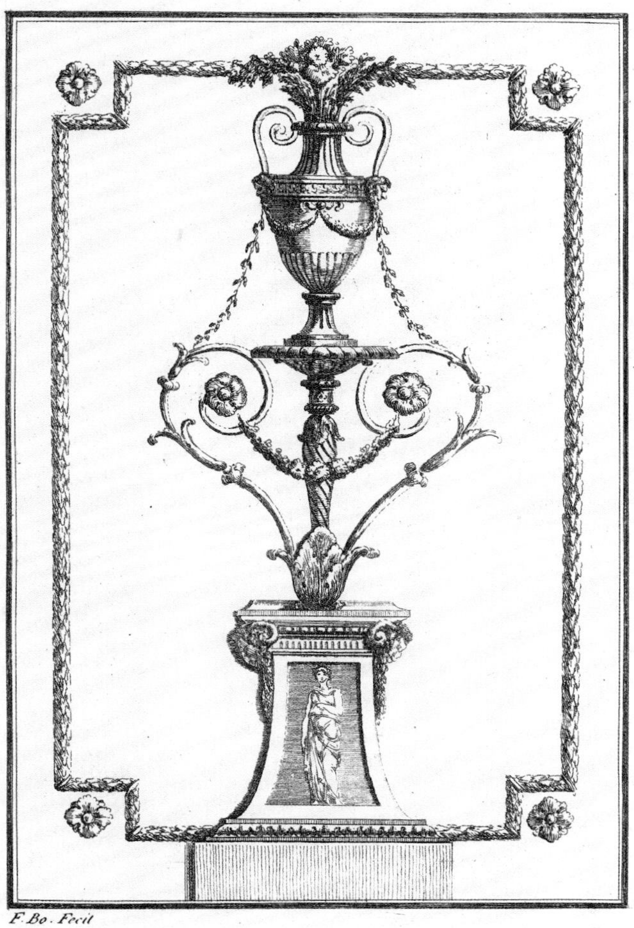

Cat. no. 80

[1] See Harris, 2001: 117, fig. 173, for a wall panel designed by Robert Adam in the anteroom of Lansdowne House, Berkeley, in which he employed similar motifs—classical pediment, arabesques, and vase in rectangular panel—to very different effect.

[2] D. Désiré Guilmard, *La Décoration au XIX siècle...* (Paris, [1872]): 230.

81. INTERIOR DESIGN OF A WALL WITH DOORS, FIREPLACE, PANELS, AND BENCHES

Michelangelo Pergolesi (act. about 1760–92, d. 1801)
November 1, 1784
Etching, plate 2 from *Designs for Ornaments*
Plate: 10 5/8 × 19 1/16 in. (272 × 485 mm)
Inscribed: [bottom center] *Pergolesi inv.t sculpt. & Published as the Act directs 1.st Nov.r 1784*
Harris Brisbane Dick Fund, 1946
46.50.2

An Italian designer of "antique" ornament active in London, Michelangelo Pergolesi was a great disseminator of Robert Adam's neoclassical style. Although bound together into a hefty volume, the many plates in his *Designs for Ornaments* were initially issued in smaller groups, probably by subscription to undisclosed patrons. The dates included in the inscriptions beneath each of the sixty-six prints in this edition reveal that they were published between May 1, 1777, and August 30, 1792.[1] Pergolesi engaged other Italians working in London, such as the noted designer, printmaker, and painter Francesco Bartolozzi, to execute some of the fine details in his etchings.[2]

This concept for a wall, complete with doorways and fireplace, is replete with sprightly decoration close to that in Adam's designs. The latter's characteristic "arabesques," or scrolling patterns and grotesques, can be seen in the frieze along the ceiling and in the rectangular spaces of the wall panels, doors, and mantle, all disposed in an orderly fashion within the elevation.[3] A variety of classical vase motifs, including three distinct designs within the door panels, occur in almost every decorative compartment. The vase in the overdoor is nearly identical to actual vessels on either end of the garniture on the mantle and remarkably similar to Josiah Wedgwood's *Vase with Lion Handles* (cat. no. 36). Their alternating shapes and sizes demonstrate how pairs and single items of such vessels could be grouped and displayed. Pergolesi was fond of showing off his versatility by including furniture and accessories in his etchings (Fig. 1).

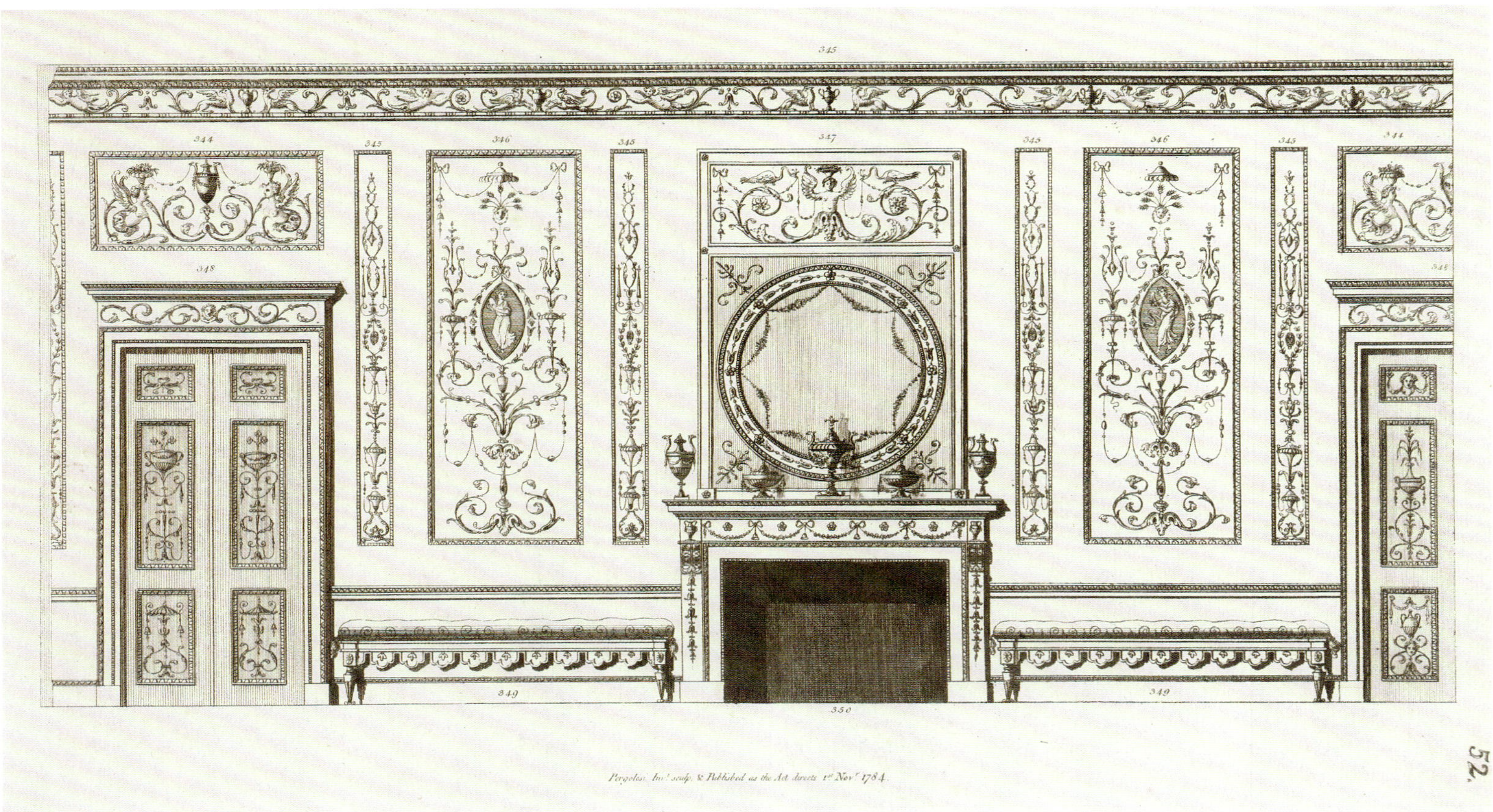

Cat. no. 81

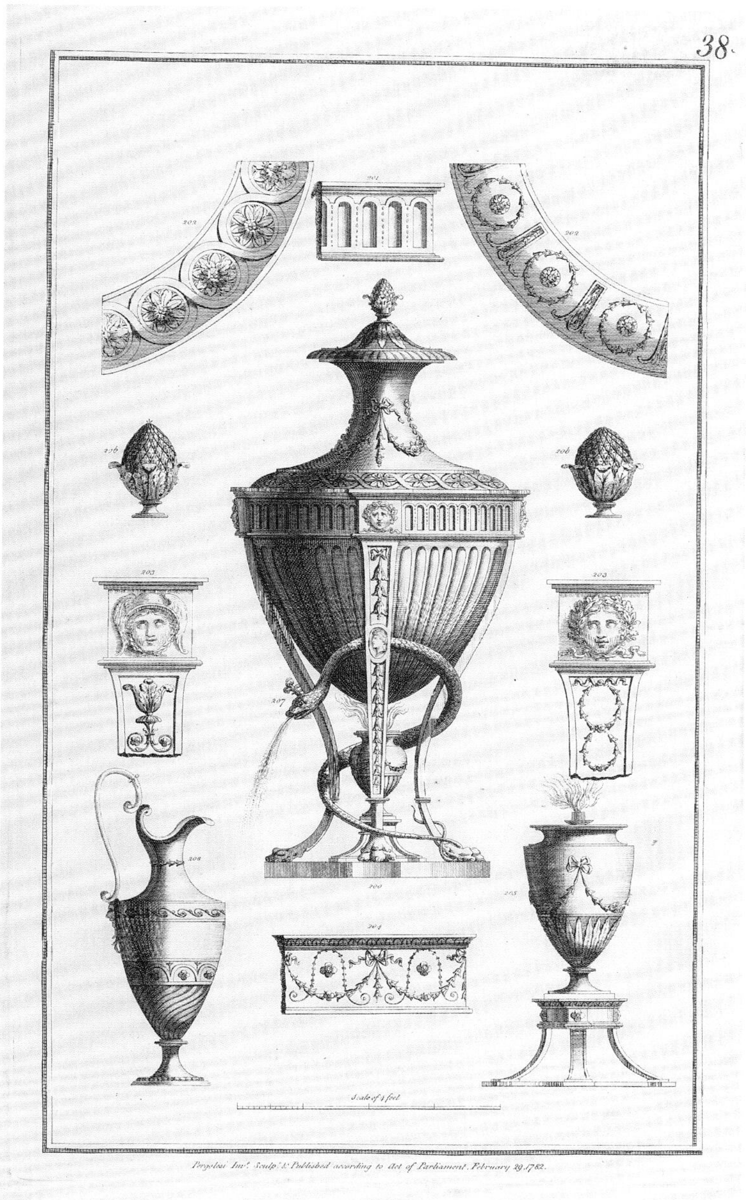

Fig. 1: *Designs for a Tea Urn and Other Objects*, 1782. Michelangelo Pergolesi. Etching, plate 38 from *Designs for Ornaments*, plate: 17 13/16 × 10 3/4 in. (452 × 274 mm). Harris Brisbane Dick Fund (1946), 46.50.2 (38).

It is no accident that this design bears a strong resemblance to the work of Adam, with whom Pergolesi had been associated from the time he arrived in England in 1760. A native of Pergola, he was raised in the artisan tradition that valued skills in a variety of media.[4] Pergolesi, a watercolorist as well as an engraver, has been credited with some of the painted decoration, as well as schemes for minor rooms, in some of Adam's houses.[5] For example, the furniture, stucco ceiling and wall panels, as well as the Ionic and Corinthian pilasters in the Long Gallery at Syon, Derbyshire, the house Adam designed for the Duke of Northumberland in the early 1760s, have been ascribed to Pergolesi.[6] Even though these were realized some twenty years prior to the publication of this print, a glance at the actual rooms reveals significant parallels to elements in the wall elevation of this etching. Most notably, the framed circular "mirror" on the wall above the mantle compares favorably to a similar construct in the anteroom at Syon.[7] Also, an elevation by Adam for a wall of a small room at Syon shows a bench similar to the one illustrated here.[8] Pergolesi's dedication in this compendium to the "memory of the late most high and puissant prince," in all likelihood Northumberland, suggests that his work for the duke at Syon was pivotal to his career.[9] Some see Pergolesi merely as an imitator of Adam, who was bothered that the Italian was offering imitations of his style when he began publishing prints dedicated to Northumberland. Others credit him with greater originality and independence.[10] Yet Pergolesi's precise role and influence with relation to his master's oeuvre remain difficult to define.[11] C.H.

1 Other known bound copies have seventy-one prints, according to a curatorial note written in the back of this volume.

2 Walter Shaw Sparrow, "Pergolesi and Robert Adam [part 2]," *Architectural Review* 50 (Oct. 1921): 13–14.

3 Each compartment of ornament is identified by a number, so that a customer or artist could realize the decoration of a room by citing the numbers corresponding to the elements they wanted to use. Identical designs bear the same number.

4 Jean Anne Vincent, "Michelangelo Pergolesi: Adam's Living Italian Source for the Lively Venetian Style," *Interiors* 114 (Dec. 1954): 79–82.

5 Emmanuel Bénézit, *Dictionnaire critique et documentaire des peintres, sculpteurs, dessinateurs et graveurs…*, 4th ed., rev. ([Paris], 1999): vol. 10, 747.

6 Constance Simon, *English Furniture Designers of the Eighteenth Century* (London, 1905): figs. 58–59; and Harris 2001: 81.

7 *Syon House: A Seat of the Duke of Northumberland* (Derby, 1992): 5.

8 Harris 2001: 83, fig. 123.

9 Plate 56, published July 26, 1791, in *Designs for Ornaments*.

10 Sparrow (note 2): 87, proposed that he was a "skilled and rapid artist who had mastered a style belonging to a school fit either to carry out the ideas for a commander-in-chief or to seek clients of his own."

11 Few contemporary Adam scholars have tried to analyze Pergolesi's contributions. In her recent study of Adam's interiors, Eileen Harris credited certain works to Pergolesi (Harris 2001: 2), but did not really address the issue of the role he played. See also Oeschslin 1982: 172, no. 208.

82. DESIGN FOR A SECTION OF A GARDEN BUILDING

German
1796
Etching and acquatint with hand coloring, plate 12 from volume 2, number 4, of *Magazin für Freunde des guten Geschmacks der bildenden und mechanischen Künste, Manufakturen und Gewerbe*
Image: 9 × 12 $\frac{1}{16}$ in. (228 × 307 mm)
Harris Brisbane Dick Fund, 1930
30.68.8

The text for this plate from the 1796 volume of *Magazin für Freunde des guten Geschmacks* describes the layout of a section through a garden building, focusing particularly on how best to illuminate a room. Other design recommendations from this "magazine for friends of good taste" are colored glass decorated with transparent paint for the three lunette windows, the use of oak or mahogany for the door, and painted candelabra stands that appear to support three-dimensional wall sconces.[1]

References to antiquity predominate here, in elements such as the reclining classical figures in the frescoes; the tall tripod candelabra; the lamps suspended from delicate blue ribbons; and an array of classical vases and tripods above the cornice and interspersed between the lunette windows. The latter perhaps refer to the often imitated, illusionistic Pompeian wall paintings.[2] According to the text that accompanies the image on this plate, most of the ornament was intended to be painted, with the

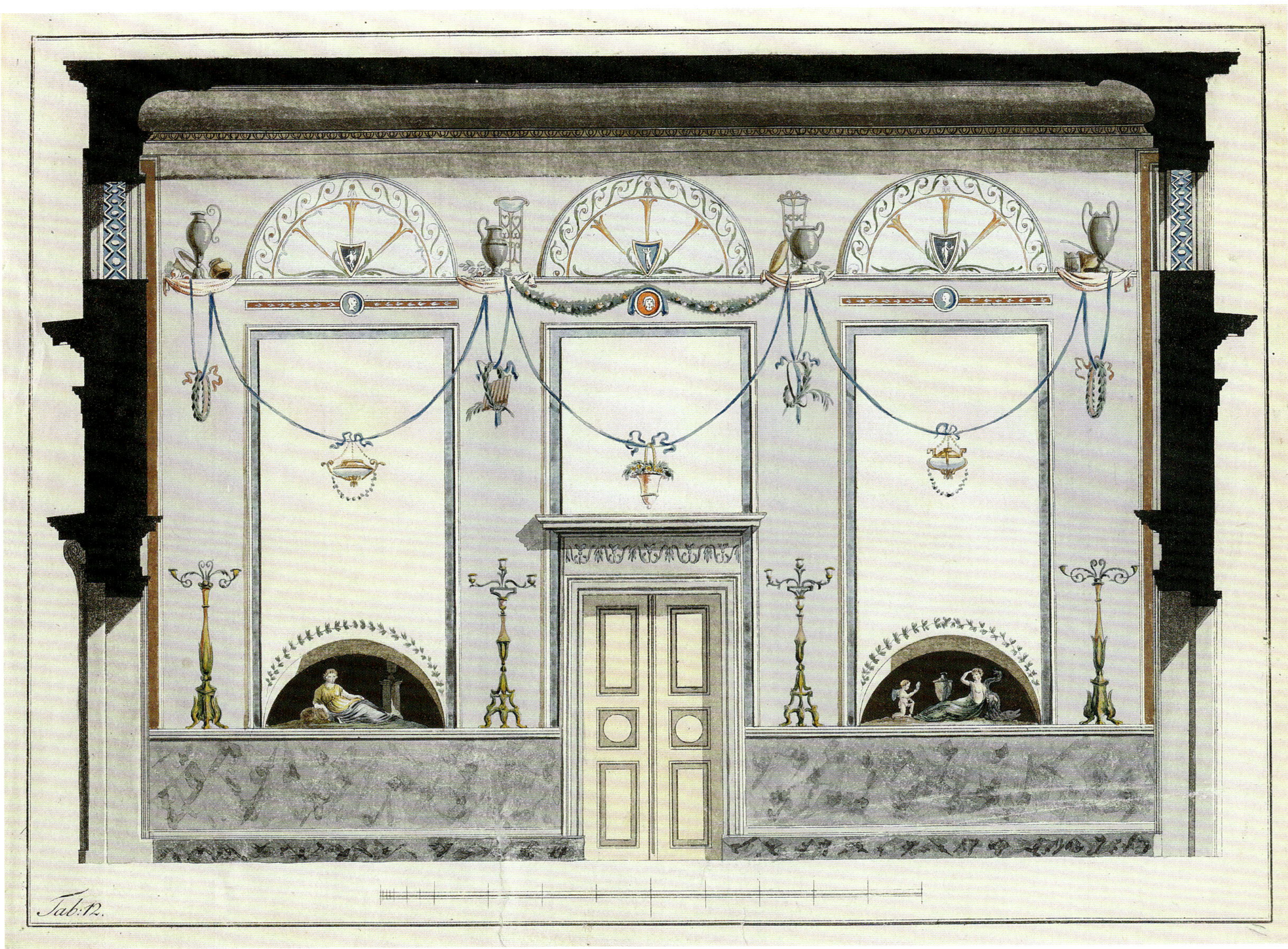

Cat. no. 82

exception of some architectural details, such as the moldings and architrave, which could alternatively be executed in stucco. Such advice indicates a shift in the second half of the eighteenth century away from relief or stucco decoration in interiors and toward plainer surfaces. Designers employed two-dimensional motifs to articulate interior space, and in this work painted vases punctuate lunette windows, simultaneously designating the different wall sections.[3] Flanking the door and mirroring the windows above them are two semicircular frescoes, showcasing female figures dressed in classical garb who recline before urns and probably represent nymphs or other personifications of nature.

The plate is one of three aquatints included in the fourth issue of the second volume of the *Magazin*, a journal that was first issued in 1794 under the direction of publisher Friedrich August Leo. In the periodical's five volumes (1794–99), Leo and his writers sought to advance "the refinement, simplification, and elevation of taste," to encourage improved production and the demand for domestic goods, and to urge a sense of national identity among the German people. The magazine focused on designs for interior decoration, furniture, and gardens and promoted good taste through images and didactic exegeses.[4]

The section on garden design encouraged readers to view gardens as a place for individual reflection, renewal, and self-improvement. In accordance with Enlightenment values, the ancient world inspired cultivation of the mind and personal refinement in the visitor while simultaneously lending dignity to garden design.[5] The pared-down ornament of this structure, which clearly was intended to allow for quiet contemplation without overpowering the visitor with busy decor or convoluted classical references, reflects above all English neoclassical ideals. In fact, the introduction to this issue of the magazine discusses the influence of English design, specifically the work of Robert Adam.[6]

The *Magazin's* beautiful, fashionable etchings appealed not only to wealthy and erudite members of the aristocracy, but also to architects, landscape and furniture designers, and their clients. The journal's readership also included members of the growing commercial class in Germany, who, having recently garnered fortunes and status, were eager to improve and modernize new properties in accordance with the most up-to-date trends. The designs the magazine featured for shop fronts, based on stores in London, and the edifying nature of the text, instructive on such subjects as the ancient world and how to implement the latest and most refined taste, attest to an audience that, at least partially, derived from a merchant class that lacked a formal, classical, and cultural education.[7]

While the journal's designers drew on English sources for inspiration, they encouraged a specifically German brand of neoclassicism. The English were freer in their interpretations of objects and buildings from the ancient world, whereas the *Magazin* advocated strict adherence to classical rules of architecture and design. In effect, comparison of the vases in this plate with some by Wedgwood (see cat. nos. 40, 42–43) shows that while German designs are not entirely historically accurate, their plainer surfaces and simpler forms correspond more closely to Greek vases than do their English counterparts. The periodical went so far as to claim that Germans' understanding of the classical world was superior to that of the English. Such an assertion not only promoted a link to ancient Greece and Rome, but also encouraged a unifying sense of nationalism at a time when Germans were divided among disparate states.[8]
K.C./M.H.

1 The original text accompanying this image reads as follows:

> Durchschnitt eines Gartengebäudes nach dessen langen Seite. Solches enthält diesen Saal, und zwei zu beiden entgegengesezten Seiten liegende Cabinets, zu deren einem hier die mittlere Thüre, von Mahagony oder Eichenholz gefertigt, führet. Der Saal ist durch halbrunde Fenster von oben erleuchtet, die auch im Durchschnitt mit ihren Seitenverzierungen zu sehen, die Thüren rechts und links führen unmittelbar in den Garten, und können Glasthüren seyn, die dann nicht allein die Erleuchtung des Saals vermehren, sondern auch die Aussicht in schöne Gartenparthien gewähren.
>
> Alle Wandverzierungen sind Mahlerei, doch werden die Arme der Leuchter natürlich ausgeführt, und dienen dann statt Wandleuchter. Die Thürverdachungen und Gewände, die Lamperie mit ihrer Zocke und Gliedern, so wie auch die Einfassungen um die grosen Füllungen können körperlich und in Stucco ausgeführt warden, die Verzierungen und Mahlerein auf den halbrunden Fenstern, werden mit transparenten Farben gefertigt. Fände man den Saal gegen dessen Höhe nicht lang genug, so kann solcher leicht durch Beifügung einer Fensterparthie mehrere Ausdehnung gegeben werden.

2 See the introduction by E. Kumsch in *Decorations-Malereien aus dem Ende des 18. Jahrhunderts...* (Dresden, 1908). The grotesque decoration of the Domus Aurea in Rome may also have been a source; see Claude Moatti, *In Search of Ancient Rome* (New York, 1993): 36–37.

3 Peter Thornton, *Authentic Décor: The Domestic Interior 1620–1920* (London, 1984): 142.

4 Sarah Richards, "A Magazine for the Friends of Good Taste: Sensibility and Rationality in Garden Design in Late Eighteenth-Century Germany," *Studies in the History of Gardens and Designed Landscapes* 20 (Fall 2000): 229.

5 Ibid.: 230, 236–37.

6 Kumsch (note 2) mentioned the popularity of English designs in other German publications, such as the *Journal des Luxus und der Moden*, published in Weimar from 1776 to 1825.

7 Richards (note 4): 233–34.

8 Ibid.: 239–40.

83. OVERDOOR WITH VASE OF FLOWERS IN A NICHE

Attributed to Michel-Bruno Bellengé (1726–1793)
About 1775
Oil on canvas
48 ⅜ × 55 in. (123 × 140 cm)
Gift of J. Pierpont Morgan, 1906
07.225.461

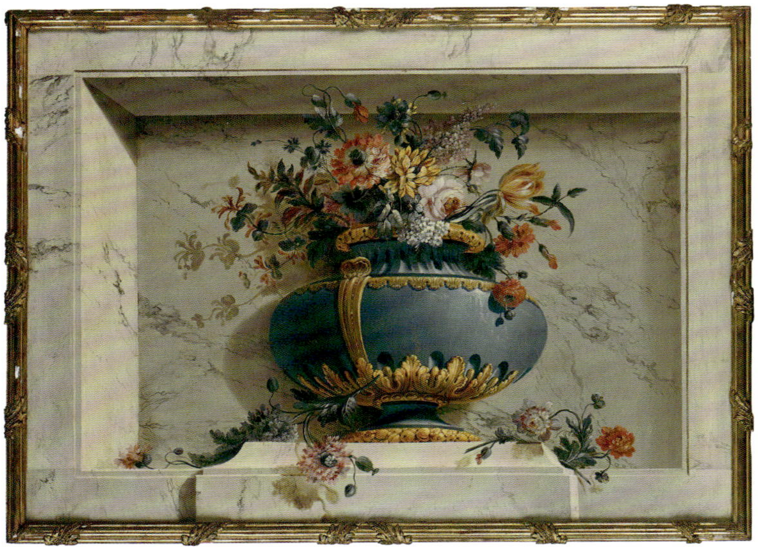

Cat. no. 83

84. OVERMANTLE PANEL WITH FLOWERS IN A BLUE VASE

Willem van Leen (1753–1825)
About 1775
Oil on canvas
55 × 29 ⅛ in. (139.7 × 74 cm)
Gift of J. Pierpont Morgan, 1906
07.225.470

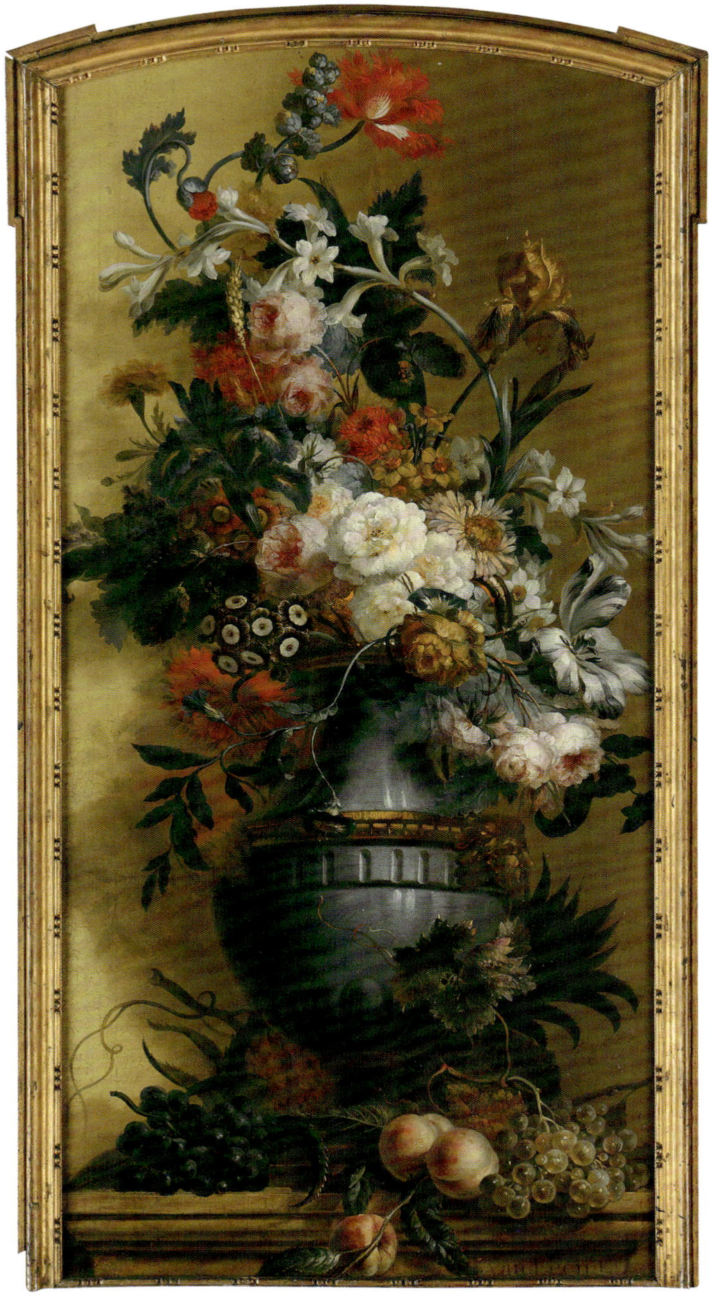

Cat. no. 84

85. PAIR OF OVERDOORS WITH PERFUME BURNERS

French
Eighteenth century
Oil on canvas
07.225.457 a (perfume burner supported by baby tritons):
21 × 26 ⅝ in. (53.3 × 67.6 cm); 07.225.457 b (perfume burner supported by snakes and putti): 21 × 26 ¾ in. (53.3 × 67.9 cm)
Gift of J. Pierpont Morgan, 1906
07.225.457a–b

For an illustration of 07.225.457a, see p. 140.

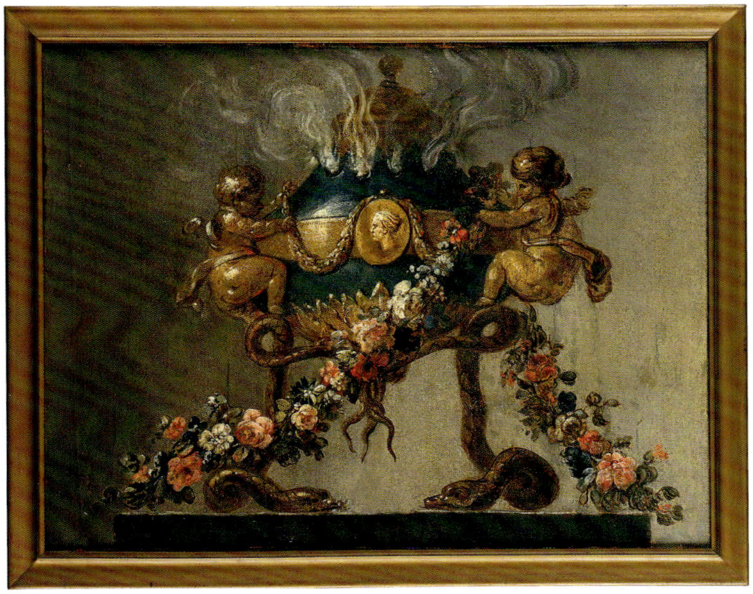

Cat. no. 85 (07.255.457b)

86. PANEL WITH A VASE OF FLOWERS DRAPED WITH GARLANDS

French
Eighteenth century
Oil on canvas
28 ⅛ × 17 ½ in. (71.4 × 44.5 cm)
Gift of J. Pierpont Morgan, 1906
07.225.267

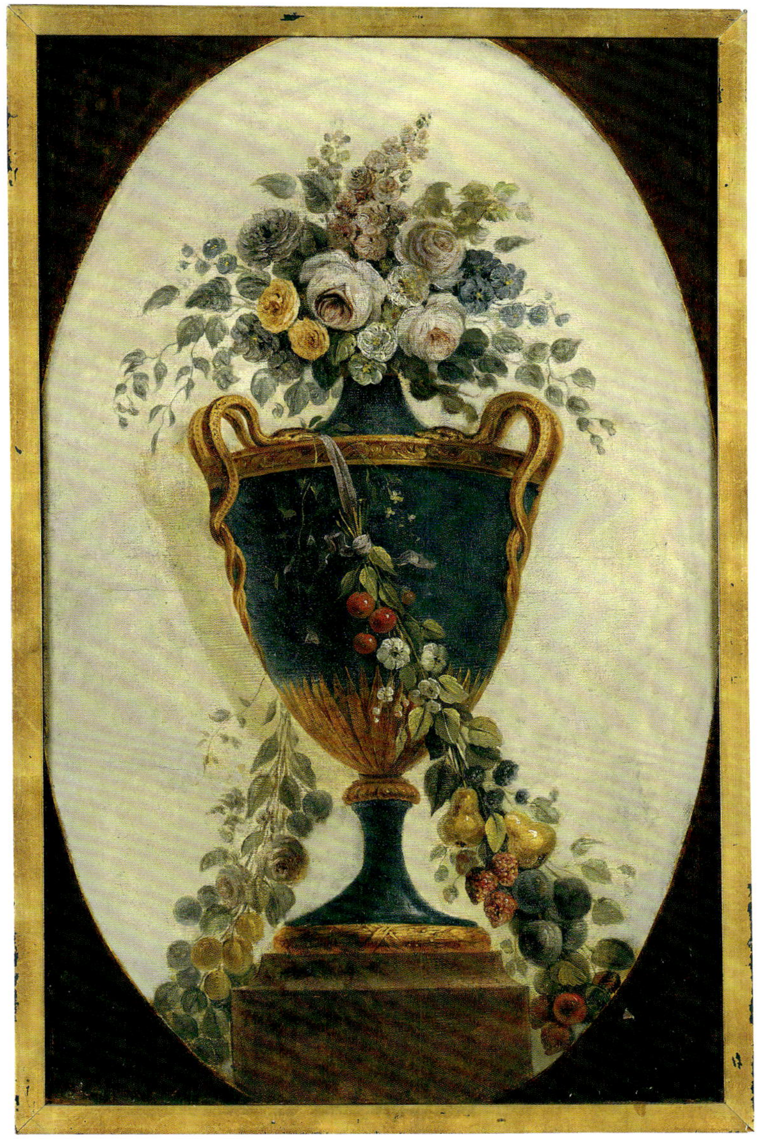

Cat. no. 86

Symbol to Satire: The Vase as Icon

Stefanie Walker

Neoclassical artists always seem to have been aware of the symbolic potential of the vase motif. Some designs are more overt than others in expressing their creator's and audience's emotional associations with the form. Depicting vessels in monumental proportions next to tiny figures or in landscapes with long neglected, overgrown ruins, Jean-Charles Delafosse, Jean-Laurent Legeay, and Ennemond-Alexandre Petitot, among many, gave vases an exalted quality that acknowledges and pays homage to them as icons of a revered ancient past.

Isolated vases rendered almost as "found objects" in landscape settings must must also be understood as representing ideas about the fundamental relationship of humanity and nature. Reviving Renaissance concepts, artists suggested that the vase can symbolize our rational, artistic abilities in contrast to nature's chaotic or haphazard creativity. In some cases, such as Petitot's *Vase with Vines* (cat. no. 91), the object seems to merge with nature, or be subsumed by it, whereas, in others, such as Legeay's *Vase with a Pestle* (cat. no. 90), it appears to triumph over nature's destructive powers. Made with elements drawn from the earth, vases can also be seen in harmony with it, as in drawings by Louis-Joseph Le Lorrain (see cat. nos. 88–89), where they coexist as symbols of timelessness and beauty.

In a more humorous vein, Petitot's *Greek Bride* (cat. no. 92) plays on the connections of the anthropomorphic aspects of the vase with a feminine ideal, although his woman costumed as a vase is virtually imprisoned by layers of neoclassical decoration. In this image, as in others, Petitot wittily acknowledged the popular obsession with all things Greek and simultaneously poked fun at overweening antiquarian scholarship and exaggerated ornament. In a similar fashion, Delafosse's two drawings from a series representing different trades (cat. nos. 93–94) show an engraver and a sculptor metamorphosizing into vases. It is no coincidence that he chose to portray these trades in this manner. Clearly, he intended his emblematic figures to represent the disciplines most closely associated with the design, production, and dissemination of neoclassical vases.

As a symbol, the neoclassical vase also served important honorific or memorial functions. Ancient funerary vessels, found in quantity and variety, served as models for a number of eighteenth-century urn shapes—usually with an ample body and a lid—that were placed on columns, pedestals, or in a larger architectural framework to celebrate or eulogize. As most such examples are usually large and site specific, few can be found in museum collections today. Included here is a charming example on a small scale: a porcelain vase commemorating the achievements of a German civic and church leader (cat. no. 95).

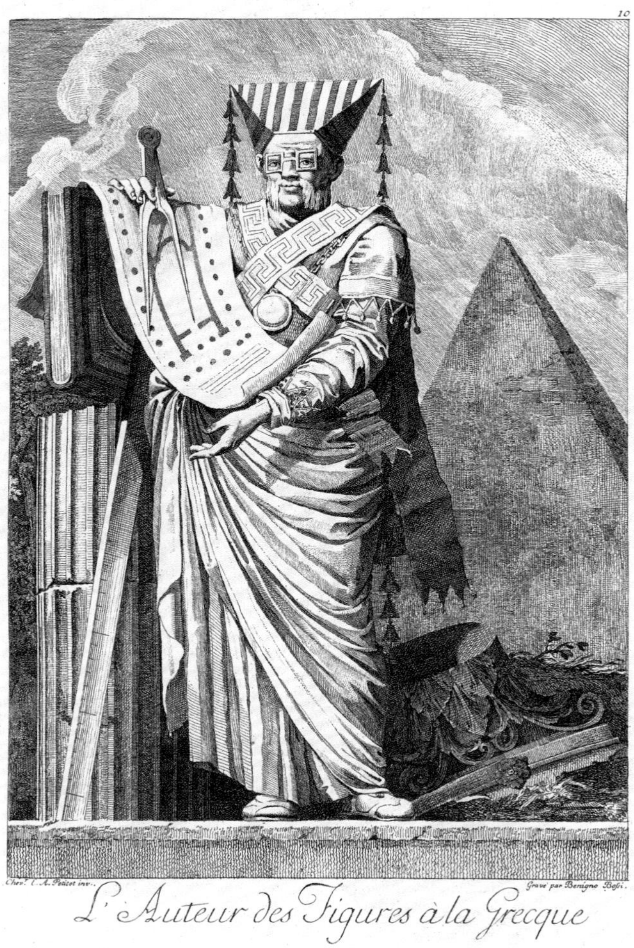

L'Auteur des Figures à la Grecque, 1771. Designed by Ennemond-Alexandre Petitot; etched by Benigno Bossi. Etching, plate 10 from *Mascarade à la Grecque*; plate: 8 ⅝ × 7 $^{3}/_{16}$ in. (220 × 183 mm). Harris Brisbane Dick Fund (1940), 40.59.1 (12).

87. COLOSSAL VASE WITH A CASCADE

Designed by Jean-Charles Delafosse (1734–1789); engraved by Jean-Dominique-Étienne Le Canu (act. 1750–70)
About 1770
Engraving, title page of the *Cinquième Livre de trophées*
Sheet: 15 ¼ × 9 ⅞ in. (388 × 252 mm)
Inscribed: [left and right, along top edge] *41.e Cahier De la Fosse/RR*; [on the base under the vase] *Cinquième Livre de Trophées/Contenant divers attributs/de Chasse et de Pêche/Inventés et dessinés par/J.C. De La Fosse Architecte/Dedié aux Artistes par l'Auteur*; [from left to right, along the bottom edge] *J C De la Fosse inv/A Paris chez Chereau rue de Mathurins aux 2 Piliers d'Or A.P.D.R./Le Canu sculp./RR/1*
Gift of Harry G. Friedman, 1964
64.682.227

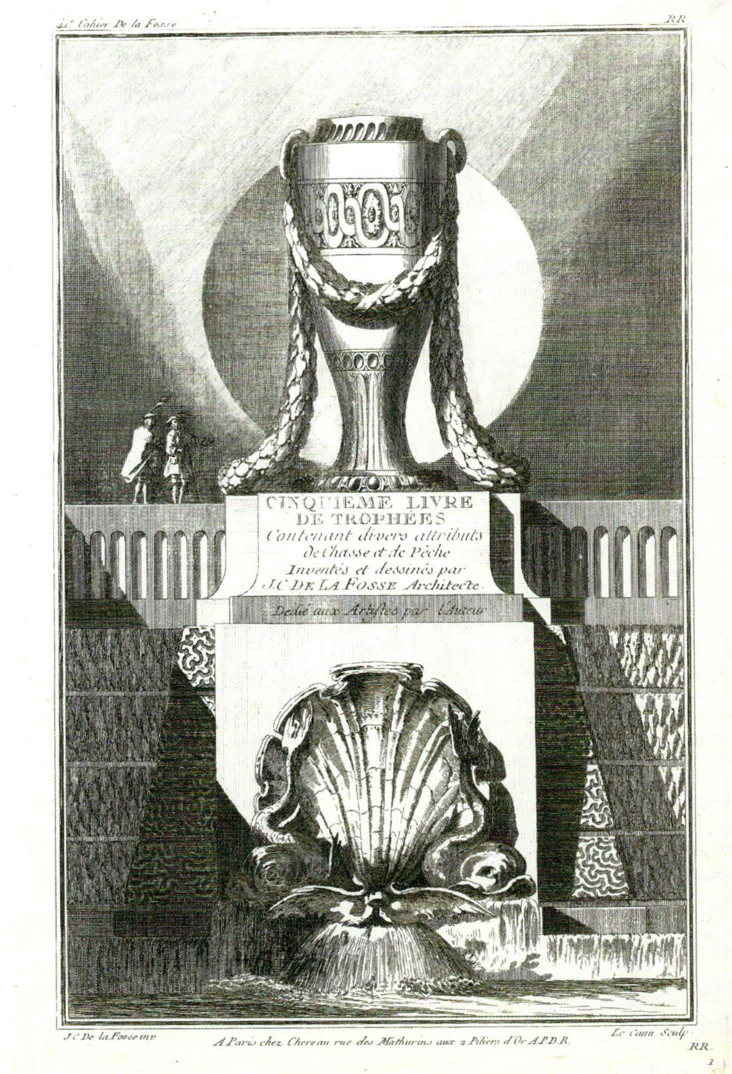

Cat. no. 87

Like other innovators of early neoclassicism in France, the Paris architect and designer Jean-Charles Delafosse explored fantastic form and abstract geometry in many of the conjectural architectural schemes and other ornamental designs that he published in the second half of the eighteenth century. First issued about 1770, this engraving—the title page for a four-plate series of trophy designs—incorporates some of these ideas. It features one of Delafosse's most characteristic vase forms, depicted as a colossal monument atop a formal cascade, silhouetted by a perfect, radiant sphere[1]. An imposing form, Delafosse's vase dwarfs the gentlemen onlookers who encounter the mammoth wonder, draped with massive laurel swags. Elevated against its halolike backdrop, with bold shafts of light and shadow falling all around, the vase becomes an exalted dramatic object. Delafosse offered an apotheosis image that reflects on, and perhaps even satirizes, the culture of the Grand Tour and the eighteenth century's reverence for classical antiquity, embodied in the vase, its most recognizable symbol.[2]

As its imprint indicates, the engraving was issued by the successful Paris publisher Jacques-François Chereau and represents the later of two known states of the plate. An earlier impression bears the name of Jean-François Daumont, a merchant who arranged the publication of many of Delafosse's designs and whose imprint Chereau replaced with his own in the present edition.[3] Throughout the 1770s, Chereau published numerous Delafosse engravings, issuing them in large volumes of the artist's work. Although it is not widely recognized, many of these plates had been published previously by other engravers and print sellers, including members of Chereau's family. The fortunate survival of this plate's earlier state provides evidence of the extent to which Chereau reused older plates, and suggests that other Delafosse designs published by Chereau may be earlier than otherwise assumed.[4]

Chereau included the series that includes this engraving in a 138-plate collection of Delafosse's designs for ornament and objects issued after 1775.[5] The publisher also sold the prints in smaller lots of five or six plates each, cataloguing the smaller sets with a lettering system. Marked "RR1," this engraving was the first plate of Chereau's *cahier* "RR," the forty-first of his small books of Delafosse designs, as the legend at the top of the plate indicates.[6]

The reissuing of engravings had been a practice of the European printing trade from its beginnings in the fifteenth century.[7] In eighteenth-century France, publishers frequently acquired plates and engravings—valued assets—from the artists and their descendents, family members in the business, or other print dealers. No exception, Chereau bought out his mother's and grandfather's stock and purchased engravings from the publisher Babuty, as well as other sources, as this engraving suggests.[8] Such traffic in used plates, along with the piracy that was equally widespread, insured broad dissemination of many admired vase designs, including the present example. H.J.M.

1 Similarly severe, attenuated vases recur throughout Delafosse's work, in numerous drawings and engravings, including plates in his influential publication, *Nouvelle Iconologie historique* (Delafosse 1768). He devised related vases for the niches of the entryway of the former Hôtel Titon, a Paris residence he designed in the 1770s (see cat. 73, n. 3); see Michel Gallet, "Jean-Charles Delafosse, Architecte," *Gazette des beaux-arts* 61 (1963): 160–61, figs. 4–5. The present sheet was separated from the other four plates in the *Cinquième Livre de Trophées* series at some point in its history. Illustrating the themes of fishing and hunting, the series is the fifth in a group of six related sets of trophy designs by Delafosse.

2 The exaggerated contrast in scale between the diminutive figures and the colossal *goût grec* vase points to what scholars propose may have been the influence on Delafosse of Giovanni Battista Piranesi's engravings. See Claire Baines, "Delafosse, Jean-Charles," in *The Dictionary of Art* (New York, 1996): vol. 8, 649; and Pia Maria Montonen, "Delafosse, Jean-Charles," in Joanna Banham, ed., *The Encyclopedia of Interior Decoration* (London/Chicago, 1997): 356. See also Sylvia Pressouyre, "La Poétique ornamentale chez Piranèse et Delafosse," in Georges Brunel, ed., *Piranèse et les français* (Rome, 1978).

3 An example of the earlier state is in the Metropolitan Museum (37.85 [2]). Numerous examples of Delafosse designs published by Daumont are recorded in Desiré Guilmard, *Les Maitres ornemanistes: Dessinateurs, peintres, architectes, sculpteurs et graveurs...* (1880): 220–22. Chereau's imprint on the present engraving—"chez Chereau rue des Mathurins aux 2. Piliers d'Or"—confirms that this is the later impression. Chereau took over these premises in 1775, moving from a shop on the rue Saint-Jacques that his publishing family had occupied since 1718. See Maxime Préaud et al., *Dictionnaire des éditeurs d'estampes à Paris sous l'Ancien Régime* (Nantes, 1987): 80–84, 98.

4 Cataloguers of Delafosse's engraved designs have not previously noted Chereau's role as a compiler. For more information on Delafosse designs that Chereau reissued, see cat. no. 9, n. 1.

5 Entitled the *IIe.[Deuxième] Volume de l'Oeuvre de J. Ch. Delafosse* (after 1775), the collection followed Chereau's first volume of the artist's work: a 1771 reprint of the *Nouvelle Iconologie historique*, which Delafosse had engraved and published himself in 1768.

6 Chereau lettered his *cahiers* consecutively from his first Delafosse volume through the second. The *Nouvelle Iconologie*, with 108 plates, comprised eighteen *cahiers*, lettered "A" through "S." The second volume comprised twenty-four *cahiers*, lettered "T" through "Z," and "AA" through "SS." "RR," therefore, is the forty-first of the combined *cahiers* of Delafosse designs offered by Chereau.

7 See David Landau and Peter Parshall, *The Renaissance Print, 1470–1550* (New Haven/London, 1994).

8 Préaud et al. (note 3): 83.

88. FRONTISPIECE FOR A SUITE OF VASE DESIGNS

Louis-Joseph Le Lorrain (1715–1759)
About 1740–48
Pen and gray ink, brown wash
7 9/16 × 5 1/16 in. (192 × 121 mm)
Inscribed: [in gray ink on front of low wall] *SUIE DE/VASE*; [signed in gray ink below] *Lelorrain*; [on mount at lower left] *Lelorrain*; [at top right] *no. 37*
The Elisha Whittelsey Collection, The Elisha Whittelsey Fund, 1948
48.148 (37)

Stylistically, this design for a frontispiece dates from sometime during Louis-Joseph Le Lorrain's stay in Rome (1740–48), and represents a transitional moment between the late baroque and early neoclassicism, or the *gout grec*.[1] Le Lorrain was a seminal figure in introducing this style in the mid-1750s, in particular through the suite of furniture he designed for Ange-Laurent de Lalive de Jully.[2] Based on an inscription on the drawing, Le Lorrain must have prepared this and at least one other related image (cat. no. 89) as part of a series intended for engravings, although no prints after these designs are known to exist.[3]

The exaggerated corpulence of the vase's body, the sculptural quality of the putti who adorn the pedestal, and the vessel's rich ornamental detail are vestiges of the baroque, while the tapering lid, thick garlands hanging from satyr masks, and large Bacchic mask flanked by grapevines all belong to the neoclassical repertory. This transitional object demonstrates that the evolution of the neoclassical style was not sudden, but rather occurred gradually, in intermediate stages.

What is distinctly new here is the lofty representation of this vase, its huge size overwhelming two figures in classical garb at the lower right. They gesture to the noble and monumental object, elevated reverentially on a pedestal that rests on a low, cracked wall

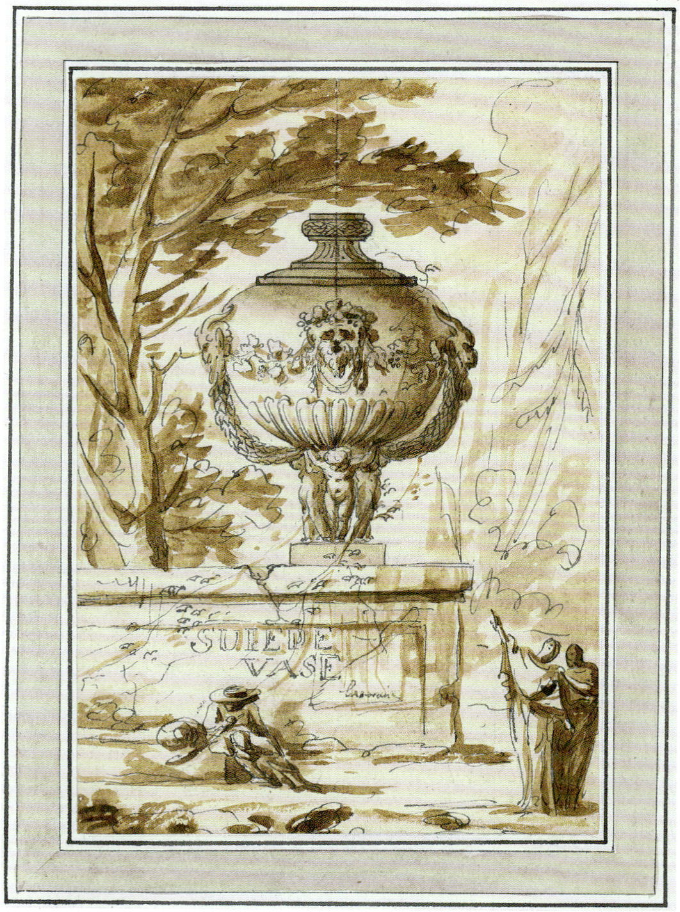

Cat. no. 88

covered with vegetation. Seated below the wall is a lone artist, his face hidden under a hat, busily sketching.

Unlike Edmé Bouchardon and other early eighteenth-century artists who depicted vases against blank backgrounds, Le Lorrain, along with Lagrenée and Legeay (see cat. nos. 29, 90, respectively), situated them in idyllic landscape settings, as if to present them as found archeological remains. Thus, we are meant to understand them as the very embodiment of antiquity. Here, a tinge of melodrama is conveyed by the deep shadow cast by a canopy of tree branches and the introspective art student in the foreground. These features reveal the influence of Giovanni Battista Piranesi, with whom Le Lorrain may have studied while he was in Rome. K.C.

1 The drawing was first published in Myers 1991: 102–05. According to Myers (104), even though Le Lorrain's prints were published in Paris, it is widely believed that he executed them in Italy.
2 Originally, the ensemble of writing desk and filing cabinet included a vase-shaped clock serving as the centerpiece on top of the cabinet. For more on the suite, see Sven Eriksen, "Lalive de Jully's Furniture *à la grecque*," *Burlington Magazine* 103 (1961): 340–47; and Eriksen 1974: 68–70. See also Ottomeyer essay in the present volume, pp. 17–18.
3 A drawing in a private French collection for a suite of eleven vases engraved by Claude-Henri Wâtelet, entitled *Raccolta di vasi Dedicata All. III. Signora GEOFFRIN Della arte amante riamanta* (1752), corresponds to the frontispiece. Four additional preliminary drawings for this suite are in the Musée des Arts Décoratifs, Paris; see Pierre Rosenberg, "Louis-Joseph Le Lorrain," *Revue de l'art* 40–41 (1978): 184, nos. 74–76.

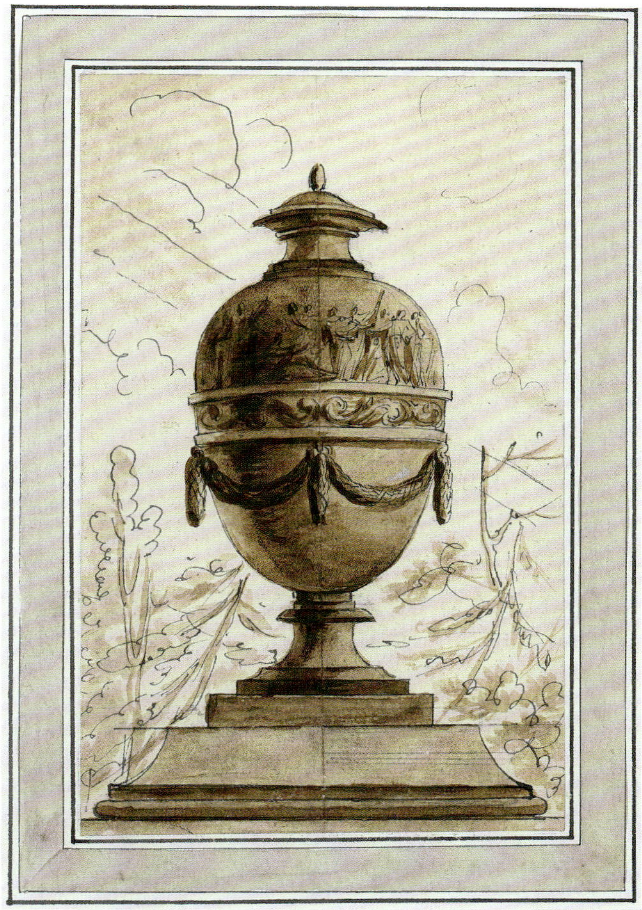

Cat. no. 89

89. VASE WITH FIGURAL RELIEF IN A LANDSCAPE

Louis-Joseph Le Lorrain (1715–1759)
About 1740–48
Pen and gray ink, brown wash
7 13/16 × 4 3/4 in. (198 × 121 mm)
Inscribed: [on mount at top right] *no. 38*
The Elisha Whittelsey Collection, The Elisha Whittelsey Fund, 1948
48.148 (38)

While in Italy between 1740 and 1748, Louis-Joseph Le Lorrain studied famous ancient Roman urns, which this drawing evokes.[1] Despite its early date, the vessel depicted here exhibits the ornamental vocabulary that would become pervasive in the *goût grec* lexicon: the frieze of classical figures around a hemispheric dome that rests on a central band of scrolling decoration, the heavy leaf swags below, and the carefully proportioned pedestal. The plain surfaces and sculptural relief allude to ancient stone carving, while the large-scale, massive ornament epitomizes the early, severe phase of neoclassicism. The way in which the vase is situated in the landscape—it towers over trees seen against a cloud-swept sky—emphasizes the object's monumentality and demands admiration for such magnificent emulations of the ancient past.

The soft, painterly drawing style, especially in the sketchy scenery, relates this sheet closely to another drawing featured here (cat. no. 88) and supports the notion that Le Lorrain had planned to use these, and other such designs, as the basis of a series of prints of vases. The presence of vertical and horizontal centering lines indicates the preparatory nature of this drawing.

Le Lorrain was one of the first artists to promote overtly neoclassical concepts. Others could have used this image to execute an ornamental garden urn or a memorial vase for a mausoleum, but most likely the artist wished to present an idea for a neoclassical vase mainly to educate and inspire his fellow artists in the study of classical design. K.C.

1 See Myers 1991: 102–05.

90. VASE WITH PESTLE IN A LANDSCAPE WITH RUINS

Jean-Laurent Legeay (about 1710–1788)
About 1770
Etching, plate 2 from *Collection de divers sujets des Vases, Tombeaux, Ruines et Fontaines, utile aux artistes, inventée et gravée par G. L. Le Jeay, architecte*
Plate: 7 ¾ × 6 ½ in. (197 × 166 mm)
Harris Brisbane Dick Fund, 1951
51.540.2 (2)

Jean-Laurent Legeay is best known today for several series of fine, small-scale etchings: *Fontane* (1767), *Rovine* (1768), *Vasi* (1770), and *Tombeaux* (1770). This print—plate 2 from the *Vasi* series (totaling twenty-four plates in all)—is part of a book that contains most of the prints from these published sets, in effect an anthology of Legeay's output of this type of graphic art. All of the images are dreamlike, romantic visions of vase forms set in ruins redolent of ancient civilizations (Fig. 1).

Like his contemporaries Louis-Joseph Le Lorrain (see cat. nos. 88–89), Ennemond-Alexandre Petitot (see cat. nos. 91–92), and Joseph-Marie Vien (see cat. no. 11), as well as other influential early practitioners of neoclassicism in Rome, Legeay used the image of the vase as an autonomous artistic template and a symbol of antiquity. In his case, his works show marked similarities to others by Giovanni Battista Piranesi, who was his most important influence.[1]

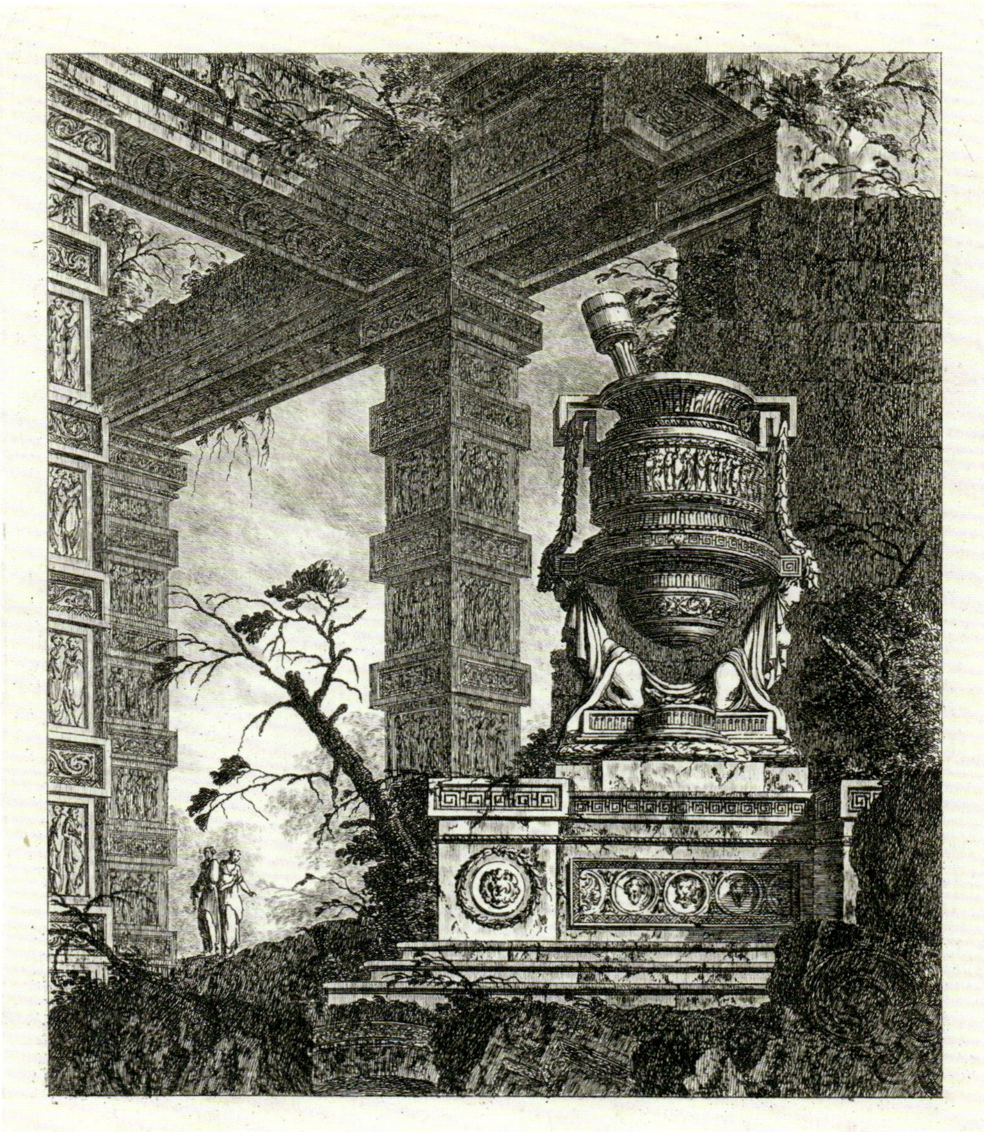

Cat. no. 90

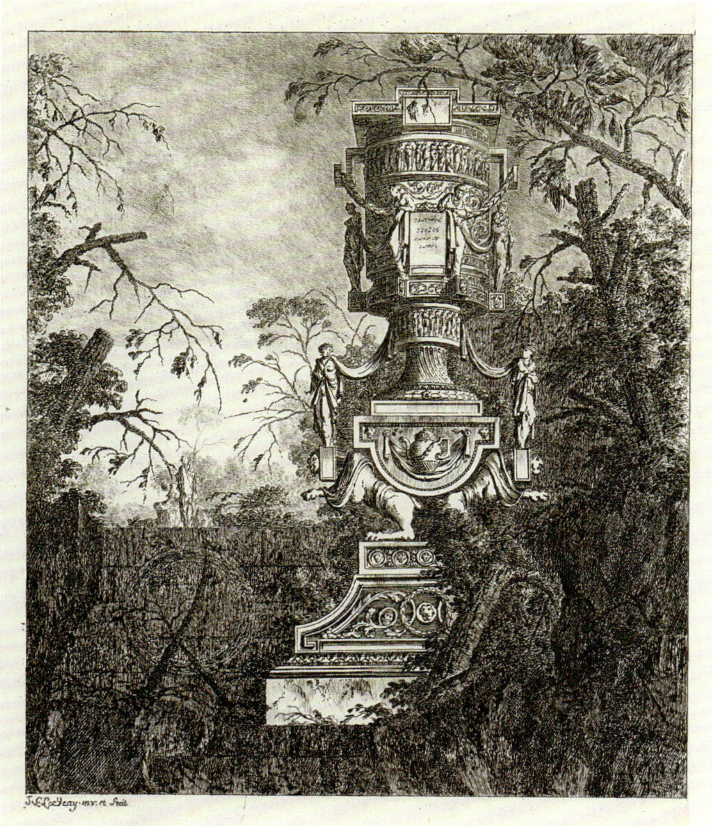

Fig. 1: *Monumental Vase on a Pedestal in a Landscape*, about 1770. Jean-Laurent Legeay. Etching, from *Collection de divers sujets...*; plate: 7 ⅞ × 6 ¼ in. (199 × 159 mm). Harris Brisbane Dick Fund (1951), 51.540.2 (4)

This vase presents Legeay's fanciful notion of the "Greek" style. The backs and heads of two shrouded sphinxes support an enormous, urnlike vase containing a pestle. Garlands dangle from the vase's angular handles, continuous meander patterns circle the body, sketchy classicizing figures constitute the main frieze. The ensemble rests on a base that repeats the meanders, in addition to featuring masks or lion heads in roundels bordered by laurel wreaths. The vessel and its support appear against a giant ruin of a strange architectural structure featuring archeologically inaccurate square columns segmented with repeated sculptural reliefs. The low vantage point from which the vase and platform are depicted contribute to the perception that the roofless ruin is immense. Two small, languid figures further reinforce a sense of monumentality.

It is difficult to assess Legeay's intentions for such compositions, but he clearly wanted to celebrate antiquity and convey its grandeur. As was true of Giovanni Battista Piranesi and others in the late eighteenth century, historical accuracy was secondary to the concept of the vase itself as the prime example of idealized antiquity. Unlike prints intended to facilitate the actual production of objects (see cat. no 53), Legeay's vessels represent romantic evocations rather than actual things. Presented as a nostalgic symbol of the classical past, the vase demonstrates the virtuosity of Legeay's imagination. Here, a neoclassical vase is invested with connotations of profundity and the sublime that recall and eulogize the ineluctable passing of time. K.C.

1 Both John Wilton Ely and Gilbert Erouart acknowledged Legeay's debt to the younger Piranesi on the basis of stylistic comparisons of etchings that are now dated to the same time period. See Denison et al. 1993: 74; Gilbert Erouart, "Jean Laurent Legeay—Recherches," and John Wilton-Ely, "Vision and Design: Piranesi's *Fantasia* and the Graeco-Roman Controversy," in *Piranese et les français, 1740–1790*, ed. G. Brunel (Rome, 1976): vol. 1, 199–213, 529–52. For the counter argument that Legeay was more influential than Piranesi, see John Harris, "Le Geay, Piranesi and International Neo-classicism in Rome 1740–1750," in *Essays in the History of Architecture Presented to Rudolf Wittkower*, ed. Douglas Fraser et al. (Bristol, 1967): 189–96. See also Emil Kaufmann, "Three Revolutionary Architects: Boullée, Ledoux and Lequeu," *Transactions of the American Philosophical Society* 42 (1952): 450–53, for the earliest mention of this debate.

91. VASE WITH VINES

Designed by Ennemond-Alexandre Petitot (1727–1801); etched by Benigno Bossi (1727–1792)
1765
Etching, plate 31 (third state) from *Suite de vases*
Plate: 8 ¾ × 6 ½ in. (223 × 166 mm)
Inscribed: [on chunk of stone at the right of pedestal] *1765*; [lower right] *Bossi Sc*
Harris Brisbane Dick Fund, 1930
30.58.7 (31)

The Metropolitan Museum owns three states of this etching, each dated to 1765. The first (Fig. 1) depicts the vase before the artist added a consistent horizontal pattern of tremulous lines, creating a subtle gray contrast to the vessel's dark contours and tonalities. The second state is bound separately at the back of another volume of *Suite de vases*.[1] While nearly identical to the first state, it contains subtle differences, the most noticeable being a greater emphasis on the shadow at the lower left. The ambiguous gray background of the first two examples is transformed into a landscape in the final state of the etching, seen here. The vase appears before a crumbling stone wall, with a hazy sky and a cluster of trees in the distance.[2]

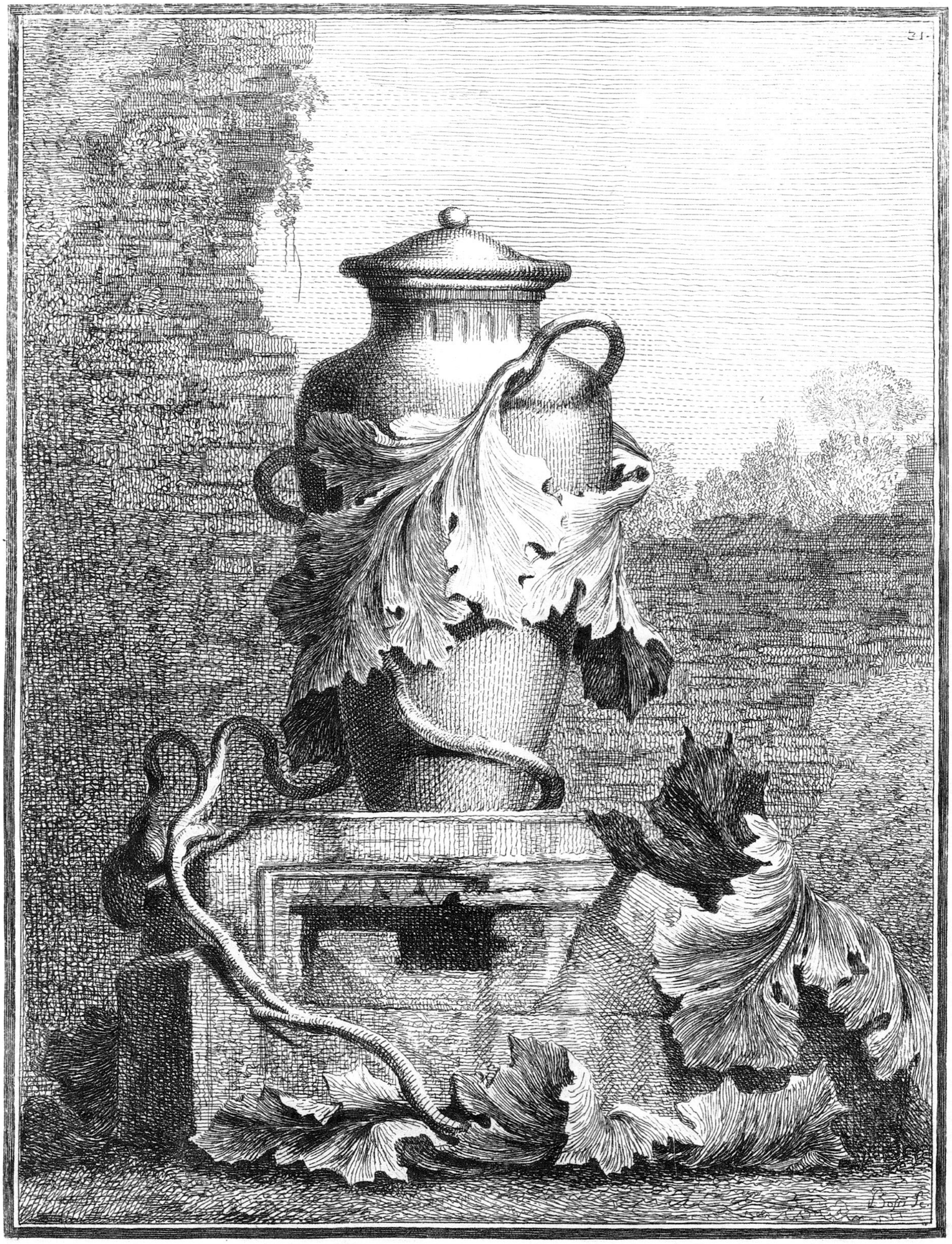

Cat. no. 91

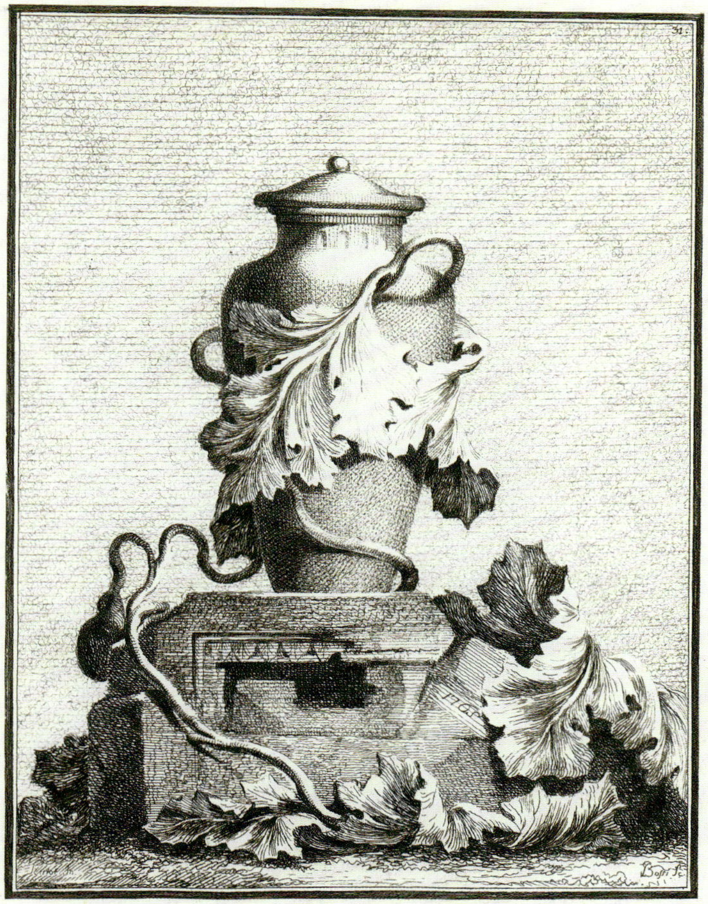

Fig. 1: *Vase with Vines* (first state), 1765. Designed by Ennemond-Alexandre Petitot; etched by Benigno Bossi. Etching, plate 31 from *Suite de vases*; plate: 8 ¹¹⁄₁₆ × 6 ⁹⁄₁₆ in. (220 × 167 mm). Rogers Fund (1985), 1985.1024 (31).

It is difficult to know the reasons for these changes; the printmaking process (from conception to execution of the image) further complicates the issue. Although Petitot did at times produce his own prints—not an unusual practice at the time—here, he designed the *Suite de vases* and Benigno Bossi etched the plates. Petitot's preliminary studies for the *Suite* were red-chalk drawings.[3] He blocked masses of color and shading in chalk, while the etcher relied on hatched lines in varying densities to express gradations of tone and shaded areas within the design. This collaborative process required a certain sensibility on the part of the printmaker to translate another's drawings. Bossi was also a painter and therefore familiar with other media, a knowledge that undoubtedly aided his thorough and perceptive interpretation of Petitot's designs. Possibly, Petitot was dissatisfied with the early states of the prints, believing the vase form might be more effective presented in a landscape of ruins than surrounded by relatively uniform and abstract series of wavy lines. Conversely, Bossi may have taken liberties with Petitot's concept, executing the changes without the latter's initiative.

Petitot's contact with Giovanni Battista Piranesi in Rome, where he studied at the French Academy from 1746 to 1750, must have influenced the dramatic environment in which the vase of the final state appears. The vessel on its pedestal remains essentially unchanged in all of the states, while its setting is completely transformed. The strong lighting, landscape, and worm's-eye perspective combine to evoke an ancient monument now overgrown with snaking vines.[4] The plant that emerges from the base and surrounds the vase recalls a legend recounted by Marcus Vitruvius in his architectural treatise.[5] According to the writer, a basket covered with a roof tile was placed on the tomb of a young girl from Corinth, inadvertently covering the roots of an acanthus plant. The following spring, the Athenian sculptor Callimachus happened upon the basket and tile, now overrun with foliage, which inspired him to create the Corinithian capital. Possibly, Petitot may have intended here a witty reference to this anecdote from ancient history, all the rage at the time. He also emphasized the long tradition of nature inspiring artists, a connection evident throughout his *Suite* and especially in this plate.

Furthermore, although the pedestal and brick wall behind the vase are chipped and crumbling, the vase is perfectly preserved as an enduring symbol of classical civilization, which withstands the destructive forces of nature. This glorification of the vase exemplifies the reverence it commanded during the eighteenth century, and its presentation here asserts that this ancient form, at least in Petitot's opinion, stands the test of time. E.W.M.

1 Second state: 40.59.1.

2 Petitot's signature, *Petitot. In.*, is clearly legible in the two earlier states at the lower left of the image, but is completely obscured in the third stage, seen here.

3 Denison et al. 1993: 208. Only one of these preparatory chalk drawings is extant, currently in a private collection in Paris.

4 At the shoulder of the vase on the right, the stem of a large leaf appears to bend in order to form a kind of handle.

5 See Marcus Vitruvius, *Ten Books on Architecture*, eds. Ingrid D. Rowland and Thomas Noble Howe (New York, 1999): 55.

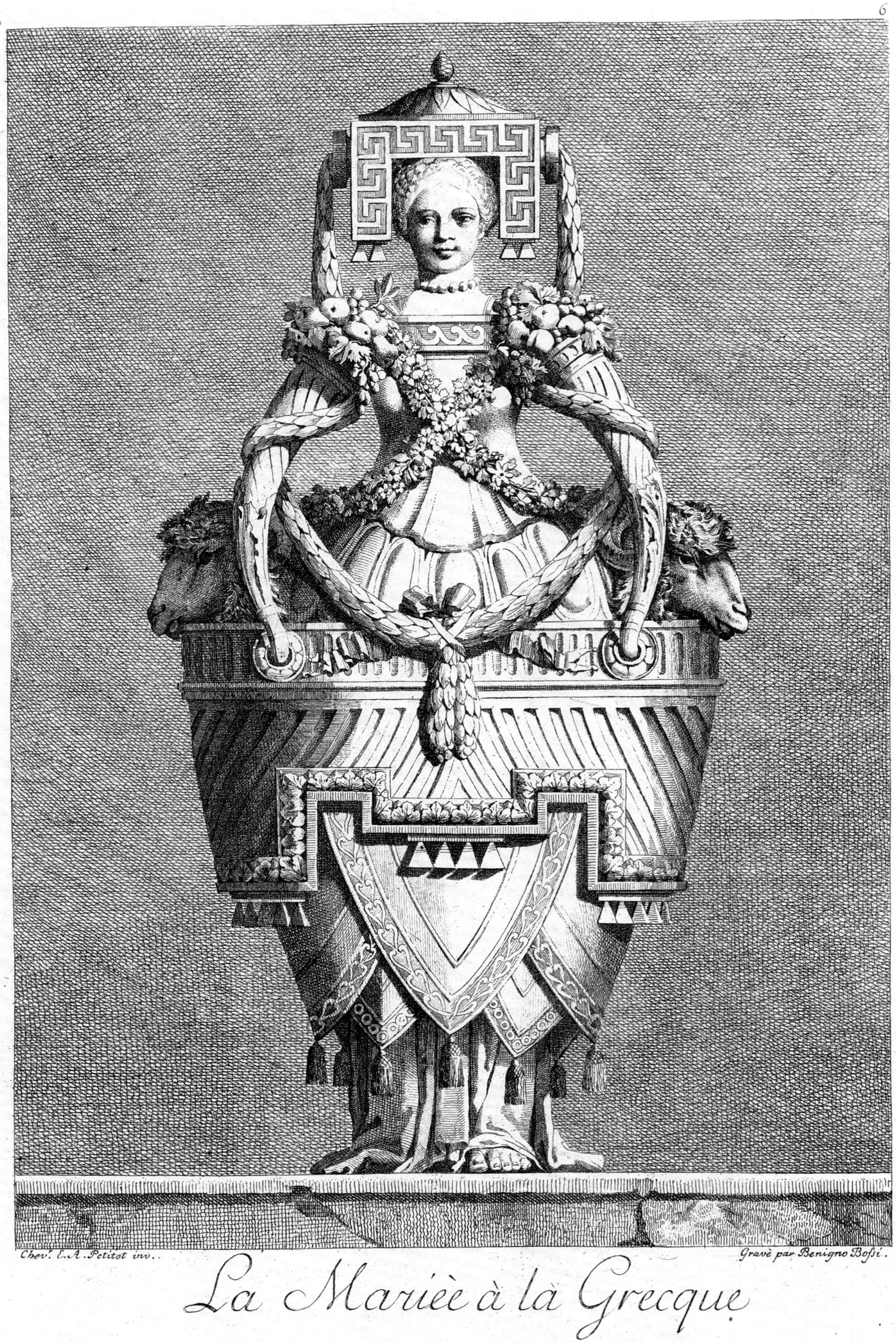
6
Chevr. E.A. Petitot inv..
Gravè par Benigno Boſsi.
La Mariée à la Grecque

Cat. no. 92

92. THE GREEK BRIDE

Ennemond-Alexandre Petitot (1727–1801)
1771
Etching, plate 6 from *Mascarade à la Grecque*
Plate: 10 ⅝ × 7 ¼ in. (270 × 183 mm)
Inscribed: [top right] 6; [from left to right, along lower edge]:
Chev.r E. A.Petitot inv. / La Mariée à la Grecque / Gravé par Benigno Bossi.
Harris Brisbane Dick Fund, 1940
40.59.1 (8)

In his *Mascarade à la Grecque*, Ennemond-Alexandre Petitot depicted a series of nine costumed personages, among them a monk, shepherd, shepherdess, bride, and groom, attending a "Greek masquerade." In this theatrical context, each figure pretends to be "Greek" by wearing a detailed, fanciful ensemble featuring a variety of classicizing design elements.

The fascination for all things Grecian, considered by some to be superior to Roman culture, played itself out in many media: architecture, decorative arts, sculpture, painting, and popular entertainment.[1] As early as 1764, Nicolas-Thomas Barthe satirized the pervasiveness of the *goût grec* in his play *L'Amateur*, which might be considered a literary equivalent of Petitot's *Mascarade à la Grecque*. Barthe wrote, "The style is Greek, our furniture, our jewels, fabric, hairstyles, riding habits, everything is Greek, except our souls."[2]

Commenting playfully on this craze in *La Mariée à la grecque*, Petitot fashioned his bride's costume into a vase ornamented with *goût grec* motifs, "wedding" her in effect to the neoclassical style. Cornucopia sleeves burst from her shoulders; floral garlands and thick laurel swags cross her bodice; ram's heads rest on her hips; and an architectural frieze, oval bosses, and fluting appear on her skirt. She balances a headdress decorated with a geometric meander and surmounted by a lid with finial. Trapped within this elaborate concoction, the bride is rendered immobile, completely consumed by the classical style, like the rest of society. Petitot's choice of a vase to make this point reinforces the form's iconic stature as the embodiment of classicism and signifier of late eighteenth-century values, beliefs, and tastes.

In the dedication printed at the beginning of the volume, to Guillaume-Léon Tillot, the Marquis de Felino, who was minister of state at the court of Parma, Petitot explained that he intended *Mascarade* as a "jest in good humor" ("plaisanterie de société"), not only about the *goût grec* but also contemporary social rituals and aristocratic behavior. Demonstrating his wit, Petitot included himself as the final personage in the series, *L'Auteur de figures* (author, or inventor, of figures), as a follower of the popular neoclassical style (see p. 158). This series of images resembles other contemporary prints and drawings, such as those of Jean-Charles Delafosse (see cat. nos. 93–94), in which the characters are composed of the tools of their respective trades.

At their height in eighteenth-century Europe, masquerades were an aristocratic activity, involving celebration, leisure, mystery, opulence, and debauchery, complete with drinking, dancing, music, and, of course, elaborate costumes and masks.[3] Such *fêtes* were simultaneously embraced and condemned for the freedom that can come with disguise. Opponents of masquerades denounced the immoral, frivolous, and often questionable behavior that the events engendered, while supporters fully endorsed the pleasurable and rare liberties they afforded.[4] Behind a mask or clad in costume, one could, for a brief time, become something one was not. As an eighteenth-century observer stated, "People dress themselves in what they have a mind to be, and not what they are fit for."[5] E.W.M.

1 On the *goût grec*, see Eriksen 1974; and Ottomeyer essay in this volume.
2 See Ericksen 1974: 50–51; and Denison et al. 1993: 214.
3 Terry Castle, "Introduction," *Masquerade and Civilization: The Carnivalesque in Eighteenth-century English Culture and Fiction* (Stanford, Calif., 1986).
4 Ibid.: 2.
5 Joseph Addison and Richard Steele, *The Spectator*, vol. 14, ed. Donald F. Bond (Oxford, 1965), quoted in Castle (note 3): 5.

93. THE ENGRAVER

Jean-Charles Delafosse (1734–1789)
After 1771
Black chalk with pen and black ink, with gray and colored washes
$7\frac{1}{2} \times 4\frac{5}{8}$ in. (191 × 118 mm)
Inscribed: [at center] *Le graveur*
The Elisha Whittelsey Collection, The Elisha Whittelsey Fund, 1960
60.576.5 (2)

Attributed to the Parisian architect-designer Jean-Charles Delafosse, this emblematic image of an engraver is one of a dispersed group of drawings of grotesque tradesmen and other figures by the artist, six of which are in the Metropolitan Museum.[1] Although unsigned, they feature Delafosse's characteristic darting line and lush ink washes enhanced by delicate transparent colors, and are inscribed with their subjects: a sculptor (cat. no. 94), tutor, copper-plate printer, print seller, and two sheets of engravers, including this one. Four similar drawings in the Ashmolean Museum, Oxford, illustrate a surgeon, an actor (*comedien*), a chemist, and a monk.[2] In 1995 the Musée des Arts Décoratifs, Paris, acquired a related uninscribed drawing of an ornament designer or architect, and one of a farrier, as complements to their large collection of Delafosse's allegorical figure drawings.[3] In theme, format, and recurring motifs, the New York and Oxford "occupational" drawings are a distinctive subgroup among the artist's many surviving images of emblematic figures. Perhaps a fragmentary series, the group may also include the two Paris examples and more that are not yet identified.

Like the other related images, *The Engraver* is a fantastic composite figure whose form, resembling an attenuated *goût grec* vase, incorporates the tools and products of his trade. His vessel-like torso holds engraved plates, one depicting an image reminiscent of the artist's *Vases antiques* (see cat. 9). Engraving tools protrude from the mouths of sheep's heads supporting the vases handles. Hanging from swagged drapery, portrait medallions illustrate the related trade of the seal or gem carver.

Only nominally a figure, Delafosse's *Engraver* might be a symbol-laden vase design but for his large, feathered mask and animal feet. On the back of the sheet, the artist experimented with a more conventional representation in lightly sketched black chalk. In this alternate design, the tradesman—tools hanging from his draped shoulders—appears to be wearing an amphora-shaped vase and conveys the effect of a man in costume more clearly than the enigmatic and more ambiguous image on the recto.

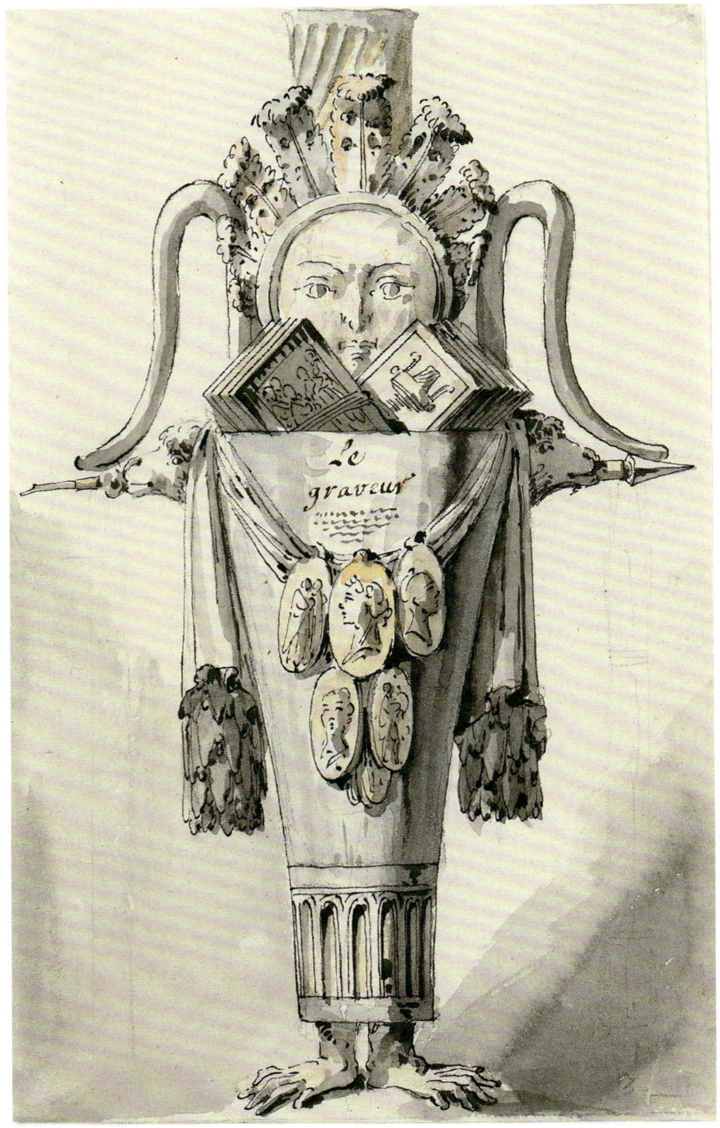

Cat. no. 93

A number of eighteenth-century artists produced fantastic costume designs. In fact, when the Metropolitan acquired the drawing, it was initially assumed to be by Ennemond-Alexandre Petitot, whose *Mascarade à la grecque*, published in 1771, lampooned the excesses of the *goût grec* by adapting forms of classical architecture and archeology, including vases, to fancy dress (see cat. no. 92). The basis for the earlier attribution to Petitot was undoubtedly the inscription on another sheet in this group which identifies the pair of figures as "Les graveurs à la grec" (engravers in the Greek style).[4] Indeed, Delafosse was almost certainly alluding to Petitot's series.[5] Yet in contrast to

Petitot's relatively straightforward satire, the full meaning of these tradesmen drawings is elusive. Like much of Delafosse's work, the drawings are dense with layered, sometimes arcane, symbolism: the sunlike mask with plume "rays," and the many animal features, even more explicit in other drawings in the group. Defying definition as simple caricatures or personifications of a single theme, Delafosse's imagery remains enigmatic, hinting at what scholars have just begun to recognize, that the originality of his designs lies not only in their visual language but also most certainly in their meaning.[6] H.J.M.

[1] The drawing was first attributed to Delafosse by Marianne Roland Michel, *Le Dessin français au XVIIIe siècle* (Fribourg, 1987): 139–40, fig. 150. Delafosse's six tradesmen drawings in the Metropolitan Museum (60.576.5 [1–6]) were acquired as a group. Four, including *The Engraver*, are included in Myers 1991: cats. 35–38.

[2] *Ashmolean Museum, Oxford, Catalogue of the Collection*, vol. 7, *French School*, cat. by John Whiteley (Oxford, 2000): cats. 666–69.

[3] Marie Noël de Gary in "Acquisitions," *Revue du Louvre* (Feb. 1995): 79. For other Delafosse drawings at the Musée des Arts Décoratifs, Paris, see *Fragments énigmatiques: Allégories de J.-C. Delafosse* (Paris, 1994).

[4] 60.576.5 (5); see Myers 1991: cat. 37.

[5] Myers 1991: 62.

[6] See Monique Mosser, "Introduction," *Fragments énigmatiques* (note 3): 5–14.

94. THE SCULPTOR

Jean-Charles Delafosse (1734–1789)
After 1771
Black chalk with pen and blank ink, with gray and colored washes
7 3/8 × 4 1/2 in. (187 × 113 mm)
Inscribed: [bottom center] *Le Sculpteur*
The Elisha Whittelsey Collection, The Elisha Whittelsey Fund, 1960
60.576.5 (1)

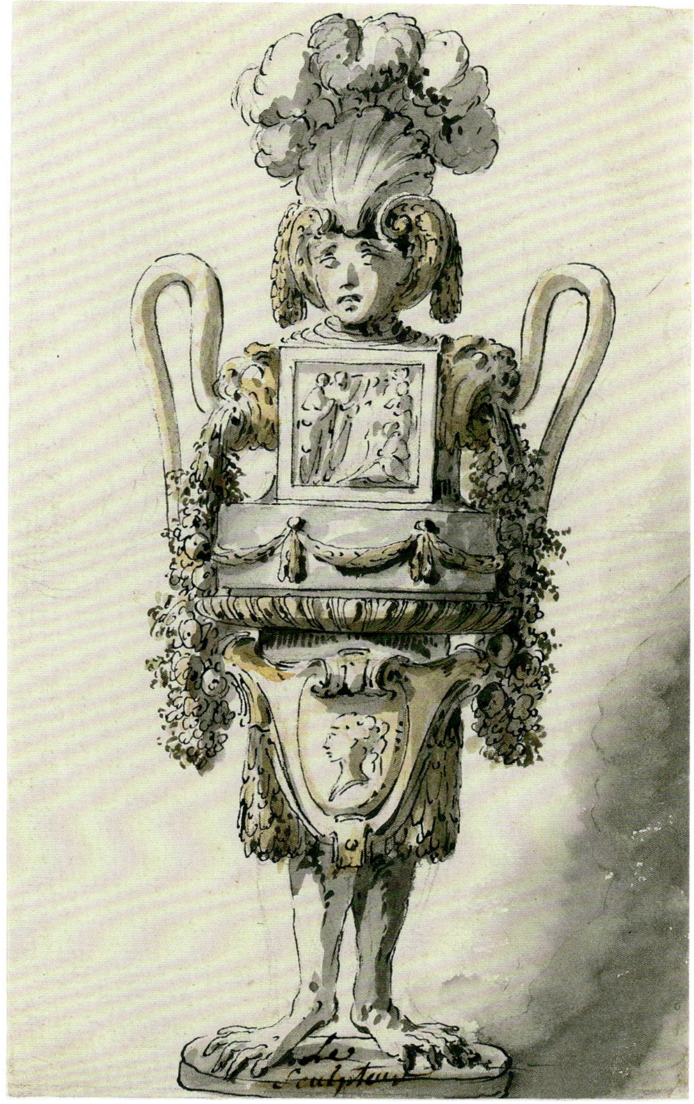

Cat. no. 94

As he did with his emblematic drawing of an engraver (cat. no. 93), Jean-Charles Delafosse conceived this fantastic ink and wash image of a sculptor in the form of a classical vase. The vase was a favorite subject of the prolific artist, whose hundreds of designs for ornament and objects shaped the development of early French neoclassicism in the decorative arts. Suggesting a vase principally with its upswept loop handles, Delafosse's image incorporates other elements, such as the paired animal masks at the figure's shoulders and draped garlands, that are typical of ancient and *goût grec* vessel forms. Additional details, including the smooth columnar belly, hanging laurel ropes, gadroonlike molding, and portrait medallion, are recurring features in Delafosse's numerous vase designs (see cat. nos. 9–10).

Adorned with leafy floral festoons and other motifs drawn from the natural world—a shell-shaped helmet with fluffy plumes, volutelike ear pieces, and mysterious clawlike feet—this apparently armless figure would be difficult to identify without the title that Delafosse inscribed on the drawing. The artist's "sculptor" is partially a composite of the products of his trade, arrayed on his form as if they were pieces of armor, or perhaps a workman's apron. A square relief panel adorns his chest and carved architectural ornament, including a frieze of festoons and egg-and-dart moldings, circles his waist, with the above-mentioned portrait relief in a sculpted cartouche hanging below.

One of six related drawings in the museum's collection devoted to symbolic depictions of tradesmen (see cat. no. 93), the present image illustrates Delafosse's extensive knowledge of the iconographical traditions that he explored in his 1768 *Nouvelle Iconologie historique*. Based on a well-established convention of depicting tradesmen as personifications of their work, Delafosse's sculptor has antecedents in images such as Nicholas II de Larmessin's seventeenth-century "grotesque trades," a series of satyrical engravings that illustrate merchants, artisans, and others as outlandish composites of their tools and wares. Delafosse's drawing may also suggest a design for a costume.

In a culture where pageantry and overt symbolism played a dominant role, Delafosse's tradesmen drawings also may have suggested costumes for processions, theatrical tableaux, masked balls, carnivals, and the festival days when the eighteenth-century corporations of artisans and merchants marched with the symbols and regalia of their guild. H.J.M.

95. COMMEMORATIVE VASE (DAHLBERG GEDÄCHTNISVASE)

Fürstenberg Porcelain Manufactory; probably designed by Franz von Piper (eighteenth century)
About 1787
Hard-paste porcelain with black enamel, glaze, and gilding
8 ¾ in. (22.2 cm)
Inscriptions: [on front of pedestal, in black] *Electus Coadjutor /Moguntiae vto Junii /Wormatiae XIXno Junii /MDCCLXXXVII*; [on the back of the vase, in lighter tones in a medallion with a sepia background] *D*; [on back of pedestal, in black] *Quod felix /faustumque sit /et Litteris et Patriae*; [on back of plinth, in black] *Franciscus a Piper /invenit*
Marks: [inside the base, in underglaze blue] *F* (factory mark); [underside of base, incised] *KK* (design mark)
Gift of Mr. and Mrs. Rafi Y. Mottahedeh, 1976
1976.71.20

By 1777 the Fürstenberg manufactory's initial burst of products in a neoclassical style, represented here by a garniture (cat. no. 25), had given way to more restrained interpretations.[1] This vase from about a decade later, however, appears remarkably reserved in comparison to both earlier and later works, its scale and sentiment in keeping with its function.

Gedächtnisvase, the German term used to describe this object, means "commemorative vase." Its central element is an egg-shaped vessel with scrolling handles, based on an antique krater. Originally, it would have had a cover with a ball knob embellished with the same flattened and gilded laurel leaves that decorate the piece's lower portion. The front of the vase features a silhouette in a medallion tied with a ribbon at the top; a similarly framed medallion, surrounding the letter "D," is located on the back (Fig. 1). The vase and columnar pedestal, minimally decorated with gold bands, painted medallions, and inscriptions, share an octagonal plinth with two standing putti in unglazed, or biscuit porcelain. Each steadies himself with an arm on the vase's foot; one extends his other arm toward his face in a gesture of exultation; the other gazes beyond the spilled cornucopia he holds; its contents—flowers, rings, boxes, watches, and chains—lie scattered at his feet.

Fürstenberg appears to have made use of known artists and familiar works of art to enhance the significance of this object. The design for this model, known by its mark, "KK," incised on the underside of the pedestal, was probably the work of Franz von Piper, whose Latinized name is painted on the back of this example.[2] However, no known documents have clarified his role at the factory. The model for the silhouette of the vase's recipient, painted on the front, has been attributed to modeler Johan Peter Melchior, who completed a similar biscuit medallion of the same subject while working at the Frankenthal porcelain manufactory.

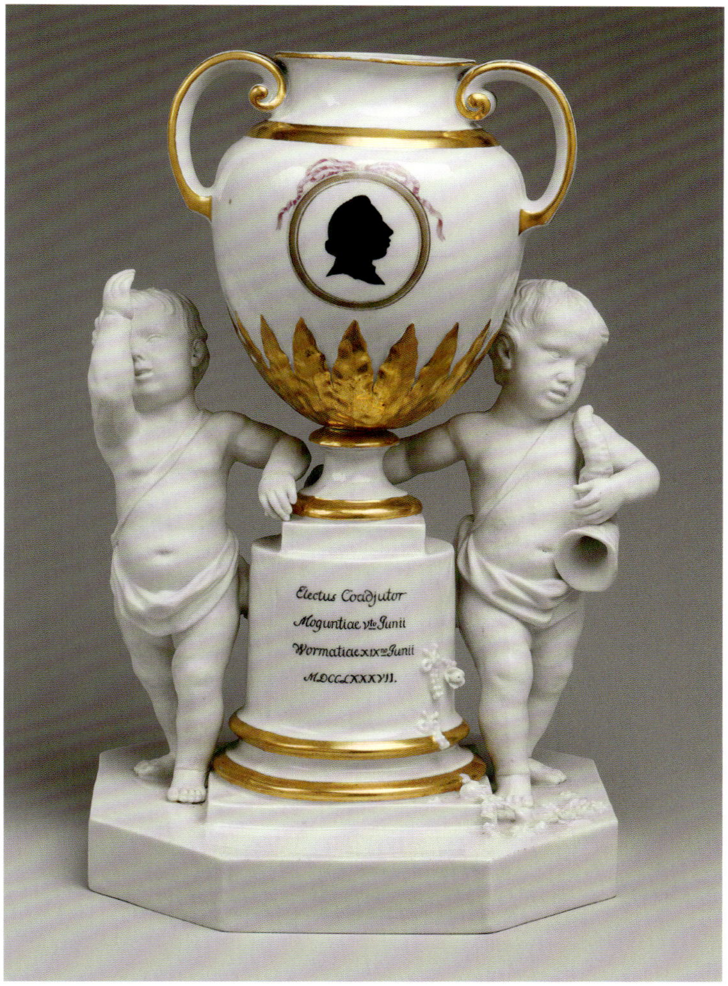

Cat. no. 95

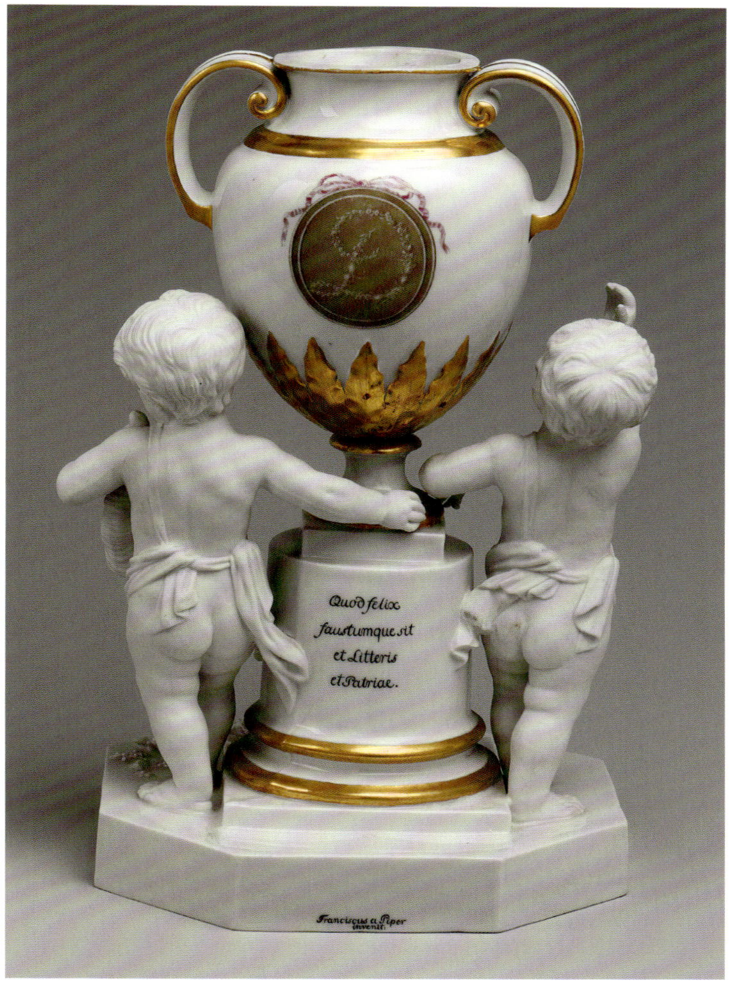

Fig. 1: Reverse of cat. no. 95.

In an advertisement in 1787, Melchior offered his model for sale; the Fürstenberg factory could easily have purchased and adapted it.[3] Visible mold marks suggest that this vase was produced in some quantity, without a great deal of hand finishing. Anna Amalia, daughter of Duke Carl I of Braunschweig-Wolfenbüttel, owned two versions of it, including one with painted putti, today in the Herzog Anton Ulrich Museum, Braunschweig.[4]

This object is dedicated to Carol Theodor Anton von Dahlberg (1744–1817), whose initial and silhouette it prominently displays on the central vase. The product of an enlightened education, Dahlberg held several high political and ecclesiastic offices. The present piece commemorates his service as coadjutant for the cities of Mainz and Worms, a position to which he was elected in 1787, according to the Latin inscription on the front. The following year, Dahlberg abruptly entered the priesthood; his outspoken liberal views, bordering on the heretical, caused the Vatican no small amount of consternation.[5]

Representing the midpoint in the development of Fürstenberg's neoclassical style, this *Gedächtnisvase* demonstrates the widespread acceptance of the new international neoclassical idiom as an appropriate way to pay tribute to significant achievements. In this instance, the antique vase form was chosen as the appropriate symbol to celebrate a man who, for many, represented selfless public service and unwavering moral values. C.H.

1 Howard Coutts, "The Fürstenberg Porcelain Factory," *Apollo* 129 (1989): 416.
2 Christian Scherer, *Das Fürstenberger Porzellan* (Berlin, 1909): 169.
3 Susanne Schroeder, *Tafelrunden: Fürstenberger Pozellan der Herzogin Anna Amalia in Weimar* (Munich, 1986): 100.
4 Inventory nos. 2736 and 2737. They are illustrated in Schroeder (note 2): 104–05, figs. 117–18. The inscriptions vary, but both are clearly dedicated to Dahlberg.
5 *Deutsche Bibliographische Enzyklopädie*, ed. Walter Killy (Munich, 1995): vol. 2, 432–33.

Selected Biographies

Robert Adam (1728–1792)

The Scotsman Robert Adam was the leading architect and designer of British neoclassicism. After training under his father, he completed his education by touring Europe from 1754 to 1758, spending most of his time in Italy. In Florence he met Charles-Louis Clérisseau, whom he employed as his drawing teacher. They traveled and resided in Rome for two years, where Adam befriended the influential neoclassical printmaker and designer Giovanni Battista Piranesi. With the help of Clérisseau, Adam began work on *Ruins of the Palace of the Emperor Diocletian at Spalatro in Dalmatia*, published in 1764 in London. This grand folio demonstrated his thorough knowledge of antiquity and validated his stature as one of the preeminent British scholar-architects.

When he returned to England from his Grand Tour, Adam set up an architectural practice in London with his three brothers. His election to the Royal Society of Art was followed by important commissions. His designs went far beyond façades and ground plans; he concerned himself with the decoration of interior walls, ceilings, furniture, and carpets. Exercising complete control over every detail of his interiors, Adam introduced a novel and extensive vocabulary of classical decorative forms and color schemes, which demonstrated his remarkable talent for ornamental invention. The broad scope of his designs inspired many artists and artisans; his vases in particular influenced the work of silversmith Matthew Boulton and potter Josiah Wedgwood. His *Works in Architecture* (1773–78) codified the Adam style with particular emphasis on his "Etruscan" idiom, derived from Greek vases, Roman architecture, and the Italian High Renaissance. The book helped disseminate his ideas and expand his influence, which ultimately was so far-reaching that the name Adam has become virtually synonymous with British neoclassicism.

Harris 2001.

Richard Askew (about 1730–1798)
Cat. no. 27

While the birth date for painter Richard Askew is not certain, he is listed as an employee of Chelsea Porcelain Manufactory from 1761 on. The Chelsea factory was owned by William Duesbury during what is termed the Chelsea-Derby period: the years between 1770, when Duesbury purchased the Chelsea factory in London and maintained it in addition to another one in Derbyshire, and 1784, when the Chelsea factory was closed and the craftsmen moved to Derby. Askew first worked at Chelsea until 1772, and then transferred to the Derby factory.

While at Chelsea, Askew primarily worked as a figure painter on painted porcelain tableware, including cups and saucers for tea and coffee, flower pots, caudle cups, vases, and plates. His vase and cup decorations include a wide variety of images: portraits of the Prince of Wales and the kings and queens of France, women with children, landscapes, mythological scenes, and allegorical images such as the Four Elements. Like other Chelsea designers, in his figural groups Askew emulated the growing neoclassicism promoted by the porcelain manufactory at Sèvres, which was highly regarded in England.

Askew may have been working as a freelance designer/painter in London by the early 1780s; in 1786 he advertised himself in a newspaper as a miniature enamel painter. He returned to Birmingham in 1793, where he remained until his death.

Gilbert Bradley, Judith Anderson, and Robin Barkla, *Derby Porcelain: 1750–1798* (London, 1990). John Twitchett, *Derby Porcelain 1748–1848: An Illustrated Guide*, (Woodbridge, Suffolk, 2002).

Henri Auguste (1759–1816)
Cat. nos. 64–67

Carrying on a longstanding family tradition, Henri Auguste succeeded his father, Robert-Joseph Auguste, as royal goldsmith in 1785. He won a gold medal at the third Industrial Exhibition in Paris in 1802 for a pair of medals commemorating a military victory. Among the many works he did for the French Crown and for the city of Paris, his most notable were a large table service for Napoleon's coronation in 1804 and a spectacular golden tiara presented to Pope Pius VII by the emperor after the coronation ceremony. Despite his fame, Auguste's business faltered because of debts and poor management. He declared bankruptcy in 1809 and subsequently left France, dying in Haiti in 1816. Auguste's oeuvre represents a transitional stage between his father's austere neoclassicism and the more monumental and embellished forms of the Empire style that would achieve full expression in the work of his rivals and successors Jean-Baptiste-Claude Odiot and Martin-Guillaume Biennais.

Honour 1971.

François-Joseph Bélanger (1744–1818)
Cat. no. 74

Architect and designer François-Joseph Bélanger was only twenty when he was admitted to the Académie Royale d'Architecture under the sponsorship of the Comte de Caylus. His instructors included Julien-David Leroy and Pierre Contant d'Ivry, both of whom encouraged him to work in a neoclassical style. In 1767 he received a royal appointment as designer of sets and decorations for court spectacles and entertainments (known as *Menus Plaisirs*). In 1777 he was promoted to inspector of the *Menus Plaisirs*, and was named first architect to the Comte d'Artois.

Working for Artois, in 1777 Bélanger created the Château de Bagatelle, considered a masterpiece of neoclassical architecture, in the record time of sixty-four days. Bélanger's many other building projects include a pavilion for the Comte de Lauraguais and a townhouse for actress Sophie Arnould, with whom he was intimately connected. Among his design for decorative objects is a jewel cabinet for Marie-Antoinette (1769) and numerous decorative bronzes, executed by Pierre Gouthière for the Duc d'Aumont and his daughter, the Duchesse de Mazarin. Though imprisoned, Bélanger survived the Revolution and, from 1809 to 1813, rebuilt the calotte of the dome of the Halle au Blé, Paris, creating the first cast-iron and glass structure, among other architecture and design commissions.

C. Arnaud et al., *La Folie d'Artois* (Paris, 1988).

Martin-Guillaume Biennais (1764–1843)
Cat. no. 62

Of all the Empire silversmiths, Martin-Guillaume Biennais produced work that perhaps best embodies official Napoleonic taste. Unlike his contemporaries Jean-Baptiste-Claude Odiot and Henri Auguste, he was not just a designer but also an entrepreneur and dealer. Biennais never actually trained as a silversmith; he originally established himself in Paris as a cabinetmaker in 1798. Taking advantage of the suppression of the guild system in 1797, Biennais extended his output to include silver. He was known especially for *necessaires*—traveling cases of silver utensils or toiletry articles—with which he attracted the patronage of Napoleon. He was appointed silversmith to the emperor in 1805. Biennais based much of his silver on designs by architects Charles Percier and Pierre-François-Léonard Fontaine, whom Napoleon favored. The silversmith eventually expanded his business to such an extent that he was able to employ six hundred workers. The key to his success was his ability to offer high-style items using pre-industrial production methods. This, plus his talent for labor organization, makes him the French counterpart to English entrepreneur Matthew Boulton. Biennais retired in 1819 and left his business to his assistant Jean-Charles Cahier.

Honour 1971. Anne Dion-Tenenbaum, *L'Orfèvre de Napoléon, Martin-Guillaume Biennais* (Paris, 2003).

Carlo Bianconi (1732–1802)
Cat. no. 14

In his formative years as a practicing artist in Bologna, Carlo Bianconi excelled at scenography (perspective painting for the stage), but he also created architectural decoration for secular and ecclesiastical settings, such as on altarpieces and for palace interiors. In 1770 Bianconi was elected a member of the Accademia Clementina in Bologna, inaugurating his successful career as an educator, especially in ornament design, for which he is best known. In 1778 he was appointed first secretary at the Accademia di Belle Arti di Brera in Milan. Although he did not have the position of director, Bianconi essentially ran the entire academy, taking on both teaching and administrative responsibilities. He served in this position for twenty-three years and remained in Milan until his death.

S. Samek Ludovici, "Bianconi," *Dizionario biografico degli italiani*, vol. 10 (Rome, 1968).

Louis-Simon Boizot (1743–1809)
Cat. no. 24

The sculptor Louis-Simon Boizot began his artistic studies at the Académie Royale in Paris. In 1762 he won the Prix de Rome, which allowed him to study in Italy at the French Academy until 1770. Boizot regularly exhibited large-scale marble portrait busts and statuary at the Salon, the government-sponsored exhibition held annually in Paris. He worked for important clients, such as Catherine the Great of Russia, and produced numerous decorative sculptures for the French court.

Perhaps through the influence of his patron Madame du Barry, Boizot was named artistic director of the sculpture studio at Sèvres in 1773. His appointment was part of the general reform at the manufactory intended to promote a new, neoclassical style, closely based on archeology. He was expected to produce and supervise the execution of at least twenty-five new ceramic models annually. Seven vase types designed between 1773 and 1780 can be attributed to him, but they must represent only a fraction of his output.

Boizot created mounts and figures for elaborate gilt-bronze vases and clocks, more than 150 biscuit groups often representing mythological scenes, and tableware that reflected ancient shapes. His neoclassical designs remained popular throughout the Revolution, and he continued to receive official commissions under the following Directory, Consulate, and Empire periods (1797–1814). Although he resigned from his factory position in 1800, he continued to supply Sèvres with models until his death.

Eriksen and Bellaigue 1987. Pierre Verlet, *Les Bronzes dorés français du XVIII siècle* (Paris, 1987). Savill 1988. Selma Schwartz, "The Sèvres Porcelain Service for Marie-Antoinette's Dairy at Rambouillet: An Exercise in Archeological Neo-classicism," *French Porcelain Society* 9 (1992): 8.

Matthew Boulton (1728–1809)
Cat. nos. 45–49, 55

Entrepreneur Matthew Boulton was born in Birmingham, England. By 1750 this city had become the largest center in England for the fabrication of "toys"—small luxury metal goods that included snuffboxes, seals, and other trinkets. In 1759 Boulton inherited his father's prosperous button and buckle manufactory, expanded it, and then relocated the firm to property he purchased in 1761 in Soho, outside of Birmingham.

Boulton's visionary plans for the new factory included utilizing a division of labor for manufacture and developing methods that would increase production of goods and keep costs down. He entered a partnership agreement with inventor James Watt, to develop and promote steam engines for use in industrial manufacture, producing and selling them to mining corporations.

Upon its completion in 1766, Soho Manufactory soon became a tourist destination for aristocratic sightseers who marveled at its production techniques and modern equipment, including machinery, driven by the factory's water mill, that was used for rolling, pressing, stamping, lathe-turning, and polishing metal and other materials, such as enamel, stone, and tortoiseshell. Boulton produced buttons, chains, buckles, boxes, instrument cases, candlesticks, cane heads, trinkets, lacquered goods, clocks, gilt-bronze objects, mechanical paintings, lamps, and coins and medals designed in the most up-to-date, neoclassical fashion. Much of his tableware was made in a highly successful type of plated silver known as Sheffield.

Boulton's wares, like those of other English designers working in the second half of the eighteenth century, had to compete for attention with stylish goods imported from continental Europe. He responded to current taste with designs that include neoclassical borders and applied ornament, that could be produced with the factory's machines, and that suited both silver and Sheffield plate.

Boulton, like Wedgwood, consciously exploited the craze for vases. During the 1760s, he sent agents to Italy to purchase prints of ornaments and vases. He also imitated mounted Sèvres porcelain vases using a local mineral, blue john, and gilt-bronze mounts. He marketed them to aristocratic patrons, culminating in sales to Queen Charlotte of England in 1770. Ultimately, however, the vases proved to be neither commercially successful nor suited to mass-production; Boulton ceased their manufacture in the 1780s.

By 1770 Boulton and his business partner John Fothergill had eight hundred employees and mercantile contacts throughout Europe. Their lucrative export business was advertised in their pattern catalogue, which promoted neoclassical ornament on a modern, industrial scale and helped to spread the taste for items in this fashion. The factory continued to operate after Boulton's death; it was sold in 1850.

Goodison 1974. Kenneth Quickenden, "The Planning of Boulton and Fothergill's Silver Business," *Silver and Jewellery: Production and Consumption since 1750* (Birmingham [Eng.], 1995): 21–28.

Jean-Claude-Thomas Chambellan,
known as Duplessis the Younger (about 1730–1783)
Cat. no. 28

Duplessis the Younger is recorded assisting his father in making models for vases at the Vincennes porcelain factory in 1752. He worked at Sèvres as a bronze founder and designer, registering as a master on June 12, 1765. After his father's death in 1774, Duplessis continued to produce neoclassical vases and gilt-bronze mounts for the manufactory. He designed hard-paste porcelain shapes for tureens, cups, bowls, and elaborate vases: for example, two presented by Louis XVI to the Comte and Comtesse du Nord in 1781. He mounted porcelains such as a toilet service for the Comtesse du Nord in 1782 and may have been involved in the design and production of bronze mounts for French cabinetmakers.

Eriksen 1974. Savill 1988.

Dominique Daguerre (act. 1772–96)
Cat. nos. 19, 30

Catering to an elite French and international clientele, Dominique Daguerre was one of the most successful French *marchands-merciers*—merchants who brokered the production of luxury goods—during the second half of the eighteenth century. They commissioned components of their decorative objects, often supplying their own, detailed designs, from various producers, including the Sèvres factory and leading cabinetmakers. Operating a shop on the fashionable rue Saint-Honoré, Paris, Daguerre specialized in furniture that incorporated Sèvres porcelain plaques as well as Sèvres vases that could be mounted with gilt-bronze ornaments of his design (*vases à monter*).

Daguerre's gilt-bronze mounts, fashioned as laurel and acanthus leaves, satyr masks, or scrolled handles, reflect the rapidly developing neoclassical style. Selling these objects to such illustrious clients as Madame du Barry and Empress Maria Feodorovna of Russia, Daguerre's firm promoted neoclassicism internationally among the fashionable elite. By the 1780s, he was exporting Sèvres porcelain directly to the English market. He developed commercial ties with Josiah Wedgwood, displaying and brokering the English potter's ceramics in his Paris shop. Daguerre moved to England in 1787, where he continued to design and market French luxury goods to such clients as King George III and other members of the English court.

Eriksen and Bellaigue 1987. John Whitehead, "George IV: Furnishing in the French Taste: The Place of Dominique Daguerre in the Royal Collection," *Apollo* 138, 379 (Sept. 1993): 153–59. Sargentson 1996.

Jean-Charles Delafosse (1734–1789)
Cat. nos. 9–10, 34, 73, 87, 93–94

Athough Jean-Charles Delafosse trained as an architect, he is best known for his work as a neoclassical ornament designer. He was educated at the Académie de Saint-Luc, Paris, under sculptor Jean-Baptiste Poullet. By 1767 he was calling himself "architect and teacher of drawing," a reference to the courses he taught in geometry and perspective at the academy.

In his designs, Delafosse developed a full range of neoclassical forms and motifs. He was particularly influential in disseminating the style's initial and most austere incarnation, the *goût grec*, through his prints of chimney pieces, architecture, and vases, particularly those in his *Nouvelle Iconologie historique* (1768). This book of 110 engravings, with designs for furniture, architectural ornament, and a variety of decorative objects, was extremely popular and was republished with additions in 1773, 1776, and 1785. Eighteenth-century French and English manufacturers of furniture, ceramics, and gilt-bronze mounts consulted Delafosse's work, incorporating his concepts for neoclassical ornament into their own designs.

Delafosse based his embellishments—heavy swags of oak leaves, twisted fluting, meanders, and stern satyr heads—on classical sculpture and architecture rather than actual ancient vessels. The Sèvres factory looked to Delafosse's designs for porcelain vases that were among the first products to reflect the rising influence of neoclassicism, introducing the antique style to clients in France and abroad.

Geoffrey de Bellaigue, "Engravings and the French Eighteenth-Century Marqueteur II," *Burlington Magazine* 108, 748 (July 1965): 357–61. Eriksen 1974. Myers 1991.

Charles-Nicolas Dodin (1734–1803)
Cat. nos. 17–18, 23

After studying military engineering, Charles-Nicolas Dodin joined the Vincennes porcelain factory, which later moved to Sèvres, as a figure painter in April 1754. At first he specialized in painting tableware and vases with monochromatic scenes of cherubs, based on paintings, drawings, and engravings by or after François Boucher that the factory owned for reference purposes. Dodin also created reserve designs, for both cuvettes and vases, inspired by seventeenth-century Dutch genre scenes.

By the early 1760s, Dodin's lexicon of themes for Sèvres pieces included chinoiseries, hunting vignettes, portraits, mythological narratives, allegories, military scenes, and pastorals. He decorated porcelain tableware, potpourris and other ornamental vases, and soft-paste porcelain plaques for mounting on furniture crafted by leading French cabinetmakers.

Despite the adoption of the neoclassical idiom at the Sèvres factory during the last quarter of the eighteenth century, in his playful designs Dodin retained the spirit, coloration, and painterly manner of the rococo period, demonstrating the enduring popularity of this style at the royal court.

Savill 1988. Marie Laure De Rochebrune, "Charles Nicolas Dodin, Miniature Painter at Sèvres," *Magazine Antiques* 158, 4 (Oct. 2000): 524–33.

Jean-Démosthène Dugourc (1749–1825)
Cat. no. 78

A designer and architect, Jean-Démosthène Dugourc studied drawing under Charles-Germain de Saint-Aubin. Around 1764–65 Dugourc went to Italy, where he met the influential art historian Johann Joachim Winckelmann, who kindled in him a lifelong interest in antiquity. Dugourc studied the Vatican Loggie decorated by Raphael and his school, which became one of the main sources of inspiration for his later work. In 1784 he was named to two royal positions, as Superintendent of Royal Buildings and as Designer of Royal Furnishings. Dugourc collaborated extensively with a leading silk manufacturer from Lyon, Camille Pernon, producing textile designs for his company from 1774 to 1790. In 1800 he left for Madrid to work for the Spanish court and did not return to France until 1814. He subsequently resumed his partnership with Pernon and was also restored to his position as royal designer. In his autobiography, Dugourc grandiosely declared himself the inventor of the *goût étrusque*. Although this is an overstatement, Dugourc was an influential figure in the development of the mature phase of neoclassicism.

Christian Baulez, *De Dugourc à Pernon: Nouvelles Acquistions graphiques pour les museés* (Lyon, 1990). Anne Forray-Carlier, "Dessins de Jean-Démosthène Dugourc pour le Garde Meuble de la Couronne," *Bulletin de la Société de l'histoire de l'art français* (1990): 133–44.

John Flaxman (1755–1826)
Cat. nos. 7, 42

The second son of a modeler who supplied plaster casts to the Wedgwood factory, John Flaxman won a prize from the Royal Society of Arts at age twelve and subsequently studied sculpture. In 1775, as a result of a recommendation from Thomas Bentley, the twenty-year-old artist received a commission from Josiah Wedgwood to design bas-relief portraits, furniture plaques, and vase decorations for production in the company's newly invented jasper ware.

Flaxman adapted some of his designs from prints and from drawings of ancient sources made by Wedgwood employees working in Italy. In addition to producing numerous portrait medallions and busts, he designed bas-reliefs for vessels, such as the Pegasus Vase (cat. no. 7), based on a scene from an ancient example owned by Hamilton.

Flaxman continued to work for Wedgwood in Rome, where he went in 1787 partially subsidized by the potter. While he continued to provide designs for the firm, Flaxman concentrated on sculpture in marble. By 1794 he had returned to England and again supplied designs to Wedgwood, as well as to Rundell, Bridge, and Rundell, silversmiths to the Prince Regent. He was elected to the Royal Academy in 1800.

Reilly and Savage 1980. Adams 1992. Reilly 1994. Young 1995.

Jean-Baptiste-Étienne Genest (1725–1789)
Cat. no. 15

Jean-Baptiste-Étienne Genest joined the Vincennes porcelain manufactory in 1752 as a figure painter, specializing in military encampment scenes. He was appointed head of the artistic studio in 1753, a position he held until his death. As such, he often provided drawings and engravings (originals and copies), for use on wares. He also executed paintings of flowers, animals, and landscapes for factory artists to copy. Genest produced designs for engraved steel dies used for jeweling, a type of decorative technique popular at Sèvres in the 1780s. Under his leadership, the factory delivered many notable commissions, including a 1777 service for Catherine the Great of Russia on which he painted reserve scenes.

Geoffrey de Bellaigue, "Sèvres Artists and Their Sources I: Paintings and Drawings," *Burlington Magazine* 122, 931 (Oct. 1980): 667–76. Linda H. Roth, "A New Attribution of Three Sèvres Vases," *MMA Journal* 37 (2002): 247–55.

Jean-Jacques Lagrenée the Younger (1739–1821)
Cat. nos. 24, 29

The prolific painter, designer, and engraver Jean-Jacques Lagrenée the Younger produced works in a variety of styles, drawing from many sources. In his obituary, he was remembered as "passionate for the works of the ancients." Accompanied by his brother, the painter Louis Lagrenée, with whom he studied, Lagrenée lived in Russia from 1760 to 1763. He studied at French Academy in Rome from 1763 to 1768, winning a second prize for his drawings of antique sculpture. He became a full member of the Académie Royale in Paris in 1775 and exhibited regularly at the Salon—the annual government-sponsored exhibition—between 1771 and 1804. Lagrenée received a number of official commissions under the Comte d'Angiviller's *travaux d'encouragement* (works intended to encourage neoclassical history painting). He completed a number of such projects, including a mural, *Apollo Surrounded by the Muses* (1775), for the ceiling of the theater in the Petit Trianon, Versailles.

In the 1780s, Lagrenée published several collections of neoclassical etchings, including *Recueil de Compositions* (1782). His prints often incorporate devices such as Roman-style military shields and helmets, and arrangements of classicizing vases and ewers with acanthus scrolls, festoons, laurel bands, and other ornament modeled after ancient sources. Interested in new techniques, he also experimented with painting on glass and printing.

In 1785 the Comte d'Angiviller appointed Lagrenée artistic codirector of Sèvres, to assist Jean-Jacques Bachelier. Lagrenée was expected to reform its designs, which were still rooted in the rococo manner, in order to make them fashionable and more commercially viable. This meant transforming the factory's products into the neoclassical idiom favored by stylish patrons. Lagrenée advocated an archeologically based version of classicism drawing from Greek friezes and Roman paintings. He supported the acquisition for the company's archives of volumes of Sir William Hamilton's vase collection published by Pierre d'Harcanville, as well as that of a study collection of ancient vases (thought to be Etruscan) from Dominique Vivant Denon, as resources for designers and painters.

Lagrenée himself produced numerous classicizing designs for Sèvres during his tenure, including breakfast services in the *goût étrusque*, ormolu mounts for vases manufactured by Pierre-Philippe Thomire, and terracotta models for porcelain vases. Under Lagrenée's direction, artists and designers at Sèvres followed a strict aesthetic that resulted in the factory's widespread adoption of neoclassicism.

Barthélemy Jobert, "The Travaux d'encouragement: An Aspect of Official Arts Policy in France under Louis XVI," *Oxford Art Journal* 10, 1 (1987): 3–14. Marc Sandoz, *Les Lagrenée: Jean-Jaques Lagrenée*, vol. 2 (Paris, 1983). Selma Schwartz, "The Sèvres Porcelain Service for Marie-Antoinette's Dairy at Rambouillet: An Exercise in Archeological Neo-classicism," *French Porcelain Society* 9 (1992): 1–35.

Claude-Nicolas Ledoux (1736–1806)
Cat. no. 75

Ledoux first studied architecture at a private school in Paris established by Jacques-François Blondel and then apprenticed under Louis-François Trouard. By the early 1760s, he was designing interiors and remodeling townhouses. One of the leading architects in Paris, he worked in a simple, classical style that attracted prominent patrons, including the Marquis de Livry, the Marquis de Montesquieu, and the Prince and Princesse de Montmorency.

Ledoux joined the Department of Water and Forests of Burgundy and the Franche-Comté, for which he designed schools, cemeteries, roads, bridges, churches, and other public works. By the 1770s, his designs, including those of the château he did in 1771 for Madame du Barry at Louveciennes, were fully neoclassical.

After Louis XVI appointed him commissioner of the saltworks of the Franche-Comté in 1771, Ledoux built the Saline de Chaux. The radical design for this factory and residential complex, which was arranged in concentric rings and included a grotto, featured simplified cubes, Doric columns, and cylinder shapes, combining Palladian references and rustication. The overall layout, and the thinking behind it, anticipated nineteenth-century utopian efforts to improve the environment for workers and their families.

From 1775 Ledoux supervised the engraving of architectural plans for major public monuments and private houses in the saltworks, inventing institutions and public buildings that would complement the multiple dwellings. First announced in 1784, this collection of engravings was ultimately published in 1804 as *L'Architecture considerée sous le rapport de l'art, des moeurs et de la législation.*

Ledoux designed over fifty guard houses or barriers at the gates of Paris to stop smuggling and to collect taxes. In the course of the Revolution, he was dismissed and imprisoned, and angry mobs tore down most of his barrier structures. After his release, Ledoux had few architectural commissions; he focused primarily on the development of ideal city plans and the publication of his engravings, today highly regarded for their revolutionary aesthetic and social vision.

Michel Gallet, *Claude-Nicolas Ledoux, 1736–1806* (Paris, 1980). Anthony Vidler, *Claude-Nicolas Ledoux: Architecture and Social Reform at the End of the Ancien Régime* (Cambridge, Mass., 1990).

Jean-Laurent Legeay (about 1710–1788)
Cat. no. 90

An architect, draftsman, and teacher, Jean-Laurent Legeay is best known today for his moody and fantastic prints and drawings of landscape and architecture. After studying at the École des Beaux-Arts, Paris, and winning the Prix de Rome in 1732, he spent six years at the French Academy. His residence in Rome coincided with that of Giovanni Battista Piranesi, who arrived from Venice in 1740. Both artists contributed to a set of engravings for guidebooks issued by Fausto Amidei, *Varie veduti di Roma antica e moderna* (1745). The association with Piranesi proved to be significant for Legeay's artistic development and his interest in neoclassicism.

In 1742 Legeay returned to Paris, where he served as a professor of architecture at the Académie Royale and became an influential teacher. He left this position by 1745 to work abroad, executing various commissions in Germany, Prussia, and England before returning to France about 1767. Shortly afterward, he began work on four sets of etchings that seem to have been inspired by his Roman sojourn: *Fontane* (1767), *Rovine* (1768), *Vasi* (1770), and *Tombeaux* (1770). In the tradition of architectural *capricci*, they are fantastic depictions of architecture, heavily ornamented fountains, vases, and tombs, inhabited by monsters, reptiles, and caricatures in landscapes with ruins. These early neoclassical etchings inspired his colleagues and students and played a role in establishing the new style in France.

Eriksen 1974. Gilbert Érouart, *Architecture au pinceau. Jean-Laurent Legeay: Un Architecte français dans l'Europe des lumières* (Milan/Paris, 1982).

Louis-Joseph Le Lorrain (1715–1759)
Cat. nos. 88–89

The multi-talented Louis-Joseph Le Lorrain was a painter, furniture designer, architect, and engraver. In 1739 he received the Prix de Rome, permitting him to spend the next eight years at the French Academy. Among his fellow students in Italy were the painter Joseph-Marie Vien and the progressive architect Jean-Laurent Legeay, both of whom were influential in the nascent stages of neoclassicism. After his return to Paris, Le Lorrain was accepted into the Académie Royale as a history painter and soon afterward elected to full membership.

In the decorative arts, he is best known for the suite of furniture he designed in 1756 for the collector and courtier Ange-Laurent de Lalive de Jully. Immediately renowned, the suite is considered the first important statement of the *goût grec* in furniture.

Pierre Rosenberg, "Louis-Joseph Le Lorrain (1715–1759)," *Revue de l'art* 40–41 (1978): 173–202.

Jean-Guillaume Moitte (1746–1810)
Cat. no. 63

A sculptor and draftsman, Jean-Guillaume Moitte collaborated extensively with goldsmith Henri Auguste, providing him with numerous drawings. Moitte originally studied in the studio of Jean-Baptiste Pigalle and won the Prix de Rome in 1768. He studied ancient sculpture in Italy from 1771 to 1773, an experience that shaped his oeuvre significantly. In 1783 he was accepted into the Académie Royale in Paris and subsequently enjoyed a successful career as a neoclassical sculptor. Moitte produced over one thousand designs for the Auguste workshop, many of which were acquired by Jean-Baptiste-Claude Odiot after the goldsmith declared bankruptcy. Moitte's drawings for the decorative arts had a significant impact on neoclassicism, anticipating the so-called Empire style.

Richard James Campbell, *Jean-Guillaume Moitte, The Sculpture and Graphic Art 1785–1799* (Ann Arbor, 1992).

Jean-Louis Morin (1732–1787)
Cat. nos. 20–22

Jean-Louis Morin was the son of an army surgeon and also trained in surgery before joining the Vincennes porcelain manufactory in 1754. Along with Charles-Nicolas Dodin, he was one of the factory's most prolific painters, completing numerous designs on vases, tableware, flowerpots, and other shapes.

From the late 1750s, Morin specialized in depicting military encampments and marine and quayside views in the reserves of many vases made at Sèvres into the last quarter of the eighteenth century. The soft, rococo-style palette of his scenes, as well as his attention to detail, stands in contrast to the neoclassicism of some of the vessels' shapes. Like other studio painters, Morin probably based his designs on engravings and paintings supplied from the manufactory's collection. He also executed trophies and grisaille cameo heads, popular neoclassical motifs employed on Sèvres vases from 1769 on.

Geoffrey de Bellaigue, "Sèvres Artists and Their Sources II: Engravings," *Burlington Magazine* 122, 932 (Nov. 1980): 751. Savill 1988.

James Neale (1740–1814)
Cat. no. 37

A merchant of ceramics and glass, James Neale acted as the London sales agent for Humphrey Palmer's Church Works pottery at Hanley, in Staffordshire, selling its neoclassical, black basalt wares. After Palmer's bankruptcy in 1778, Neale took over the business, added Robert Wilson as a partner in 1786, and renamed the factory Neale and Wilson. They continued to produce cream ware, black basalt, and jasper ware vases in imitation of vessels made by Josiah Wedgwood. The firm was absorbed by Ridgeway pottery during the nineteenth century.

Edwards 1987. Edwards 1994.

Jean-Baptiste-Claude Odiot (1763–1850)
Cat. no. 61

A silversmith and cabinetmaker, Jean-Baptiste-Claude Odiot came from a family of silversmiths who had risen to prominence in the early eighteenth century. Fearing persecution during the Terror, Odiot joined the Republican army and returned to Paris around 1800 to establish his business. The jury for an 1802 industrial exhibition awarded the gold medal jointly to him and Henri Auguste, his chief competitor. Capitalizing on Auguste's subsequent bankruptcy, Odiot bought all of his rival's models, tools, and drawings, a valuable asset housed in the archives of the still-flourishing Maison Odiot.

During the Empire period, Odiot's output was so compelling that he challenged the primacy even of Napoleon's personal silversmith, Martin-Guillaume Biennais. Odiot, for example, furnished the emperor with a ceremonial silver service in 1804. He enjoyed substantial foreign patronage from the courts of Naples, Poland, and Russia. Odiot is credited as the inventor of a screw system that secured applied ornaments to silver bodies. Under the Restoration, he produced works in the Empire style for Louis XVIII. He retired in 1827, and his descendants have continued the family business.

Jean-Marie Pinçon and Oliver Gaube du Gers, *Odiot l'orfèvre: Iconographie, légendes, préface, introduction et notice technique* (Paris, 1990).

Michelangelo Pergolesi (act. about 1760–92, d. 1801)
Cat. no. 81

An architect, painter, and designer of ornament and furniture, Michelangelo Pergolesi grew up in the town of Pergola, in the Roman Marches, where he received his early training. He met Robert Adam during the latter's sojourn in Italy between 1754 and 1758. Two years later, Pergolesi joined the Scottish architect's office in London as a draftsman. His activities there included producing designs for paintings, engravings, ornament prints, calling cards, and furniture. He was responsible for the painted decoration of sixty-two pilasters in the Long Gallery at Syon House, designed by Adam in the early 1760s. From 1777 to 1792, Pergolesi published installments of *Designs for Ornaments*, a collection of etchings illustrating concepts for decorative arts and interiors.

Jean Anne Vincent, "Michelangelo Pergolesi: Adam's Living Italian Source for the Lively Venetian Style," *Interiors* 114 (Dec. 1954): 79–82.

Ennemond-Alexandre Petitot (1727–1801)
Cat. nos. 12–13, 91–92

Although his father wanted him to pursue an ecclesiastical career and enrolled him in a seminary, Ennemond-Alexandre Petitot soon turned to art and design. After training in Lyon under Jacques-Germain Soufflot, he entered the Académie Royale d'Architecture in Paris in 1741 and won the Prix de Rome four years later. From 1746 to 1750, he studied classical art and architecture in Rome, sketching ancient buildings and recording original ideas inspired by his surroundings. His talent was soon recognized by the Comte de Caylus, for whose publication *Recueil d'antiquités* he provided engravings in 1751–33. Caylus referred the artist to the duke of Parma, who promptly appointed Petitot court architect in 1753. He designed the city's new palace, gardens, and library, as well as supervising many other projects. In 1764 he published his *Suite de vases*, comprising engravings of thirty-one vessels, some of which were realized and placed in Parma's ducal gardens. He issued another another print suite, *Mascarade à la Grecque*, in 1771.

Giorgio Cusatelli, *Petitot: Un Artista del settecento europeo a Parma* (Parma, 1997).

Giovanni Battista Piranesi (1720–1778)
Cat. no. 2

The Venetian engraver and designer Giovanni Battista Piranesi first learned drawing from his uncle, a designer and engineer, and proceeded to study both stage design and architecture. In 1740 he went Rome to work as a draftsman for Marco Foscarini, the Venetian ambassador to Pope Benedict XIV, and was introduced to antiquarians who were promoting early neoclassical ideas. Within the year, Piranesi was working as a freelance printmaker, producing souvenir views of the city for the Grand Tour market. He also specialized in the restoration and redesign of classical antiquities, including vases, to sell to wealthy visitors.

Piranesi's first published folio of twelve etchings, *Prima Parte di Architetture e Prospettiva* (1743), features imaginatively reconstructed Roman buildings and landscapes with ruins which influenced the work of other designers, such as Louis-Joseph Le Lorrain and Jean-Laurent Legeay. Championing the supremacy of Rome over Greece, he was convinced that the Etruscans were the sole founders of Roman civilization and an older people than the Greeks. Many scholars, including Johann Joachim Winckelmann, disagreed.

However, such opposition had virtually no effect on the dissemination of Piranesi's popular prints or his importance for both neoclassical designers and their clients, who were moved by the printmaker's dramatic combinations of classical motifs, especially his romanticized views—infused with a sense of grandeur—and illustrated guidebooks to Rome. By the 1770s, he had enlarged his print business and was also creating objects, for example chimney pieces that incorporated classical fragments and motifs in an imaginative manner. The artist continued to publish Roman scenes and depictions of Herculaneum and Pompeii.

By the end of his life, Piranesi had produced over one thousand individual etchings with multiple impressions and revisions, which were copied and collected by advocates of the antique style, especially in England and France. Influential designers, such as Robert Adam, Pierre-François-Léonard Fontaine, Thomas Hope, Charles Percier, in addition to Le Lorrain and Legeay, further promoted Piranesi's approach to neoclassicism in their designs for interiors and objects, including vases.

John Wilton-Ely, *Piranesi* (London, 1978). Denison et al., 1993. John Wilton-Ely, *Giovanni Battista Piranesi: The Complete Etchings* (San Francisco, 1994).

Jacques-Nicolas Roettiers (1736–1788)
Cat. no. 50

Jacques-Nicolas Roettiers succeeded his father, one of the finest makers of rococo silver, as royal gold- and silversmith in 1772. His best-known surviving work is the enormous dinner service ordered by Empress Catherine the Great for Count Grigory Orlov, which was completed and sent to Russia in 1775. Roettiers's silver is heavily decorated with rosettes, gadroons, oak, and laurel leaves, a style of ornamentation typical of the initial neoclassical idiom. Exhibiting a certain affinity with designs by Jean-Charles Delafosse, the forms and decoration are lavish and weighty, conveying architectural grandeur. Having made his fortune early, Roettiers seems to have retired and turned to sculpture after 1777.

Honour 1971.

Paul Storr (1771–1844)
Cat. no. 55

Paul Storr was one of the most important English silversmiths of the late eighteenth and early nineteenth centuries. He was apprenticed from 1785 to 1792 to Andrew Fogelberg, a Swedish-born silversmith, in Soho, London. After being made a freeman in 1792, he joined the workshop of William Frisbee, a plateworker. In 1807 he began to manage a subsidiary of the famed silversmiths Rundell, Bridge, and Rundell known as Storr and Co., and became partner in 1811. Eight years later, Storr established his own business, and by 1822 opened a retail outlet in partnership with John Mortimer. Storr's business eventually declined, and he retired in 1839.

Although he worked in a number of modes, Storr's neoclassical pieces have been considered his most successful. Influenced by Robert Adam, he accented his designs in this style with motifs such as floral swags, vases and urns, beaded edges, and fluting.

N. M. Penzer, *Paul Storr 1771–1844, Silversmith and Goldsmith.* (New York, 1954). John Culme, *Nineteenth Century Silver* (New York, 1977).

Joseph-Marie Vien (1716–1809)
Cat. no. 11

While Joseph-Marie Vien's career as a painter, draftsman, and engraver is often overshadowed by that of his most famous pupil, Jacques-Louis David, he was one of the earliest French painters to adopt the neoclassical style, and as such exerted a significant influence on the subsequent generation of artists.

A student of Charles-Joseph Natoire in Paris, Vien won the Prix de Rome in 1744, studying both classical and baroque art in Italy until 1750. Returning to Paris, he was elected to the Académie Royale in 1751; the following year, he declined an invitation to serve in Copenhagen as court painter to Frederick V of Denmark.

Vien based his earliest paintings and engravings on baroque and rococo antecedents. Over time the artist's work became more classical. An enthusiast of archeology, he revitalized the age-old technique of wax encaustic to copy or model designs after ancient examples. The most famous example of Vien's neoclassical work is his painting *Cupid Seller* (1763; Musée National de Fontainebleau), modeled after a wall painting excavated in Naples in 1759.

In 1775 Vien returned to Rome to become director of the French Academy. Upon the death of Jean-Baptiste-Marie Pierre in 1789, he was appointed royal painter to Louis XVI. Despite this monarchist connection, Vien's association with David and other neoclassical artists who championed the republican cause granted him protection during the Revolution. He died in 1809, one year after being honored with a title from Napoleon.

Thomas W. Gaehtgens and Jacques Lugand, *Joseph-Marie Vien, peintre du roi, 1716–1809* (Paris, 1988).

Francis Vivares (1708–1780) and Thomas Vivares (1735–1821)
Cat. no. 34

Francis Vivares was a French engraver and print publisher active in England from 1739 until his death in 1780. Trained in the family tailoring business in Geneva, he moved at age eighteen to England, where he studied with the painter Giacomo Amiconi and Swiss engraver Joseph Wagner. Vivares began to publish books of rococo ornament and produce landscape engravings, many of which he based on paintings by French and Dutch Old Masters, as well as contemporary English artists such as Thomas Gainsborough. By 1749 Vivares's London shop was selling paper and chalk as well as Italian, Flemish, and French prints; in 1753 he published his own collection of engravings of Yorkshire views.

Vivares became a member of the Society of Artists of Great Britain in 1761, and exhibited his landscape prints with the society in 1766 and 1768. Subsequently, he engraved ornamental borders for use in print rooms and published a variety of books copying French engravers whose work helped introduce the neoclassical style to England. After the publication of Jean-Charles Delafosse's *Nouvelle Iconologie historique* (1768), Vivares adapted some of the architect's ideas in a collection of six vase designs, engraved by his son Thomas, entitled *A New Book of Vases.* Vivares, aided by his son, continued to publish other copies of French neoclassical-style drawings and engravings, including vase designs by Edmé Bouchardon that had been originally engraved and published by Gabriel Huquier in Paris before 1731. Although Vivares's publications often literally copied such prototypes, he heightened their dramatic effect with shading and contours, enhancing the monumental and sober *goût-grec* style of the 1760s.

Thomas Vivares continued in his father's footsteps and ran the business with considerable success. He contributed plates to Robert Adam's *Works in Architecture* (1773–78) and exhibited paintings at the Royal Academy in the 1780s. Six landscape prints dating to 1800 that are documented as by him are now at the Victoria and Albert Museum, London.

H. Vivarez, *Pro domo mea: Un Artiste graveur au XVIIIe siècle, François Vivares* (Lille, 1904). Elizabeth Miller, "Landscape Prints by Francis Vivares," *Print Quarterly* 9 (1992): 272–81.

Josiah Wedgwood (1730–1795)
Cats. 5, 8, 31–33, 36, 39–44

Born in Burselm, England, Josiah Wedgwood belonged to the fourth generation of a family of potters. After an internship with his brother, he apprenticed with Thomas Whieldon, using this time to experiment with glazes and ceramic formulas. By 1759 Wedgwood had started his own pottery, producing different kinds of earthenware decorated with a variety of glazes, often resembling ancient hardstones and fashioned in the new neoclassical idiom.

Wedgwood eventually perfected a formula for a cream-colored glazed earthenware. Operating a showroom in London, the potter received an important commission in 1765–66 to make a cream ware tea service for Queen Charlotte. As a result of the queen's admiration for the set, Wedgwood was appointed royal potter; his cream ware took on the moniker Queens ware, catapulting his work into the limelight.

In 1767 Wedgwood formed a partnership with Thomas Bentley, an experienced merchant. Wedgwood moved much of his pottery production to a new, expanded factory in Staffordshire, which he called Etruria. He immediately began producing neoclassical vases in cream ware, variegated earthenware of mixed dyed clays, and basalt (an unglazed black-colored stoneware).

In 1775 Wedgwood created jasper ware, made from fine-grained, unglazed stoneware stained or dipped in metallic oxides and decorated with white applied or sprigged designs. This hugely successful material further expanded the repertory of neoclassical vase types in England. The bas-reliefs on jasper vases were based on wax models, sketches of ancient vessels, and actual antiquities that Wedgwood often borrowed from museums and private collections. The Wedgwood firm collected published engravings in sets or books that constituted a rich source for factory designers and artisans.

Wedgwood's genius extended to technological innovations for mechanized ceramic production. By 1763 the potter had begun to utilize the engine-turned lathe for cheaper and more efficient production of cutting and shaping vases. His aggressive marketing strategies, including advertisements, public showrooms, and trade catalogues, attracted an ever-growing stream of wealthy clients seeking to buy vases based on classical designs. Wedgwood expanded his business to France, Italy, Russia, and North America.

Neil McKendrick et al., *The Birth of a Consumer Society: The Commercialization of 18th Century England* (London, 1982). Reilly 1989. Reilly 1994. Young 1995.

James Wyatt (1746–1813)
Cat. no. 55

James Wyatt, one of the most respected British architects and designers of his time, studied under Antonio Viscentini in Venice and then lived in Rome from 1764 to 1768. After completing the critically acclaimed Pantheon on Oxford Street, London, in 1772, Wyatt became an established architect at age twenty-six. In the 1770s, he provided Matthew Boulton with designs for silverware decoration and objects. There is no evidence, however, that he contributed to any of the gilt-bronze pieces for which Boulton is most famous. Although Wyatt is best remembered as one of the founders of the Gothic revival in England, his early work constituted an important contribution to neoclassical design.

A. Dale, *James Wyatt, Architect*, 1746–1813 (Oxford, 1936). J. Frew, "James Wyatt and the Gothic Revival," (Ph.D. diss., University of Oxford, 1975).

Glossary

Acanthus
Plant of the Mediterranean region with deeply cleft and scalloped leaves that was freely imitated on the capitals of Corinthian and composite orders. It was often used in varying degrees of abstraction on ornament moldings, friezes, and so forth, or as a scrolling vegetal motif.

Adam style
Named after leading British architect and designer Robert Adam. He introduced into Britain neoclassical ideas inspired chiefly by Roman antiquity, but also from Greek, baroque, and Renaissance sources. Expressing a new unity of architecture and interior design, his ornamental schemes feature elegantly flattened and refined classical motifs in a delicate and subdued palette.

Agate ware
See Variegated ware.

Amphora
Literally means "carried on both sides." Often called the "Greek vase archetype," this two-handled vessel was the most common storage form used in ancient Rome and Greece. It likely stored wine and/or oil, but was versatile and undoubtedly multifunctional. Each region produced its own style and type of jar, some with a pointed base (to facilitate pouring), narrow neck (to prevent spillage), and/or variously shaped handles.

Anthemion
Stylized vegetal motif widely used in classical architecture and ultimately derived from the honeysuckle or a branch of the date palm. See Palmette.

Aquatint
Etching technique involving a specially treated ground of minute particles of resin that act as resists to acid. These create tiny depressions which, when printed, give the effect of a soft grain and look very much like watercolor wash. The demand for printed facsimiles of drawings stimulated the development of aquatint and other tonal printmaking processes in the eighteenth century. See also Manière de lavis.

Arabesque
Originating in Roman art, a form of flowing and fanciful painted decoration incorporating scrolls, animals, and flowers. It developed into the complicated interwoven geometric patterns of Middle Eastern ornament that eventually were re-imported to Europe and, by the eighteenth century, included strapwork, foliate scrollwork and grotesques, generally symmetrically aligned. See Grotesque.

Athénienne
A piece of neoclassical furniture consisting of a vessel on a tripod support. It could be made in a variety of sizes and materials and served different functions, such as a wash basin, flower stand, or a perfume burner. The type was invented by the English architect James "Athenian" Stuart in 1757 and depicted in Joseph-Marie Vien's 1763 painting *La Vertueuse Athénienne* (The Virtuous Athenian Maiden), which gave the form its name.

Biscuit (porcelain)
Fired porcelain that has been left unglazed deliberately.

Black basalt
Improvement of the traditional Staffordshire stoneware "Egyptian Black," which Josiah Wedgwood introduced in the summer of 1768. It comprised clay, calcined ocher, iron stone slag, manganese dioxide; coloration occurred naturally in the sediment from the drainage of local coal deposits, called "car." Experimenting with various chemicals, Wedgwood improved the color and produced a denser, finer-grained, glossier ceramic material. He dubbed these unpainted black vases "Etruscan." In his first catalogue of ornamental wares (1773), he called them "Basaltes" and used the name "Etruscan Encaustic" for his imitation of red-figured Greek vases.

Bleu lapis
Underglaze ground color that often resulted in uneven intensity, in use at the royal manufactory at Sèvres by 1749. Inconsistent contemporary documents and subsequent scholarship have created confusion around the color terminology used at Sèvres.

Bleu célèste
Developed in 1753 by Jean Hellot, chemist and technical director at Sèvres, this was a turquoise-blue overglaze enamel ground color also known as *bleu de Sèvres*.

Bleu nouveau
Dark blue overglaze ground color launched by Sèvres in December 1763 and called bleu nouveau at the factory by about 1773. Its uniform tone and relatively light shade were seen as improvements upon the preceding *bleu lapis*.

Blue john
Type of fluorspar most commonly known as blue john. It has a radiating crystalline structure and is found only in Derbyshire. There are fourteen main varieties, often all called blue john in the eighteenth century, although the dominant color varies. It is difficult to handle since it is soft and brittle; because of this propensity for flaws, it is usually bound by resin or shellac.

Bonheur de jour
Literally meaning "luck of the day," this is the French term for a lady's writing desk.

Bright-cut engraving
So-called for the sparkling brilliance that results from lifting out short slivers of silver with a sharp, beveled graver, this method was introduced in London about 1770.

Bucranium
Derived from ancient art, an ornamental motif involving an ox-skull, usually garlanded.

Cassolette
Eighteenth-century term for perfume burner.

Chasing and engraving
Two different methods of working metal to finish a surface and produce a decorative design. Chasing uses various shaped punches and a chasing hammer to model the surface of a metal object. In flat chasing, the surface is only incised to form lines. This can look very similar to engraving, but there is no actual removal of metal. Engraving, a subtractive process, involves carving or cutting lines into the surface using a graver. See Bright-cut engraving.

Cippus
Ancient Roman circular altar form that Josiah Wedgwood reinterpreted as a stand for vases.

Ciseleurs-doreurs
French term for those who finished bronzes by tooling, chasing, and gilding.

Cream ware
Lead glazed earthenware with a cream-colored body composed of clays, flint, and Cornish stone that Josiah Wedgwood perfected in the 1760s as an alternative to porcelain. He based his recipe on formulas developed by John and Thomas Astbury, Enoch Booth, and others in the early eighteenth century. The type was renamed Queens ware after Queen Charlotte of England ordered a cream ware tea service in 1765–66.

Diced ware
Josiah Wedgwood used an engine-turned lathe, a technical innovation, to pare away discrete sections of the ceramic body. The lathe employed a system of weights and pulleys that regularly brought the cutting tool into contact with the surface, creating a diced or checkered pattern. Dipping the solid white jasper ware body in colored slip before cutting the design enhanced the pattern by creating an interplay of contrasting colors. After turning, bas-relief ornaments in white jasper or a third color could be applied.

Ébéniste
French term for a cabinet-maker who specialized in veneered furniture.

Egg and dart
Ornamental molding, derived from classical architecture, of alternating oval and pointed forms.

Enamel
Pigments of a vitreous nature, colored with metallic oxides, and applied to ceramic wares over the glaze. Enamel decoration requires a second firing at lower temperatures. The range of enamel overglaze colors is much greater than those used under the glaze, which must withstand the greater heat used for the initial firing.

Encaustic
Term coined by Josiah Wedgwood to describe his method, patented in 1769, of imitating the decoration of Greek red-figure vases. A black basalt body was fired to its biscuit state, and then red figures were painted onto the surface in a pigment composed of enamel and slip, which had a matte rather than glossy finish. Although Wedgwood knew his technique was entirely different from that employed by the ancient Greeks, he claimed to have rediscovered it. As described by Pliny and eighteenth-century writers like the Comte de Caylus, the term referred to painting with heated, pigmented wax. This is not the method Wedgwood used.

Engraving
See Chasing and engraving.

Etruscan style
Ornamental manner derived from the shapes, motifs, and palette of ancient vessels, initially misunderstood as Etruscan and later recognized as Greek. The style also drew inspiration from Renaissance grotesque ornament and the discoveries at Pompeii and Herculaneum. Josiah Wedgwood's black basalt pots, based closely on William Hamilton's collection of Greek vases, were said to be in the Etruscan style, as were a number of interiors designed by Robert Adam. See also Grotesque.

Étui
French word for case. This can mean a protective case for a larger object or, more specifically, a small container for ladies' personal items (sewing implements, for example); it was often made of precious materials.

Fête galante
Outdoor court entertainments in the seventeenth and eighteenth centuries that became extremely popular as a subject for artists in the rococo period.

Fondeurs-ciseleurs
Founders who specialized in finishing the surface of a cast object by chasing it before it was gilt; often hired by *marchands-merciers* to create mounts for luxury objects and furniture.

Gadrooning
Lobed ornament, usually employed as edging on a curved surface, consisting of a regular series of convex forms; essentially, the opposite of fluting.

Garniture
Matching or related set of three or five decorative vessels, usually in two differing forms or sizes; in the eighteenth century, such ensembles often included a clock. A *garniture de cheminée* was meant for display on a fireplace mantle.

Glaze
Layer of silicate glass applied to the ceramic vessel's surface in order to seal it and decorate or change its surface. Glazes are made from essentially the same ingredients as the ceramic body but with more silicates (glass formers), such as feldspar or quartz, and with fluxes (substances that promote fusion), which reduce the melting point of the silicates. Glazes are classified by the type of flux used. Low-fire glazes include lead, alkaline, and tin. Glazes are colored with metallic oxides such as iron, cobalt, copper, and so forth, and are usually applied as a liquid suspension.

Grotesque
Grotesque ornament was inspired by the late fifteenth-century discovery of Nero's Domus Aurea (Golden House) in Rome. The interiors of these underground rooms (by that time, they were buried under accumulated layers of earth and so were known as *grotte* or caves) were painted in a light and playful manner distinct from the classical vocabulary familiar from more accessible ancient architectural ruins. They featured fanciful juxtapositions of scrolling foliate forms with human and animal subjects sometimes disposed within frames or arranged symmetrically on large, plain backgrounds of solid color.

The discoveries and excavations of Pompeii and Herculaneum in the eighteenth century renewed interest in the classical grotesque and Renaissance interpretations of it, particularly the work of Raphael in the Vatican Loggie. Grotesque ornament became an element in the Etruscan style.

Guilloche
Ornamental motif comprised of a decorative band or border of interlocking circles derived from classical architecture.

Hard-paste and soft-paste porcelain
Hard paste is also called pâte dure, true porcelain, china, or at Sèvres, *porcelaine royale*. The word porcelain was originally applied to Chinese wares imported into Europe. They are characterized by a fine, hard, white, and resonant body that is more or less translucent. True porcelain requires both kaolin (china clay, a non-fusible silicate of aluminum) and petuntse (china stone, a fusible silicate of aluminum), which vitrify when fired at temperatures of 1300 to 1400 degrees centigrade. Hard paste was first used in Europe around 1710 under the auspices of Augustus the Strong, King of Poland and Elector of Saxony, who founded the ceramics manufactory at Meissen. The discovery in 1768 of kaolin deposits in Limousin led to the introduction at the Sèvres factory of true hard-paste porcelain. Also called artificial porcelain, *porcelaine de France*, or *pâte tendre*, soft paste dominated European production until late in the eighteenth century. Instead of china clay or petuntse, a powdered glassy material or frit was added to the clay. When fired at approximately 1100 degrees centigrade, this material vitrifies, creating a glassy body that imitates true porcelain. Recipes for soft paste vary and can include bone ash or soaprock. Soft-paste porcelains were decorated with lead glazes that required a second, lower temperature firing. Due to the inherent instability of the paste and kiln losses, production costs were high. Once the secrets of hard paste were known, and deposits of china clay located, production of soft-paste porcelain dwindled. Sèvres produced hard- and soft-paste wares until soft-paste was abandoned in 1804.

Hydria
Ancient Greek vessel with rounded body, small neck, and handles—one vertical for pouring, and two horizontal for lifting—that was generally used for water.

Japanning/japanware
European decorative finish intended to imitate Asian lacquer work.

Jasper ware
Hard, fine-grained, vitrified stoneware invented by Josiah Wedgwood and named after a type of opaque quartz. This entirely new clay composition was obtained by adding sulfate of barium to semi-porcelain clay. The result was slightly translucent and pure white when fired. Jasper ware could be tinted in a variety of colors by mixing metallic oxides with the white body. After 1785, instead of mixing metallic oxides into the body itself, color was applied in surface washes in a technique called jasper dip. Jasper ware is capable of holding very sharp details and can be fired with relief decoration in another color in order to achieve a cameo effect. Wedgwood's experiment book records over five thousand attempts to perfect the body between 1772 and 1774.

Krater
Vase for mixing wine and water that figured importantly in Greek ritual banquets and entertainments. It was stored in an adjacent room or connecting courtyard for use during these gatherings. Dances and ceremonies often took place around this libation vessel. Some examples supported an inner container (psykter) that held the wine or water, with the larger krater functioning as a cooling liner filled with snow or cold well water. Kraters appear in many forms, including volute, bell, and column types, depending on body shape, rim, and handles.

Kylix
Term employed generically in antiquity for "vessel," but in modern usage it usually refers to a shallow, stemmed drinking cup with two handles. The cup was actually held by the foot, as documented in ancient vase paintings. Also known as a tazza: a footed drinking vessel with a wide, shallow bowl or cup.

Lancet leaf
Lance-shaped motif employing leaves tapered at each end often combined side to side to create a running border.

Lead glaze
See Glaze.

Manière de lavis
French printmaking term that generally refers to any method used to achieve the effect of a wash drawing, including aquatint. A tonal process related to aquatint is *lavis,* in which the plate is brushed directly with acid. This is useful for giving tone to small areas or creating tonal variations within an aquatint ground, but is not easily controlled over large areas.

Marchand-mercier
French mercers, incorporated in the twelfth century, traded especially in imported goods. In the late seventeenth century, Jean-Baptiste Colbert adopted an economic policy designed to foster the market in luxury goods; as a result, mercers were given rights to trade more freely than any other Parisian corporation. From the seventeenth into the eighteenth century, they became active in the distribution of luxury goods made in Paris and elsewhere in France. Although they did not actually produce goods, they had access to designers, manufacturers, as well as imported wares denied to other trades, and the right to finish goods. This unique position enabled them to manipulate and sell to the most fashionable and high-end markets.

Meander
Geometric ornamental pattern consisting of a series of stylized spirals sometimes in the form of the Greek key,

Oenochoe
Ancient Greek jug with a handle at the neck, used for pouring wine into kraters for mixing with water.

Ormolu
Literally, *or moulu,* ground gold in French. Gold is made into a powder so that it can be easily amalgamated with mercury; this mixture is smeared onto bronze, but sometimes copper, silver, or brass, which is heated until the mercury evaporates, leaving the gold alloyed to the surface. Also known as mercury gilding.

Overglaze decoration
Decoration or marking that is added to the surface of a ceramic vessel after it has been glazed. This decoration is in enamel colors that are fixed by a second, low-temperature firing.

Palmette
Palm-shaped ornament, often combined with a stylized lotus, and used on ancient vases and architecture.

Patera
Round or oval architectural ornament derived from a Roman sacrificial plate or saucer. It can feature radiating petals around a central point.

Porphyry ware
Josiah Wedgwood's porphyry glaze involved applying ground mineral oxides to a translucent colored glaze, which achieved a speckled effect in imitation of the purplish or greenish hardstone found in Egyptian and Roman sites. In 1770 Wedgwood produced a green porphyry and later a "true" or purplish porphyry. The minerals used to create the reddish glazes were difficult to control in the kiln. See also Variegated ware.

Red-figure wares
Ancient Greek vases made of red-ware clay. The figures were depicted by reserving the reddish color of the clay ground for the bodies and painting slip around them, which would turn black in a complicated three-part firing. The method was unknown in the eighteenth century but was imitated by encaustic painting.

Répareurs-modeleurs
Craftsmen charged with molding, assembling, and readying a porcelain object for firing.

Sheffield plate
Achieved by a technique invented in the early eighteenth century by Thomas Bolsover in Sheffield, England, involving a copper-sheet core sandwiched between two rolled sheets of silver fused together by heat. Electroplating, introduced in the 1830s, replaced this process.

Slip
Liquid clay.

Soft-paste porcelain
See Hard-paste porcelain.

Stamnos
Large ancient Greek storage jar with a small mouth that typically was covered.

Thyrsus
Staff topped by a pinecone or bunch of vines or ivy leaves with grapes or berries that Bacchus and his retinue carried.

Tourneurs
French term for lathe workers.

Trophies
Originally, a carved or painted representation of a group of arms or armor derived from the real arms taken in battle and displayed as monuments of victory in ancient times. By extension, the term is used for any decorative group of congruous items, such as musical instruments, objects associated with the four seasons, and so forth.

Underglaze decoration
Glaze decoration, usually cobalt (blue), which is applied directly to the ceramic body prior to the initial firing. Generally, a clear lead glaze is applied over the decoration at the same time.

Urn
Vase form featuring a cover and usually provided with a foot or pedestal. In ancient times, it was connected with funeral rites.

Variegated ware (agate ware)
Process used by Josiah Wedgwood from 1768 on to imitate a variety of stones such as granite, agate, and various marbles. He developed three methods. For striated stones like marble and granite, he mixed together different colored clays in a technique called "wedging," the patches and different stripes of the clay remaining distinct. The clays were tinted by the addition of minerals such as iron oxide, manganese oxide, or oxides of copper or cobalt. The ware was then covered with a clear lead glaze. "Veining" and "marbling" could also be achieved by painting the surface of cream-colored ware with slips of various colors. The slips were tinted with the same minerals used in solid agate ware. Agate ware was first developed in Staffordshire around 1725. To imitate grained stones such as granite, one sprinkled powdered mineral colorants on the vases and then covered them with a lead glaze, or dusted the liquid lead-glazed surface with the powder before firing. See also Porphyry ware.

Vermeil
Gilded silver.

Vitrify
Transformation of the clay body into a glassy substance through firing in a kiln.

Frequently Cited Works

Adam and Adam 1773-1822. Adam, Robert, and James Adam. *The Works in Architecture.* 3 vols. London.

Adams 1992. Adams, Elizabeth Bryding. *The Dwight and Lucille Beeson Collection at the Birmingham Museum of Art, Birmingham, Alabama.* Birmingham.

Beazley 1984. Beazley, J. D. *Attic Red-Figure Vase Painters.* 2nd edition. Oxford.

Brunet and Préaud 1978. Brunet, Marcelle, and Tamara Préaud. *Sèvres: Des Origines à nos jours.* Fribourg.

Caylus 1752–67. Caylus, Anne-Claude-Philippe, Comte de. *Recueil d'antiquités Égyptiennes, Étrusques, Grecques, Romaines, et Gauloises.* 7 vols. Paris.

Clayton 1985. Clayton, Michael. *The Collector's Dictionary of the Silver and Gold of Great Britain and North America.* About 1871. Reprint. Woodbridge, Suffolk.

Clifford 1978. Clifford, Timothy. "Some English Ceramic Vases and Their Sources, Part One." *English Ceramic Circle Transactions* 10, 3 (1978). Pp. 159-173.

Clifford 1992. Clifford, Timothy. "The Plaster Shapes of the Rococo and Neo-classical Era—Britain." *Journal of the History of Collections* 4, 1. Pp. 39–66.

Dauterman et al. 1964. Dauterman, Carl, James Parker, and Edith Appleton Standen. *Decorative Art from the Samuel H. Kress Collection at The Metropolitan Museum of Art: The Tapestry Room from Croome Court, Furniture, Textiles, Sèvres, Porcelains, and Other Objects.* London.

Delafosse 1768. Delafosse, Jean-Charles. *Nouvelle Iconologie historique, Ou, Attributs Hiérogliphyques Qui Ont Pour Les Objets, Les Quartre Élémens, Les Quartre Saisons, Les Quartre Parties Du Monde Et Les Différentes Complexions De l'Homme.* Paris.

Denison et al. 1993. Denison, Cara D., Myra Nan Rosenfeld, and Stephanie Wiles. *Exploring Rome: Piranesi and His Contemporaries.* New York.

Edwards 1987. Edwards, Diana. *Neale Pottery and Porcelain: Its Predecessors and Successors 1763–1820.* Foreword by Geoffrey Godden. London.

Edwards 1994. Edwards, Diana. *Black Basalt: Wedgwood and Contemporary Manufacturers.* Woodbridge, Suffolk.

Emmerson 2000. Emmerson, Robin. "Wedgwood's Oven Books." Ars Ceramica 16. Pp. 10-35.

Ennès and Ducrot 1997. Ennès, Pierre, and Brigitte Ducrot. *Un Défi au goût: 50 ans de la création à la manufacture royale de Sèvres (1740–1793).* Paris.

Eriksen 1968. Eriksen, Svend. *The James A. Rothschild Collection at Waddeson Manor: Sèvres Porcelain.* Fribourg.

Eriksen 1974. Eriksen, Svend. *Early Neo-Classicism in France: The Creation of the Louis Seize Style in Architectural Decoration, Ormolu, Gold and Silver, and Sèvres Porcelain in the Mid-Eighteenth Century.* Translated and edited by Peter Thornton. London.

Eriksen and Bellaigue 1987. Eriksen, Svend, and Geoffrey de Bellaigue. *Sèvres Porcelain: Vincennes and Sèvres, 1740–1800.* Translated by R. J. Charleston. London and Boston 1987.

Finer and Savage 1965. Finer, Ann, and George Savage, eds. *The Selected Letters of Josiah Wedgwood.* New York.

Frégnac 1966. Frégnac, Claude. *French Master Goldsmiths and Silversmiths from the Seventeenth to the Nineteenth Century.* New York.

Goodison 1974. Goodison, Nicholas. *Ormolu: The Work of Matthew Boulton.* London.

Grant 1967. Grant, Maurice Herold. *The Makers of Black Basalts.* 1910. Reprint. Woodbridge, Suffolk.

Grimwade 1982. Grimwade, Arthur C. *London Goldsmiths: 1697–1837: Their Marks and Lives from the Original Registers at Goldsmith's Hall and Other Sources.* 2nd edition. London.

Hamilton/d'Harcanville 1767–76. Hamilton, William, and Pierre d'Hancarville. *Collection of Etruscan, Greek, and Roman Antiquities from the Cabinet of the Hon. W. Hamilton His Britannic Maiesty's Envoy Extraordinary and Plenipotentiary at the Court of Naples.* 4 vols. Naples.

Harris 2001. Harris, Eileen. *The Genius of Robert Adam: His Interiors.* New Haven and London.

Hepplewhite 1969. The Cabinet-Makers and Upholsterer's Guide [by] George Hepplewhite. 3rd edition. London, 1794. Reprint, with new introduction by Joseph Aronson. New York.

Honour 1971. Honour, Hugh. *Goldsmiths and Silversmiths.* New York.

Jenkins and Sloan 1996. Jenkins, Ian, and Kim Sloan. *Vases and Volcanoes: Sir William Hamilton and His Collection.* London.

Johnson 1990. Johnson, Howard A. "Books Belonging to Wedgwood and Bentley, the 10th of August 1770." *Ars Ceramica* 7. Pp. 13–23.

Johnson 1991. Johnson, Howard A. "Books Belonging to Wedgwood and Bentley the 10th of August 1770." *Ars Ceramica* 8. P. 34.

Meteyard 1970. Meteyard, Eliza. *The Life of Josiah Wedgwood from His Private Correspondence and Family Papers.* 2 vols. London, 1865–66. Reprint. London.

Miller 1999. Miller, Elizabeth. *16th-Century Italian Ornament Prints in the Victoria and Albert Museum.* London.

MMA Bulletin. The Metropolitan Museum of Art Bulletin. 1906–.

MMA Journal. The Metropolitan Museum of Art Journal. 1968–.

Myers 1991. Myers, Mary L. *French Architectural Ornament Drawings of the Eighteenth Century.* New York.

Oechslin 1982. Oechslin, Werner, et al. *Die Vase.* Zurich.

Percier and Fontaine 1801–12. Percier, Charles, and Fontaine, Pierre-François-Léonard. *Recueil de décorations intérieures: Comprenant tout ce qui à rapport à l'ameublement....* Paris. Reprint. Farnborough, Hants.

Reilly 1972. See Reilly 1994.

Reilly 1989. *Wedgwood.* 2 vols. New York.

Reilly 1994. Reilly, Robin. *Wedgwood Jasper.* 1972. Reprint. New York.

Reilly and Savage 1980. Reilly, Robin, and George Savage. *The Dictionary of Wedgwood.* Woodbridge, Suffolk.

Roth and Le Corbeiller 2000. Roth, Linda H., and Clare Le Corbeiller. *French Eighteenth-Century Porcelain at the Wadsworth Atheneum: The J. Pierpont Morgan Collection.* Hartford.

Rowe 1965. Rowe, Robert. *Adam Silver: 1765–1795.* London.

Sargentson 1996. Sargentson, Carolyn. *Merchants and Luxury Markets: The Marchands Merciers of Eighteenth Century Paris.* London.

Savill 1988. Savill, Rosalind. *The Wallace Collection Catalogue of Sèvres Porcelain.* 3 vols. London.

Whitehead 1992. Whitehead, John. *The French Interior in the Eighteenth Century.* London.

Young 1995. Young, Hilary, ed. *The Genius of Wedgwood.* London.

Index of Names

Page references precede catalogue number references. Bold type indicates a work by the listed artist.

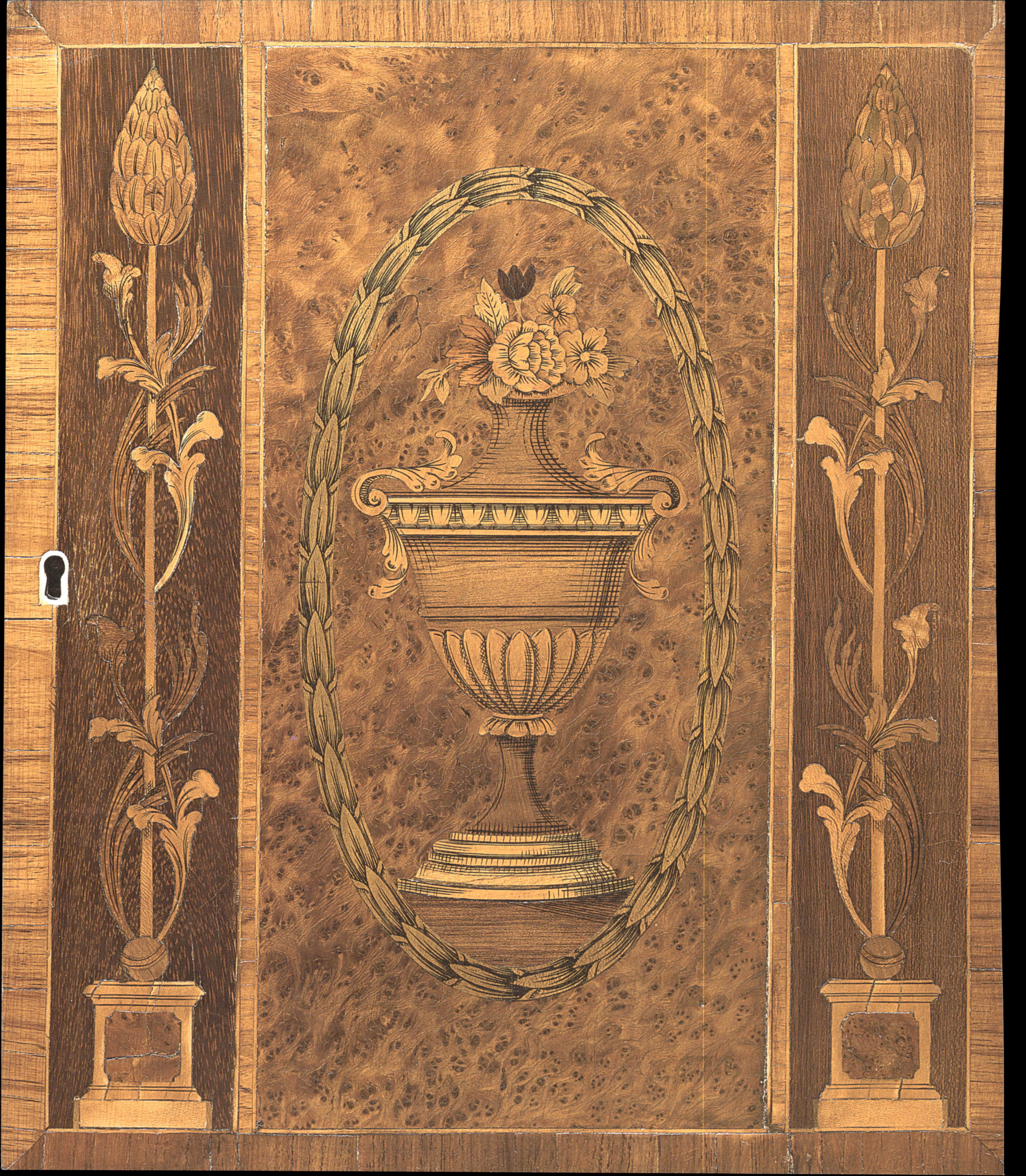